BASIC TECHNICAL WRITING

BASIC TECHNICAL WRITING

Third Edition

HERMAN M. WEISMAN

National Bureau of Standards

Charles E. Merrill Publishing Company
A Bell & Howell Company
Columbus, Ohio

Published by
Charles E. Merrill Publishing Co.
A Bell & Howell Company
Columbus, Ohio 43216

ISBN: 0-675-08833-X

Library of Congress Catalog Card Number: 73 – 90882

1 2 3 4 5 6 7 8 9 10 11 12 13 14 15 — 80 79 78 77 76 75 74

Printed in the United States of America

75-1243

To **MARGARET** my wife

CONTENTS

3   Technical Correspondence   27

**4   Report Writing — Reconstruction of an
        Investigation   79**

PART TWO

8   Semantics and the Process of Communication   182

9   Special Expository Techniques in Technical Writing — Definition   201

**10   Special Expository Techniques in
          Technical Writing — Description,
          Explanation, and Analysis   223**

11   The Technical Article and Paper   276

PART THREE

12   Technical Style   307

PREFACE

Basic Technical Writing, since its first appearance in 1962, has been accepted as a standard textbook for courses in technical writing and report writing. It is intended for

— students in colleges and technical institutes,
— semiprofessionals and professionals taking in-service training courses, and
— scientists, engineers, and others who in the practice of their profession must communicate through the written medium.

I have based this book on the premise that if you, the reader, can understand how the process of technical communication operates and can understand the problems inherent in the operation, such understanding will better enable you to communicate information more effectively and efficiently. I have designed *Basic Technical Writing* to provide you with the understanding of the principles of communication and with instruction to master the techniques of the craft of technical writing. *Basic Technical Writing* is not a "how-to" book in the usual sense. Its emphasis is as much on the background leading to the "how-to" as it is on the "how-to" itself.

Now, a few words about this Third Edition. All experienced writers know that writing is not written but *rewritten*. Mark Twain once observed that the difference between the right word and nearly the right word is the difference between lightning and the lightning bug. The Third Edition of *Basic Technical Writing* is my continuing effort to improve the earlier versions. Not only have some lightning bugs been eliminated for the real thing, but some sections of the text have been reorganized for greater cohesion and clarity. To help the reader deal with common problems of grammar, mechanics, and usage, I have added for his convenience of use a reference index and guide.

The passage of time aged some illustrations and examples in the text. Fresher and more appropriate samples have been selected. Hindsight also called for some additional material. Novelty, however, for the sake of novelty did not determine my substitutions or additions. My criterion for inclusion of any matter, illustration, or example always was and remains its capability for demonstrating the principle or technique involved.

I am greatly indebted to the many students who have both used the book and contributed to it. To my many colleagues, teachers, and professionals who have been users of the book and who have by their comments and suggestions enabled me to strengthen the treatment at many points, I wish to express publicly my thanks. I want to acknowledge also my gratitude to the authors and copyright holders for permission to reprint their material and to share their insights with the readers of *Basic Technical Writing*. These permissions are inherent in the citations within the text to the References noted at the end of each chapter. Finally, let me mention Mrs. Ellen Ring whose very capable typing eased my chores in this revision.

While my present affiliation is with the National Bureau of Standards, my work on the Third Edition of *Basic Technical Writing* has been done independently of my duties at the National Bureau of Standards.

Now, let *Basic Technical Writing* speak for itself.

Herman M. Weisman
Bethesda, Maryland

PREFACE

to the Second Edition

The second edition of *Basic Technical Writing*, like the first, is intended for two types of readers:

1) Students in college and technical institutes,
2) Scientists and engineers who in the practice of their profession must communicate through the written medium.

Technical writing is a craft; it is an expository form of factual writing about technical and specialized matters. Factual writing adheres to the disciplines of accuracy, order, precision, and reasoning based on facts. *Basic Technical Writing* was designed to present instruction in the principles and techniques of communication and technical writing in general, and in the report, the technical article, the letter and memo, and in the expository elements of definition, description, explanation, and analysis in particular.

Since the appearance of the first edition in 1962, both the publisher and I have extended sensitive antennas to receive comments, criticism, and suggestions from the book's users so that improvements could be incorporated in revisions. Before I began work on the second edition, the publisher surveyed some of the major users among colleges and institutes who had long experience with the text. Their reaction was solicited to several approaches I was considering for both enlarging the scope and revising the treatment. The users responded that they were highly satisfied with the text's scope and treatment. There were comments from some that in following the order of arrangement of chapters in the first edition some difficulty was experienced in encompassing all of the material within a single academic quarter or semester course. Any additional chapters would aggravate their situation. Nor was there a recommendation for deletion of any of the chapters or material in the first edition.

The decision became clear that the revision should center mainly on the reorganization of the chapters around two major cores; each appropriate — if the academic user wished — to a semester's work in technical writing. The second edition has been reorganized accordingly. The first core, Part I, concentrates on the principles and techniques of writing reports; the second core, Part II, concentrates on advanced elements of technical writing. Suggested course outlines for each of the cores are found in the Appendix.

Other than the reorganization of the material of the book, the revisions in the second edition bring the text up to date, sharpen the focus of important principles and techniques, and correct typographical errors and the few errors of inadvertence. The bibliographical and advanced reading references at the end of each chapter have been fully updated to include new references and other sources of value to the reader.

The second edition, like its predecessor, provides summaries of principles and techniques and illustrative examples and models. Discussion problems and exercises are integrated as teaching aids for exemplifying and underscoring the principles and techniques within the text.

While my present affiliation is with the National Bureau of Standards, my work on the second edition of *Basic Technical Writing* has been done independently of my duties at the National Bureau of Standards.

H. M. W.
May, 1968

PREFACE

to the First Edition

Basic Technical Writing is an outgrowth of my experience as a technical writer and editor and as a teacher of technical writing. The book gives major consideration to the language and thinking processes in science and technology. Since no person can write more clearly than he can think, I have concentrated on clarifying the thought processes involved in technical communication.

I do not believe that a technical writing text which has been adapted from the college composition text by the technique of adding a chapter on correspondence and substituting a section on the research report for the standard section on the library research paper is adequate to the requirements of the field. Students and practitioners of technical writing — whether technical writers, scientists, or engineers — need to understand the significance of the research investigation and the relationship of this investigation to its end product, the report. They require a background in the *what* and *why* — that is, in the scientific methods — and in the *how*.

The *how* necessitates not only the obvious mechanics of grammar, style, and structure but also an understanding of the communication process — language and semantics.

To make the explanation of these principles and techniques clear and readable, I have introduced at the appropriate places summaries of principles and guide-rules for applying techniques. Also, I have made extensive use of models which show principles and techniques in action. These have been taken from both professional writing and student writing.

Discussion problems and exercises have been structured at the end of each chapter, not as busy academic work, but as a teaching device. The exercises have been interrelated throughout the book to make for a core of projects that will give the student mastery of technical writing and better enable him to meet the variety of reporting situations encountered by the technical writer.

BASIC TECHNICAL WRITING

PART ONE

1

What Technical Writing Is

What is technical writing? Though an established and important pursuit, technical writing as a term is not found in present dictionaries. The words *technical* and *writing* are defined separately. Webster's *New World Dictionary* defines *technical* as: "having to do with the practical, industrial, or mechanical arts or the applied sciences." *Writing* is defined as: "the occupation of the writer . . . the practice of composition." We might combine these two definitions to define technical writing as: "writing about science and technology." People engaged in the profession might want to restate this as: "Technical writing is a specialized field of communication whose purpose is to convey technical and scientific information and ideas accurately and efficiently."

History of Technical Writing

Technical writing is a new term for a very old and very common communication process. Archaeologists tell us that man has inhabited this planet for more than 1,000,000 years. For about 975,000 years, man lived like other animals inhabiting the earth. Only in the last 25,000 years have civilization and progress come to man. Man's invention of writing has contributed immeasurably to this advance. Writing gave him the ability to record and communicate his experience and knowledge. Each generation thereafter did not have to begin all over again. The writing of predecessors became an available source of information and ideas that stimulated further thought and progress. One technical writer has put it this way: It took man almost one million years to arrive at the agricultural revolution; only 25,000 years to get to the industrial revolution; but a mere 150 years to come to the atomic edge of the space age; and but one year to bring prednisone (for arthritis) from idea, to research stage, to commercial production and sale. The process of technical writing has given continuity to human endeavor and has accelerated the tempo for increasing knowledge and progress.[1]

Technical writing might be traced to the prehistoric cave paintings in France and Spain which describe pictorially primitive man's techniques for hunting buffalo. More directly, technical writing had its origin in the earliest cuneiform inscriptions of the Akkadians and Babylonians. Modern science had its beginnings in Babylonia. The ancient Babylonians achieved many accomplishments in astronomy and mathematics; they were the first to divide the year into 360 days, and the circle into 360 degrees. They were also our earliest agricultural scientists. We know this from their technical writing which has survived in the form of clay tablets. In the New York Metropolitan Museum, for example, is a clay tablet of about 2000 B.C., a technical manual, which gives instruction on the proper making of beer.

The Egyptians invented paper. We have remains of Egyptian technical writing in the fields of medicine and mathematics, written on papyrus, dating back to about 500 B.C. Among the more prolific technical writers were the ancient Greeks. Their writings in mathematics, the physical sciences, astronomy, biology, and psychology are more than a historical curiosity. Present-day mathematics, physics, and medicine show the influence of Euclid, Archimedes, and Hippocrates. Aristotle's writings on physics, astronomy, biology, and psychology were used as texts far into the fifteenth century. Writers today may learn much about principles of writing from Aristotle's *Poetics* and *Rhetoric*.

A major area in technical writing is the writing of instruction books or manuals. Today, there are more than one hundred thousand technical writers engaged in this activity alone. Modern instruction books may be traced to the sixteenth century when the first manual on the use and care of military weapons was written. The early instruction books covered the complete range of military weapons of a branch of service. Until about one hundred years ago, the artillery man learned "the book." A single book covered all that his branch of the service was supposed to know — types of weapons, how to use them, maintenance, elements of drill, and tactics. Similar books were prepared for the infantry. The first technical manual for a specific weapon was written in 1856 by Commander John Dahlgren to cover an

---

[1] Dr. Israel Light, formerly of the U.S. Public Health Service, in a personal communication.

advanced type of naval gun. Shortly thereafter, other specific instruction books on such weapons as the Gatling machine gun were produced by the military (2).[2]

From the middle of the nineteenth century to beyond World War I, military weapons were comparatively simple in design. When instruction manuals were inadequate, time was usually available for the gun crew to learn the needed answers through trial and error. It was not until World War II, when there were many advances in war machines with elaborate electrical and hydraulic systems, that instruction books became of primary importance.

World War II brought a tremendous speed-up in research and technology. A quick and efficient way to explain new scientific devices and weaponry to nonscientists and soldiers who were going to use them was urgently needed. Industrial mobilization produced vast quantities of complex war equipment. Manpower mobilization created a huge armed force. How could these men, in the shortest period of time, be taught to operate and maintain the new equipment? Instruction manuals were the answer. Every piece of equipment that was used in World War II required an instruction manual. As a result, the field of technical writing grew up almost overnight.

Today's scientific and technological society needs the services of qualified technical writers not only for writing about electro-mechanical instrumentation and products but also for writing requirements in other important fields such as agriculture, forestry, medicine, pharmaceuticals, biology, chemistry, and straight consumer goods.

What's in a Name

Technical writing is a cloudy term. It covers indistinctly a large communications territory. In some organizations and companies, writers who write manuals of instruction, progress reports, specifications, and proposals may be called *technical writers*; in others, they may be known as *publications engineers;* in still other organizations, *technical editors* or *industrial writers.* Yet all of them may be doing essentially the same thing. These writers may be working on reports and technical manuals, and at the same time, their duties may call for them to write articles for magazines and news releases for newspapers in order to explain the significance of scientific and technological advances which their organizations have developed. In this activity, they are doing much the same thing that journalists for newspapers and magazines do; yet, journalists do not consider themselves technical writers, but journalists. Writers on newspapers and magazines whose major assignment is in the area of science and medicine consider themselves science writers or medical writers, but their activity is no different from that of the technical writers working for Westinghouse or Smith Kline & French Lab-

---

[2] The number in parentheses refers to a reference listed in sequence in the back of the chapter. If other numbers follow a colon after the reference number, the numbers refer to appropriate pages in the cited reference. See pp. 141-142, chapter 6.

oratories who translate or interpret technical material for use in stockholder reports or company publications for the general reader.

Not only do the physical and life sciences require the services of technical writers, but also so do the social sciences and the humanities. For example, a discussion of suprasegmental phonemes or metathesis in linguistics or an explanation of the vagaries of the business cycle or the economics of technological obsolescence can become very "technical." Demography offers a variety of aspects involving complex statistical data and mathematical formulas. Population analysis, sampling racial typology, cohort analysis, etc., are certainly technical. In psychology, the physiology of stimulus-response, the techniques of learning to read faster, and analyses of processes like cognition or association pose problems of explanation, interpretation, and simplification which are no different from the problems facing the writer in explaining scientific processes like dielectric absorption or frequency modulation. The problems and demands involved in structuring and explaining a sociometric diagram of an individual's interactions with other members of his group are similar to the demands involved in explaining the electronic circuit of a new search radar.

For definition purposes, we might look at these diverse situations to see whether we want to divide the writing activity into narrow areas or whether it would be better to examine the complete spectrum of demands for the communication of specialized problems.

Some people would like to distinguish between the writing done by scientists and engineers for their colleagues and the writing done by technical writers who serve as "middlemen"—interpreters, chroniclers, or translators of scientific and technological matters researched, developed, created, or produced by another person. However, though there may be a distinction in the source of the writing and in the purpose for which it is written, the principles of the writing process are the same, whether the reporter is the developer of a new electronic device, a technical writer producing an instruction manual for it, or a journalist telling about it in the Sunday supplement of a newspaper. We are concerned in this text with the principles common to the requirements for technical writing, no matter who does the writing or by what means it is being disseminated.

Factual Communication of Experience

Let us begin with principles. Moliére, the great French comic playwright of the seventeenth century, has an amusing play about a rich shopkeeper who wanted to become a gentleman and attain culture. This Monsieur Jourdain hired a pedagogue to teach him the social graces. "Teach me to talk in prose and poetry!" he ordered, "But, sir," replied the teacher, "you do talk in prose." "I do?" asked Monsieur Jourdain disbelievingly. "Yes, you do. You have been talking in prose all your life." "I have? Fancy that! I have been talking in *prose* all my life!" exclaimed the delighted shopkeeper.

Many of us might be similarly astonished when we discover that we are scientists on occasion. Part of human experience is to observe and to induce or deduce valuable conclusions which we have derived from our observations, just as the professional scientist does. For science, as the dictionary defines it, is concerned with observation and classification of facts for the discovery of general truths and the operation of "natural" laws.

Let me illustrate this with a personal example. One day several years ago, I took my family for a drive into the country. We drove past a farm where cattle were grazing. "See the cow, Daddy!" my then two-year-old daughter, Lise, exclaimed. "Cow is hungry." She observed the cows, saw them browsing, and induced from her own experience the generalization that the cows were eating because they were hungry. Here is a basic example of the scientific principle of induction based on observation and past experience. Of course, Lise's reasoned inference is based on a limited observation, but it has the same probability of being correct as some accepted scientific "laws" based on a similar proportion of observations; for example, metal expands when heated, or altitude affects the boiling point of water.

To discover a fact—I am using the scientific definition of fact: a verifiable observation—we need not peer through a microscope nor trace a pattern on an oscilloscope. To reach a general truth, it is not necessary to collect and examine a whole notebook of data. Science is the knowledge of the physical world we see, hear, touch, taste, and smell.[3] In other words, science is learning from experience. Technical writing is the factual recording of that experience or knowledge for the purpose of disseminating it.

Do you mean, you might ask me, that if I record whatever I perceive through my five senses I am doing technical writing? In a very limited way, yes—provided you are accurate both in your observation and in your writing. What does that mean? you may ask.

First, let me quickly say that most single-sentence definitions are oversimplified (as you might infer from footnote 3 and as you will see in chapter 9), but they can serve a useful purpose. For the moment, I need to provide you with a background for understanding the process of technical writing. Many elements within the process are common, but some have aspects which are specialized, not easily apparent, and often difficult.

Now, to return to your question: let me explain what I mean by accurate observation. Do you remember the story of the six blind men and the elephant? These blind men had often heard of elephants but had never "seen" one. One morning an elephant was driven down the road where they stood.

---

[3] According to psychologists, perception and apprehension of one's environment come about in a number of ways. The human organism has developed a variety of receptor organs of amazing sensitivity and complexity with which it receives and interprets "messages" from its environment. While it is traditional and logical to divide sensations on the basis of five senses, psychologists say that there are at least ten distinct sensory systems, and a typical sensory reponse involves a "global" response of the organism to a message of the environment which can almost never be limited to the response of just one of the traditional five senses. Perception is a complex process involving the sensations specific to a sensory modality interacting with past experience and other ongoing sensations from other systems.

When they were told that the elephant was before them, they asked the driver to let them see him. They thought that by touching him they could learn just what kind of animal he was.

The first blind man felt the elephant's side and called him a wall; the second grabbed the animal's tusk and labeled him a spear; the third touched the trunk and believed the beast to be a snake; the fourth grasped one of the elephant's legs and judged the animal to be like a tree; the fifth, a tall man, took hold of the animal's ear and was certain the elephant was like a fan; the sixth seized the beast's tail and was convinced the elephant looked like a rope.

I am sure you will recognize that this old story has a moral. The oversimplified moral is: Even people who have eyes and all their senses sometimes do not observe carefully. But as we think about the story of the six blind men and the elephant, we realize something more significant: *No matter what we see and report, there is always something left out in our reporting.* When what we omit, for whatever reason, is more significant than what we include, our observations are inaccurate and distorted.

Problems in Factual Communication

The accurate recording of what we perceive through our senses is complicated further because of the means we use to do the recording—language. Language is, as you know, a body of *words and methods of combining words* to communicate to others our experiences, feelings, and thought. Man uses artificial symbols to represent things and ideas. Complications follow because words are not precisely things or ideas but representations of them. This artificiality is the occupational hazard of language. You or I may intend a word to convey a certain meaning, but the reader or listener, because he has a different background, experience, and personal make-up, receives a different meaning.

I shall explain this "hazard" in greater detail in a later chapter, but for the moment, let me illustrate what I mean with a simple situation. Your neighbor's wife has come in to show off her new Easter bonnet. Your wife exclaims, "How exquisitely chic! I love that feather! It's fabulous! It gives you that come-hither, dangerous look!" Now, would your own description of the hat be the same as your wife's? You probably take one look at it, mumble to yourself that the feather looked better on a duck, and turn your thoughts to the fine, expensive cigar you have been saving to enjoy with the relaxation of the evening paper. You no sooner light it and begin to savor the aromatic flavor of it, when your wife exclaims, "Dear, how can you smoke that *thing!* It smells like burning wool. The living room has that foul odor for days!"

It is obvious that what are purportedly described in this situation are not an Easter bonnet and a superbly mellow cigar. Your wife and yourself are describing only the impressions created in your minds by a combination of present sensations and past experiences. Neither description is objective.

In this distinction between subjective and objective lies the crux of the matter. The subjective lies within our minds and is colored by our make-up and experience; the objective is independent of our minds and has its source in external things, apart from us. This important distinction corresponds to the difference between literature and technical writing. Literature is an interpretive record of man's progress and is based on imaginative and emotional experiences rather than the factual record of man's climb from his primitive cave to his flights into outer space. Technical writing, however, is not history. (History is a record of man's past and of his institutions; history subjectively interprets the past and offers value judgments about it.) Literature is concerned mainly with ourselves, with our thoughts, feelings, and reactions to experiences. Its purpose is to give us insight. Technical writing concerns itself with factual information; its languages does not appeal to the emotions nor the imagination, but to the intellect. Its words are exact and precise. Its primary purpose is to inform. Its information is the activity and progress of science and technology.[4] Technical writing is the term given to this type of informative reporting because much of the writing deals with highly technical matters. Technical writers report technical and scientific progress in books, reports, magazines, and professional and trade publications for the expansion and utilization of knowledge.

Discussion Problems and Exercises

1. (a) Who reads and uses technical writing? Before you read further, list at least four types of users and comment on how they might be benefiting from the process.

   (b) Now consider the home owner who opens up a large crate and begins to assemble the various components of a power lawn mower; the maintenance man searching for the sputter in the engine roar of the F-111 as it warms up on the airstrip; the mother who reads a pamphlet on nephritis because the consulting physician has tentatively diagnosed her son's ailment as being just that; the local councilman perusing a communication outlining objections to the fluoridation of his town's water supply. These four people are reading an explanation of a particular scientific fact or condition. What generalizations come to your mind?

2. Many of the large metropolitan papers have columnists and feature writers who specialize in reporting on scientific, medical, and technological advances. These writers object strongly to the appellation of technical writer and call themselves science writers or medical writers. Is this aversion justified? Explain. If your answer is yes, differentiate between science writing, technical writing, and medical writing.

---

[4] Some technical writing, because of the excellence of its expression and the imaginative and creative insights it brings to the elements written about, is regarded as literary. Examples are the writings of T. H. Huxley, Charles Darwin, Albert Einstein, Rachel Carson, and Harlow Shapley.

3. *Chambers Technical Dictionary,* revised edition, defines science as: "The ordered arrangement of ascertained knowledge, including the methods by which such knowledge is extended and the criteria by which its truth is tested." Does this definition disagree with the definition on page 6 which says, "Science is learning from experience"? Explain.

4. One textbook has attempted to distinguish "scientific writing" from "science writing" and these two categories from "technical writing":

> The scientific writer is first a scientist and secondarily a writer; the science writer is primarily a professional writer. Scientific writing stems from the scientific tradition, science writing from the journalistic tradition. However, since some science writers are well-trained in science and some scientists devote part of their time to journalism, the categories do at times overlap. Scientific writing done strictly by and for specialists may be termed technical writing (1:131-32).

These distinctions are useful but often academic because writers like Isaac Asimov, George Gamow, and Lancelot Hogben have done scientific writing, science writing, and technical writing simultaneously. Consult the library and find examples from these three authors and others indicating this triple categorization.

5. Technical writing falls into the larger category of factual writing. Factual writing concerns itself with verifiable observations and reasoning based on facts. In its purest form, factual writing adheres to the disciplines of accuracy, order, precision, and reasoning based on facts. Literature is concerned with a writer's imaginative and subjective world as he sees and interprets it. Poetry, drama, and fiction belong to literature. In the category of factual writing are some of the following: news stories, biography, history, informational articles, reports, essays, reviews, and technical writing of all types. In what category would you place the ensuing types of writing? Explain why.
    (a) Tonight's front-page news story
    (b) The lead editorial in tonight's paper
    (c) A biography of Madame Indira Gandhi; a biography of Lord Byron
    (d) The book of Genesis
    (e) *The Outline of History,* by H. G. Wells
    (f) The Annual Report of General Motors
    (g) *The Odyssey*
    (h) *Origin of Species,* by Charles R. Darwin
    (i) "The Chambered Nautilus," by Oliver Wendell Holmes
    (j) A book review of a best seller
    (k) An instruction sheet on how to use a can opener
    (l) *Critique of Pure Reason,* by Immanuel Kant
    (m) *On the Nature of Things,* by Lucretius

6. Can you write objectively, factually, scientifically using the first person singular? Give reasons for your answer.

7. Would the following qualify as technical writing? Why? The terza rima is a three-line stanza form borrowed from the Italian poets. The rhyme scheme is aba, bcb, cdc, ded, etc. One rhyme sound is used for the first and third lines of each stanza and a new rhyme introduced for the second line; this new rhyme, in turn, is used for the first and third lines of the subsequent stanza.

The meter is usually iambic pentameter, but in the example shown below from Browning's "The Statue and the Bust," the meter is iambic tetrameter:

There's a palace in Florence, the world knows well,	*a*
And a statue watches it from the square,	*b*
And this story of both do our townsmen tell.	*a*
Ages ago, a lady there,	*b*
At the farthest window facing the East	*c*
Asked, "Who rides by with the royal air?"	*b*

8. Which of the following three statements is scientifically most accurate?
    (a) Water freezes at 32 degrees Fahrenheit.
    (b) The freezing point of H_2O is 0° Celsius.
    (c) At sea level, pure fresh water starts to freeze when its temperature falls to 32 degrees Fahrenheit.

9. If words cannot mean the same thing to two people, how is communication possible, especially technical writing?

10. Are recipes in cookbooks to be considered as a type of technical writing? What are the similarities and differences in the following pieces of writing?

    (a) "Ballad of Bouillabaisse" by William Makepeace Thackeray
    > This Bouillabaisse a noble dish is —
    > A sort of soup, or broth, or brew,
    > Or hotchpotch of all sorts of fishes,
    > That Greenwich never could outdo;
    > Green herbs, red peppers, mussels, saffern,
    > Soles, onion, garlic, roach and dace:
    > All these you eat at Terre's tavern
    > In that one dish of Bouillabaisse.

    (b) Bouillabaisse

1 small lobster	2 tomatoes
2 crabs	1 bay leaf
1 medium eel	1 cup white wine
3 pounds fish, assorted	½ cup olive oil
3 chopped onions	1 piece orange peel
3 garlic cloves, minced	pinch of saffron
1 sprig thyme	bread for toasting
1 sprig fennel	salt and pepper

Choose small fish of the following varieties: bass, striped bass, porgy, black fish, and a slice of cod or halibut to make a total of three pounds. Clean and cut up. Throw the heads into cold water, and bring to a boil; then add eel, lobster, and a few minutes later, the crabs. Boil fifteen minutes, take out eel, lobster, and crabs, strain the liquor and throw away the fish heads. Heat the strained liquor to boiling point.

Place raw fish, vegetables, and all seasonings except saffron, in a saucepan. Pour olive oil over and cook, adding the boiling strained liquor. Boil rapidly over hot fire for 5 minutes. Cut the eel into 2-inch lengths, cut lobster into pieces and remove meat from the crabs. Add to the boiling sauce. Stir the saffron in a little cold water and put that in. Stir up, bring to boiling point, strain liquor into a tureen over triangles of toast. Place meats on hot platter, so diners may choose what they want to add to the liquor served in hot soup plates.

(c) Pararosaniline

Pararosaniline is a triamino-triphenylcarbinol of the formula:

It is obtained by the oxidation of a mixture of *p*-toluidine (1 mol.) and aniline (2 mols.). As oxidizing agents, arsenic acid or nitrobenzene are used. Acids effect the elimination of water and the formation of a dyestuff with a quinonoid structure; e.g., the hydrochloride of pararosaniline or parafuchsine has the formula:

References

1. Emberger, Meta Riley and Marian Ross Hall, *Scientific Writing*, New York: Harcourt, Brace and Company, 1955.

2. Marschalk, Henry E., "Our Economy Depends upon Good Technical Manuals," *Technical Writing Review*, October 1959.

2

Scientific Method and Approach

It is traditional for the scientist to report on the results of his work: research is not completed until it has been recorded, reported, and disseminated. The report is the chief end product in much scientific activity. Since the act of reporting is part and parcel of the scientific method, it would be useful for us to pause and examine the background, the meaning, and the characteristics of the scientific method.

Science, including the activity which characterizes it—research—has been defined in various ways. One approach to an appropriate definition would be operational. The purpose of scientific research is to discover facts and ideas not previously known to man. According to many scientists, science always has two aspects. On the one hand, it gives us knowledge and understanding of what was previously unknown, mysterious, or misunderstood; on the other hand, it increases our control over nature and the forces at work in nature (5:508). The implication in this definition is that science is verifiable and communicable knowledge. The word *communicable* is significant, for as Benjamin Lee Whorf has pointed out, scientific knowledge is, in large part, symbolic knowledge or knowledge expressed in symbols. Because man has the ability to communicate, he is able to apprehend and describe the world (6:221).

Historic Origins of Science

Science, as we know it today, dates back to the introduction of the experimental method during the Renaissance. However, there were earlier noble beginnings.

How Man Learns

People learn new facts and ideas in several ordinary ways: through chance, trial and error, and experience. Primitive man probably learned about cooking through chance or accident. One of our prehistoric ancestors came across a stag burnt in a forest fire started by a bolt of lightning. Being hungry, this primitive bit into the roasted carcass. It tasted surprisingly good—better than raw meat. This experience made him want to have all of his food acted upon by fire. Man learned about cooking by chance but had to learn the art through trial and error. Primitive man soon found that if he were not watchfully careful, his meat might burn into ashes or, at the other extreme, remain raw.

Primitive man acquired many types of knowledge by deliberately trying things out. Watching some animals eat herbaceous plants, he tried to do the same. Through trial and error, he was able to discover that some plants were satisfying pleasant and some were bitter and harmful. These discoveries through chance and through trial and error were communicated to other men so that the discoveries became common knowledge. In this way, apprehension and comprehension of the world began. Such apperceived information was passed on by whatever means of communication primitive man had, and it became a permanent body of useful knowledge.

T. H. Huxley, in his essay "On Improving Natural Knowledge," surmises how this type of activity started the primitive savage on the path toward civilization:

> I cannot but think the foundations of all natural knowledge were laid when the reason of man first came face to face with the facts of nature; when the savage first learned that the fingers of one hand are fewer than those of both; that it is shorter to cross a stream than to head it; that stones dropped stop where they are unless they are moved, and that they drop from the hand which lets them go; that light and heat come and go with the sun; that sticks burn away in a fire; that plants and animals grow and die; that, if he struck his fellow savage a blow, he would make him angry, perhaps get a blow in return, while, if he offered him fruit, he would please him, perhaps receive a fish in exchange. When men had acquired this much knowledge, the outlines, crude though they were, of mathematics, physics, chemistry, biology, of moral, economic and political science were sketched. (4:22)

Information generalized from experience was added to the body of useful information derived from chance and from trial and error. Thus, heavy, clouded skies presaged rain; it was best to seek shelter quickly. Certain shellfish made one sick; it was best not to eat them. Primitive man advanced a notch in civilization by arriving at useful conclusions based on experience. Generalizing from experience involves the faculty of logic, that is, reasoning things out, not only from one's own experience but also from the experience of others. Man thus began to use an important tool or process of modern scientific inquiry. This process enabled man to acquire more knowledge by examining his own experience and that of other members of his tribe. It helped him to meet potentially harmful situations and advance his well-being. He thus was able to build up a considerable body of knowledge as a common heritage.

But let us not think that primitive man was happily on his way to discovering vacuum cleaners, compact cars, and vitamin tablets. There were many things about him and his environment which he could not understand and which offered him difficulties. That which he could not explain satisfactorily by his experience, primitive man attributed to magic and the influence of some supernatural power. When his tribe was attacked by fierce beasts or other tribes, when it was struck by plague, or when a thunderstorm flooded him out, he wanted to know why such a thing had happened. Primitive man would turn to his priest, to his witch doctor, or to the tribal wise man. The priest might account for the conflagration as a punishment for sins or failure to propitiate the tribal gods. The priests or tribal wise men frequently offered good advice, and men came to trust their chosen authorities. Thus there arose a tradition of reliance and trust upon authority, based on the belief that great thinkers of the past were able to discover truth for all times and that man could best learn about himself and his world by studying their words. This subservience to authority was most influential during the medieval period when the teachings of the ancients—Plato, Aristotle, and the early church leaders—dominated all learning and received more credence than did first hand observation and analysis of facts.

Greek Contributions — Logic

Aristotle and other Greek philosophers and scientists controlled much of world thought for a period of some twelve hundred years. The Greeks were responsible, more than anyone else, for developing mental tools by which man has been able to solve many of the mysteries of the universe. By developing logic and mathematics, the Greeks were greatly responsible for bringing explanations by magic to an end. Greek philosophers were the first to arrive at the concept of the universe as an ordered cosmos in which everything happens according to definite laws of cause and effect. The Greek philosophers believed in the natural order of things, and they set out to learn the characteristics of this natural order.

Aristotle believed that one could know the unknown by examining the known through a deductive process of which the syllogism is the basis. The syllogism is a device for testing the truth of any given conclusion or idea. This may best be illustrated by Aristotle's classic example:

Major premise: 1. All men are mortal (must die).
Minor premise: 2. Socrates is a man.
Conclusion: 3. Therefore, Socrates is mortal (therefore will die).

If the first two assertions, or *premises* as Aristotle called them, can be shown to be true, it must follow that any reasonable person will agree that the third statement (the conclusion) must be true. By these principles of classical logic, a conclusion properly deduced or arrived at from reliable premises is necessarily reliable. This, of course, takes for granted that the premises are true and that every element in the argument has been carefully examined and all unrelated or false materials have been discarded. Hence, if it is certain that all acids turn blue litmus paper red, then acetic acid, CH_3COOH, will turn blue litmus paper red. This is certainly true. Given a valid major premise and a related valid minor premise, we can deduce a valid conclusion. The process can be an intellectually stimulating and valuable exercise.

Deductive reasoning, however useful, offers the hazard that it may interest the user more in the mental processes and skillful argumentation than in the facts of the case. Scholars in medieval times were influenced more by Aristotle's methods of deduction than by his caution to follow direct observation of nature. As a result, they arrived at their conclusions and generalizations by means of reasoning or logic alone. By the use of syllogisms, they deduced conclusions and principles from selected statements of approved authorities without checking their accuracy and reliability. Through skillful selection of premises, medieval scholars were able to prove almost anything, even the number of angels that could dance upon the head of a pin. The syllogistic approach is limited in that only if the premises are true and properly related can we know that the conclusion is true. However, the truth of premises and their proper relation can be determined only from observation and experience. Take, for example, a common American myth which might be considered a major premise, gentlemen prefer blondes. Our syllogism would be:

Gentlemen prefer blondes.
John Doakes is in love with a blonde.
Therefore, John Doakes is a gentleman!

Or this syllogism:

All creatures that fly with their wings are birds.
The bat flies with its wings.
Therefore, the bat is a bird.

Not all creatures that fly with their wings are birds. Insects, some fish, some mammals, like the bat or lemur, and an extinct order of reptiles, the Pterosauria, fly.

Unless each of the premises is carefully tested and its relationship shown, the conclusion may be wrong. Of course, the basic weakness of the syllogism is that it can guarantee the truth of the conclusion only when it begins with true and properly related statements of fact, but it can provide no assurance of the truth of its premises.

Deductive reasoning held scholarship in its grasp from the time of the ancient Greeks to the sixteenth century when Francis Bacon attacked this method of reasoning as limited. Bacon advocated the need for basing general conclusions upon specific facts arrived at through direct observation. Bacon urged scholars to ignore authority; to observe nature closely; to perform experiments; to draw their own inferences; to classify facts in order to reach minor generalizations; and then to proceed from minor generalizations to greater ones. In other words, Bacon advocated inductive reasoning, generalizing from examples—that is, going from the particular to the general, rather than from the general to the particular, as is the case in deductive reasoning. Induction is based on the assumption that examples we encounter in the future will be like those we experienced in the past:

Water freezes at 32° Fahrenheit.
Metal expands when heated.
The sun will rise tomorrow.
Honesty is the best policy.

However, this is not to say that research problems can be solved by the inductive approach alone. Induction is incomplete until the accumulated data have been analyzed and classified and related to a guiding principle or problem. Charles Darwin found this out when he followed Bacon's advice. He collected fact after fact in his biological researches, hoping that the facts themselves would lead to an important generalization, but, as he noted:

My first notebook [on evolution] was open in July, 1837. I worked on true Baconian principles, without any theory, and collected facts on a wholesale scale, more especially with respect to domesticated productions, by printed inquiries, by conversation by skillful breeders and gardeners, and by extensive reading. When I see the list of books of all kinds which I read and abstracted, including a whole series of journals and transactions, I am surprised at my industry. I soon perceived that selection was the keystone of man's success in making useful races of animals and plants. But, how selection could be applied to organisms living in a state of nature remained, for some time, a mystery to me.

In October, 1838, that is fifteen months after I had begun my systematic inquiry, I happened to read for amusement, Malthus' "On Population," and being well-prepared to appreciate the struggle for existence which everywhere goes on for long-continued observation of the habits of animals and plants, it at once struck me that, under these circumstances, favorable variations would tend to be preserved, and unfavorable ones to be destroyed. The result of this would be formation of new species. Here, then, I had at least found a theory by which to work.... (2:68)

Having gathered considerable data, Darwin made a guess or assumption derived from his reading in Malthus as to what the data might mean. He formulated a hypothesis built on the facts he already knew. Using this hypothesis as a guide, he further investigated to see whether it would be supported or proved wrong by additional evidence which could be gathered. Darwin thus stumbled upon a possible solution to the problem of how evolution takes place and began to test it. By making deductions from it, he was able to see how his facts could be put together to form a workable theory. In this way, Darwin used both inductive and deductive reasoning to arrive at his final conclusions. The foregoing is a good example of how modern research is frequently carried on.

The Scientific Method

What are the distinguishing characteristics of the scientific method? Science begins with the observation of the selected parts of nature. Although the scientist uses his mind to imagine ways in which the world might be structured, he knows that only by careful scrutiny can he find out whether any of these ways truly correspond with reality. He rejects authority as an ultimate basis for truth. Where primitive man ascribed anything unusual that he might see or hear to the special intervention of the gods, modern man looks for "natural" causes. Though he is compelled by necessity to consider facts and statements issued by other workers, the scientist reserves for himself the decision as to whether these other workers are reputable; whether their methods are good; and whether, in any particular case, the alleged facts are credible.

Until Galileo's time, it was commonly assumed by thinking men that Aristotle was right in saying that heavy objects fall to the ground faster than lighter objects. This assumption appears perfectly logical and reasonable to anyone who thinks about it without taking the trouble to test it by experiment. Galileo, however, refused to accept either authority or logic as the basis for his conclusions until he experimented by dropping cannon balls of various sizes and weights from the leaning tower of Pisa in 1589. Perhaps even to his own surprise, he found that the cannon balls he dropped, except for minor differences caused by the resistance of the air, all fell at the same rate of speed.

The Problem Concept and Its Significance to Scientific Investigation

Man has been called the two-footed animal, the talking animal, the tool-using animal, and, very appropriately, the curious animal. From earliest history, one of man's greatest urges has been to *know*. The deep human impulse to learn and to share knowledge with others has probably been one of the most potent factors in man's rise from his primitive cave to his yearnings for travel into space.

What the the lights in the sky? What makes grass grow? Why is grass green? How long is a river? How deep is the sea? What makes me sick? How far away is the moon? What is the moon? Questions such as these were no doubt asked by our earliest ancestors. The basic elements of our own "new" science are contained within the primitive's ability to identify and articulate the unknowns of life and his crude but fundamental attempt to find solutions to those basic questions. It was the primitive savage's recognition of an obstacle, difficulty, or problem and the urge to know its answer that started man on his way to civilization.

As primitive man conquered his obstacles, difficulties, and problems, he let others of his tribe know about his successes and experiences. Communication of the primitive savage's research aided the clan. The more "research minded" the primitive clan, the greater was its survival prospect. Civilization grew as man was able to observe the phenomena of his world, recognize and isolate his problems, investigate them, and then arrive at answers. As man approached more difficult obstacles, he found it necessary to get others to cooperate with him as a team. For example, one man might not be able to get a large log across a raging stream, but two, three, or four might be able to do it. The clan could then cross the rapid stream without hazard.

The recognition of the function of the problem in scientific research is a dominant feature of our modern world. The role of the problem in scientific advancement might be illustrated by the following episodes.

In the fifteenth century seaport city of Genoa lived a young boy who loved to visit the docks and watch the ships come and go. He was fascinated by the seamen who sailed the galleons. What fascinated him most were the ships leaving the docks and sailing out into the horizon. This boy would watch the ships as they grew smaller and smaller and smaller, and suddenly they would disappear entirely. It was as if they dropped into the ocean. "What caused the ships to disappear?" he would ask himself. "Is it that the ships reach the end of the earth and then fall straight down? It couldn't be," he said to himself, "because the ships that disappear return after a year or two." As the boy grew older and became a sailor himself, he was intrigued by an added phenomenon—the disappearance of land from the sailor's view as the distance between ship and port increased. In talking with a fellow sailor who had been blown off his course and had claimed that he had reached a land in the west and returned, Christopher Columbus was strengthened in his hypothesis that the reason why ships disappear to the viewer on land, and why as the ships go west beyond the curvature of the earth they are no longer visible to the watchers on the shore, is that the earth is not flat but round. The "beyond" in the west was not the edge of the world, but the Orient. This hypothesis Columbus sought to prove. If the earth were round, he ought to be able to reach India by following the curvature of the earth. The fact that you and I are here on the North American continent attests to the most fundamental accuracy of his hypothesis.

Another boy, who lived in Scotland about three hundred years later, was fascinated by the sight of his mother's teakettle boiling over the hearth fire. When the water boiled, the lid of the teakettle was lifted and steam came out from under the lid as well as through the spout. "What causes the lid to lift?" James Watt would ask himself. "It must be the steam," he hypothesized. "If the steam in the teakettle has the power to lift the lid, could steam in a larger vessel have power to lift heavier things?—Or even move heavier things a great distance? Could steam be put to work?" he asked himself. History has provided the answer to this problematic question, also.

Paul Ehrlich, the father of chemotherapy, hypothesized that since certain dyes had a special affinity for certain tissues in the body, it might be possible to find substances which would go one step further and seek out not only special tissues but also germs. His faith in this hypothesis made him persist in his investigation despite long-continued frustration, repeated failures, and attempts by friends to dissuade him from an apparently hopeless task. Eventually, on his six hundred sixth attempt, he came upon his "magic bullet" or salvarsan which could track down the germ of syphilis and destroy it without harming any of the surrounding healthy tissues and cells.

In our time, after it was discovered that poliomyelitis was caused by a virus, a doctor in Pittsburgh asked himself, "If polio is caused by a virus, why can't we use immunization techniques against polio and come up with a protective inoculation?" Dr. Jonas Salk and others have fortunately proved this possible.

Recognition of a problem, then, is the starting point of inquiry. In his book, *Logic: The Theory of Inquiry*, John Dewey states:

> It is a familiar and significant saying that a problem well put is half-solved. To find out what the problem and problems are which a problematic situation presents to be inquired into, is to be well along in inquiry. To mistake the problem involved is to cause subsequent inquiry to be irrelevant or to go astray. Without a problem, there is a blind groping in the dark. The way in which the problem is conceived decides what specific suggestions are entertained and which are dismissed; what data are selected and which rejected; it is the criterion for relevancy and irrelevancy of hypotheses and conceptual structures (3:117).

Morris R. Cohen and Ernest Nagel, in their book *An Introduction to Logic and Scientific Method* point out, "It is an utterly superficial view that the truth is to be found by 'studying the facts.' It is superficial because no inquiry can even get underway until and unless *some difficulty is felt* in a practical or theoretical situation. It is the difficulty, or problem, which guides our research for some *order among facts* in terms of which the difficulty is to be removed" (1:199). The end result after the investigation is the answer to the problem.

Various problems faced by scientists can be categorized into three major types: problems of fact, problems of value, and problems of technique.

Problems of Fact. As the term implies, problems of fact seek answers to what the facts are. Is the earth flat or round? At what temperature will water freeze at an altitude of 10,000 feet? What causes cancer? How much salinity is in Farmer Brown's well? Who killed Cock Robin? How effective is streptomycin against staphylococcus? Problems of fact involve questions of what occurs, when it occurs, how it occurs, and why it occurs.

Problems of Value. Problems of value deal with what is more valuable, what is to be preferred, or what ought to be the case (rather than what is the case). Problems of value are involved in setting up standards or criteria. Examples of problems of value that affect scientific research are the determination of criteria or standards of safety, health, efficiency, tolerance, economy, etc., within a situation. Many problems of value are related to problems of fact.

Problems of Technique. Problems of technique are concerned with the methods by which a desired result can be accomplished. How, for example, can a space station be launched in the mesosphere? Problems of technique are usually the concern of applied science or engineering. The problem of technique combines elements of fact and value. The solving of a significant problem frequently entails all three categories. For example, in the polio situation, there was first a problem of fact—what causes polio? After the fact (that a virus was the cause) was discovered, a problem of value came into being: What known techniques of immunology against viruses are effective against polio? After Dr. Jonas Salk perfected his vaccine, there arose a technical problem—how to manufacture the vaccine in adequate amounts to supply the need.

Apprehending the Problem

Every field of endeavor or study has problems requiring investigation. New areas of knowledge open every day. New discoveries uncover the need for further research. Today, our state of knowledge might be compared to that of the period of exploration opened up by Columbus' discovery of America.

Much time and effort often go into apprehending or identifying the problem for research. This is especially true when the researcher is inexperienced or lacks knowledge of his field. Frequently, the reseacher is not in a position to make an unrestricted choice; a client or a supervisor thrusts a situation into his hands. The researcher may be a staff member of an industrial concern, of a government laboratory, or of a research unit in a university. In such a case, the field will probably be designated for him with the preliminary work involved in making the selection already carried through by others. At academic institutions, students are frequently required to pursue selected fields for research. The field of choice may be further limited for the student by the requirement of majoring or specializing in a particular department. These restrictions are frequently necessary and desirable since it is certain that after graduation the young researcher will be confronted with similar restrictions of choice by his industrial or government employer. So, the freedom to select the problem for research is frequently limited by the exigencies of the researcher's position or affiliations. Personal interest often is compromised with the demands of actual circumstances. However, there are research workers who are aggressive, alert, imaginative, and quick to see fields which are fertile for investigation, and there are also those who are passive and unimaginative, content to work out problems suggested by others.

Guides for Making a Choice

The first and most important step in selecting a problem for research is a thorough understanding of the known facts and accepted ideas of the field of interest. The researcher who is familiar with his field and knows what research has already been completed in it will also know something of the many problems that remain. His acquaintance with the literature of his field often leads to long lists of areas or problems requiring additional research. Too, the more experienced researcher, from his own work, can readily observe which phenomena are not satisfactorily accounted for by existing knowledge and therefore need additional investigation.

The researcher or student who decides to investigate a problem because it interests him has a natural incentive because he wants to know the solution. Sometimes an academic advisor or a supervisor may make a suggestion which opens the researcher's eyes to new possibilities which had not previously occurred to him.

Not to be overlooked are the actual and pressing problems which arise out of the researcher's everyday experience, problems which are related to his everyday work in business and industry. When serious problems arise, solutions are often found as a result of research. Thus actual experience frequently suggests many topics that need investigation.

Here are some guides you will find helpful in defining and formulating your problem:

1. Can the problem be stated in question form? Stating your problem in the form of a question is an excellent way of making your problem clear and concise. Stated in question form, your problem obviously requires a specific answer. The quest for the answer then becomes the objective of the study.

2. Can the problem be delimited? By deciding in advance the boundaries and considerations of your problem, you will be saving yourself much useless work.

3. Are the resources of information available and the state of the art practical? The previous two points become immaterial if the present state of knowledge is inadequate or the means for research is unavailable. For example, certain research material in the field of aerodynamics may not be available to a researcher because of the classified nature of this matter. Security restrictions by our gov-

ernment may prevent a researcher from obtaining all necessary resources of information for his investigation.

4. Is the problem important or significant? This is not to imply that your problem must be earthshaking in its importance, but the question should be worthwhile for the expenditure of time, energy, and funds involved. No one in our day can afford to pursue the problem, as did medieval scholars, of how many angels can dance on the head of a pin.

The Hypothesis

The observation that leads the scientist to recognize the problem also suggests the tentative answer. This answer is called a hypothesis or a working guess. Hypothesis is derived from a Greek verb, *hypotithenai*, meaning "to place under." The hypothesis, then, is an explanation placed under the known facts of a problem to account for and explain them. Scientists test their explanation or hypothesis by experiment. If the hypothesis does not meet the test, it is revised in accordance with the new facts brought about by the testing. If the hypothesis is proved by sufficient testing, it becomes one of the accepted generalizations of science.[1] Let us see how this works.

A hypothesis, we said, was an explanation placed under the known facts of the problem to account for and explain them. Recognizing and stating the problem clearly move the researcher along the path where the answer may lie. Directions are pointed toward an answer by identifying clues and recognizing roadsigns of the problem. The hypothesis is tested by following the directions and picking up the clues. As the road of the hypothesis is traveled, facts about the problem are found. Some do and some do not corroborate the hypothesis. Constant examination along the way enables revision to correspond with the facts of the researcher's investigation.

The Researcher as a Detective

We are all familiar with the process just described from having watched TV mysteries. A man is found murdered in a hotel room; there has been nothing unusual to call attention to the murder—no one heard any noise

---

1 The terms *hypothesis, theory, law, generalization,* and *conclusion* are closely related in meaning. They relate to the solution which the investigation has disclosed for the problem studied. The term *hypothesis* is most frequently considered as a provisional working assumption. As a rule, *theory* is much broader than *hypothesis*. A *series* of hypotheses may be investigated, all of which lead to a theory. *Theory* is an umbrella term suggesting a basic condition common to a number of hypotheses. A particular theory about psychology of learning evolves from a series of hypotheses relating to a variety of ways of learning which must be validated. Because knowledge arrived at through the scientific method is subject to revision in the light of new data, a theory is frequently composed of a number of working assumptions. *Law* is a term applied to a *theory, generalization,* or *conclusion* purportedly proved as conclusively as the state of knowledge or art permits.

or saw any suspicious characters. The problem is easily recognized: Who killed this man? The detective on the scene examines the facts of the problem. The man has been shot through the heart. From the size of the wound, the murder weapon appears to have been a small caliber revolver. There are no powder burns on the man's clothes; so, he must have been shot from the distance of across the room. There is no sign of struggle; so, the murdered man apparently trusted or did not suspect the individual who shot him. There are cigarette butts in the ash tray—some of a brand found in the pocket of the murdered man's coat and some of another brand. The cigarette butt of the other brand has lipstick on it. Here is the first clue. At some time there must have been a woman in the room with the man who was murdered. She certainly could tell much about it if she were in the room at the time of the shooting. The detective moves into the kitchenette from the room where the body was found. Two empty cocktail glasses are on the serving table. One contains lipstick on the rim—the same color as the lipstick on the cigarette butt. This fact reinforces the beginnings of a hypothesis—that a woman who smokes a certain brand of cigarettes and who uses a certain brand and color of lipstick should be suspect. From the question—who killed this man?—certain evidence has been gathered to point toward a possible solution to the problem—a woman is involved! *Cherchez la femme!* If he finds the woman, he may find some further data which may lead to the solution.

This problem has been made familiar by innumerable television programs and movies. It serves to illustrate how hypotheses are derived from a recognition of the problem and a recognition of data which potentially point to the answer. The detective's provisional conclusion that a woman was involved in the murder was reached from the available clues—the data. Similarly, educated guesses lead a researcher down the pathway of his investigation toward an answer. Answers are not always simple. There are many complexities which throw the researcher off. For example, in the situation just described, the detective might find the woman in the case also murdered and stuck in a closet. All previous clues have been false starts, and he must begin all over. All of us have seen evidence of such complications involving Perry Mason and other crime solvers. Frequently, the most obvious answer is the one furthest from the truth. Researchers follow all clues until the answer to the problem is found.

The problem and its hypothesis are the starting points in report writing —whether the report answers the question, Who killed the man found in the hotel room? or problems like, How can clover mites be controlled? What causes cancer? What is the best design for a photoelectric amplifier for a given gain factor?

Role of Hypothesis Summarized

1. The investigator, after preliminary gathering of data or evidence and an analysis of it, uses inductive reasoning to see what the data, evidence,

and clues add up to. This adding up of evidence leads to a preliminary conclusion as to what the answer might be. The provisional answer or hypothesis is the first or trial hypothesis.

2. With this trial hypothesis in the background, the researcher next makes use of deductive reasoning to decide what kind of data, evidence, or facts he will need to test the trial hypothesis. In other words, he determines what should logically follow from the provisional conclusions he is testing.

3. By this analysis, the researcher then proceeds to apply his hypothesis by gathering all possible data and testing them to see whether the actual evidence accumulated and his hypothesis are in agreement.

4. If the evidence fails to support the hypothesis, the investigator either rejects it or modifies it to conform with the evidence he has. If he rejects it, he will analyze the facts and search further until he arrives at a second hypothesis which he again tests by comparing it with the new total evidence he has amassed.

5. In the type of search which calls for finding facts or information alone, there may sometimes be little use for a hypothesis. For example, if the search is historical in nature or if the search is a request for information—e.g., What industries use computers? How many computers are in use throughout the country?—then a hypothesis is rarely necessary. However, most legitimate research involves interpretation of facts. For example, the knowledge of the number of computers in use in the United States may raise the problem: Why isn't the computer of our company's manufacture in greater use? After the facts have been discovered, they must be analyzed to find out what they mean, what conclusions should be drawn, and what should be done about those facts. The conclusions and recommendations are usually the chief object of such research.

Summary of The Scientific Method

The scientific method has these features:

1. It concerns itself with problems to be solved; therefore, the scientific method is purposeful, with specific goals directing the research activity.
2. When the problem is clearly identified and stated, a working hypothesis or hypotheses concerning the explanation of the phenomenon or solution to the problem are formulated.
3. The establishment of a hypothesis or theory is followed by observation and/or experiment to test the hypothesis.
4. And finally, the data of the research testing of the hypothesis are recorded, analyzed, and interpreted. The conclusions are reported, published, and disseminated.

Relations of Research to Report Writing

Report writing is the reconstruction of a purposeful investigation of a problem in written form. Leonard Bloomfield[2] has succinctly observed this relationship by noting that scientific research begins with a set of sentences which point the way to certain observations and experiments, the results of which do not become fully scientific until they have been turned back into language, yielding again a set of sentences which then become the basis for further exploration into the unknown.

Discussions Problems and Exercises

1. The origin of the word *science* is the Latin term *scientia,* meaning "to know" or "to learn." Discuss the appropriateness of this terminology. Early civilizations possessed a science of a sort. Modern science had its origins in ancient Babylonia. Science as we know it today is represented by the difference between Babylonian *astrology* and modern *astronomy.* Through library research, prepare a paper comparing the ancient Babylonian science of astrology with today's science of astronomy.

2. List and explain examples of scientific impulses in children. Is the curiosity of a puppy or kitten scientific?

3. Read an account of a significant scientific investigation. To what extent did the procedure of the scientist correspond to that outlined in this chapter? The contributions of the following scientists would be appropriate to the study: Copernicus, Galileo, Kepler, Newton, Franklin, Lavoisier, Pasteur, Boyle, John Dalton, Faraday, Harvey, Becquerel, Mendel, Roentgen, Michelson, Pavlov, Salk, Watson and Crick.

4. What is the relationship between the scientific method and ordinary logical thinking?

5. Explain the significance of experimental verification and its relation to the scientific method.

6. Explain the process of induction. How does it differ from deduction?

7. In the last twenty-five years the sale of raw popcorn has increased 900 percent. The National Association of Popcorn Poppers initiated research to find the reason. They found that the sale of raw popcorn to commercial poppers, such as candy stores and movie theaters, dropped 350 percent during the period in question. They also found that sales of raw popcorn in retail outlets, such as drugstores and supermarkets, increased 1200 percent in the same period, accounting for nearly all of the overall increase. How do you account for the increased sale of raw popcorn? How would you go about testing your theory?

---

[2]See chapter 8, p. 197.

8. Explain the differences among hypothesis, theory, law, generalization, conclusion, results.

9. What is the role of the hypothesis in problem solving?

10. In what ways are scientific method and technical writing related?

11. Do you agree with Leonard Bloomfield's generalization about scientific research? Explain.

12. Choose a broad field or area of study for investigation. Narrow or localize a topic within that field. Pinpoint a problem within the delimited topic in the form of a question in accordance with the principles discussed in this chapter. Submit the interrogative statement of your problem in memo form to your instructor for approval. Explain why you have chosen the problem by reason of your interest, experience, and ability to follow through toward its solution by means of a report of the investigation.

References

1. Cohen, Morris R. and Ernest Nagel, *An Introduction to Logic and Scientific Method,* New York: Harcourt, Brace and Company, 1934.

2. Darwin, Frances, Editor, *The Life and Letters of Charles Darwin,* 2 Vols., New York: D. Appleton and Company, 1899, Vol. 1.

3. Dewey, John, *Logic: The Theory of Inquiry,* New York: Henry Holt and Company, 1938.

4. Huxley, Thomas H., *Method and Results,* New York: D. Appleton and Company, 1893.

5. Urban, Wilbur M., *Language and Reality,* New York: The Macmillan Company, 1951.

6. Whorf, Benjamin Lee, *Language, Thought and Reality, Selected Writings of Benjamin Whorf,* John B. Carroll, Editor, New York: John Wiley and Sons, Inc., and the M.I.T. Press, 1956.

3

Technical Correspondence

Correspondence is the basic communication instrument in business and industry. Much of the activity of science and technology is conducted through letters; many technical reports are written in letter form. It is not only the technical administrator or his subordinate but also the man in the laboratory who daily must use correspondence to accomplish his work. Even the basic science researcher in the relative isolation of his laboratory is called upon to write letters. He sends out inquiries and requests. He answers inquiries and requests. He orders equipment, sends acknowledgments, writes letters of instruction, and, on certain occasions, writes sales letters and letters of adjustment. When he wants to change jobs, he will write an application letter.

Because the technical man's successful activity depends very much on social interaction, it would be well to examine the mechanics of correspondence, the form constituting a major portion of such interaction. Moreover, many of the principles of letter writing apply to report writing. While letters are intended for a single reader and reports for a wide range of readers, letters, like reports or any piece of organized information, receive the classical structure of a beginning (which indicates the purpose), a middle (which elaborates on or develops the purpose), and an ending (which completes the purpose). Neither the report nor the business letter is written for the pleasure of the writer or reader. It is intended for some practical objective. The distinguishing dissimilarity between the report and the business letter is the definite intrusion of personal elements in the letter. Modern business letters are reader-centered. Their style is based on the premise that a business letter communicates an attitude as well as a message. Reports, on the other hand, though directed toward a specific audience are impersonal and objective. The letter tends to establish a personal relationship between the writer and the reader. The style of the report, on the other hand, usually demands subordination of any personal relationship between the writer and the reader. The stress in the report is on fact; in the letter it is on rapport.

The intention of this chapter is not to replace a text or handbook on business correspondence but to offer fundamentals, principles, and techniques of modern business correspondence which will be useful to the technical person. A list of some of the better texts on business correspondence will be found in the Bibliography.

The Reader-Centered Letter

The modern business writer is much concerned with his reader. The "You Psychology" plays an important role in letter composition. Put yourself in your reader's shoes is the maxim. When you compose a letter, therefore, remember you are writing to a specific reader who is a human being. The reader of your letter is interested only in how your message will benefit *him*. He will not buy your product or service merely because you want his business. He will not hire you merely because you want to work for him. He will not accept delay in the delivery of your product only because you are having procurement problems. He will not buy for cash just because you ask him to. If your letter is to appeal to him, it must be constructed in terms of benefit to him. You must convince him that, by hiring you, he will get financial returns. You must show him that the material you want to use in the product he ordered is worth his waiting. You must prove that his buying for cash is the best thing for him. Visualize your reader; tailor and personalize your letter specifically for him. Begin your letter with something which will be of interest to him. For example, do not say, "In order to help us simplify our problems in processing orders, we require customers to include our work order number in their correspondence to us." Would it not be better to say, "So that your order may be promptly serviced, please include our assigned work order number in your correspondence."

Write simply, write naturally. Make your letter sound as if you were talking directly to your reader. Business English is not pompous English; it is a clear and friendly language. Stilted, formal letters build a fence between the writer and the reader. Compare "Please be assured, kind sir, of our continued esteem and constant desire to be of service in any capacity" with "We are always glad to help in any way we can."

Be sure that the general appearance of your letter creates a favorable impression. We like certain people because their appearance impressed us favorably the first time we met them. The same principle holds for letters. If you receive a letter that is neatly and evenly typed, well-centered, correctly punctuated, and free from typing, spelling, and grammatical errors, your impression of the writer is likely to be a favorable one. Your letters can be one of the best public relations and advertising mediums because they reveal the quality of the service which can be expected from the writer.

Planning

No one can write a good letter without being exactly sure what he is after. Good letters are based on planning. Before starting to dictate or write, ask yourself, What do I want this letter to do? What action do I want the reader to take? What impression do I want to leave with him? If you have to answer a letter, read carefully the letter you have received; underline questions or statements to be answered; jot down comments in the margin. In composing a reply, it is often helpful to look through past correspondence. When you have gained all the background and facts, ask yourself, What is the most important fact to the reader? Usually the most important fact should be dealt with first; let the rest follow in logical sequence.

Paragraphing the Letter

Set a number of paragraphs—one for each main thought or fact—before you start. This will force you to order your thoughts and prevent confused ramblings. Make use of 1-2-3 or a-b-c lists wherever possible to streamline your message. This will help to clarify and emphasize.

Use the first paragraph to tell your reader what the letter is all about. Link it with any previous correspondence, but do not repeat the subject of the other person's letter as a preliminary to your reply. The result will be something trite and clumsy like, "In reply to your letter of April 25 requesting. . . ." Neither is the participial type of opening, like "Regarding your letter of April 25," any better. A simple "Thank you for your letter of April 25" or "Thank you for your quotation request of April 25" is very effective. Businessmen do not read letters for pleasure. They want letters to be brief and specifically to the point. Begin your letter directly. For example, "Production tells me we can now promise delivery of the 48 items of your order No. P-1465 on August 18, or a day or two sooner." Or "Here is some information about our fuel pumps, which we are very glad to send you."

Try to see the closing paragraph before you begin; keep moving toward it when you set down your opening sentences. Use the last paragraph to make it easier for your reader to take the action you want him to take, or use it to build a favorable attitude when no action on his part is needed. When your message is completed, stop. For example, "Please send us the completed forms by May 15. We'll do the rest." Or "If you have a special measurements problem, our engineers might be able to help you. Just call us, and we'll be glad to send someone to see you."

Keep sentences and paragraphs short. Business letter experts recommend a sentence averaging twenty words, but vary sentence length. Opening and closing paragraphs normally should be brief. The longest paragraphs usually come in the middle of the letter. Paragraphs keep thoughts together which belong together. They enable the reader to get your meaning as easily and as certainly as possible.

Read your letter carefully before you sign it. Once it goes out, *you*, the signer, are responsible for any errors or confusion.

Format of the Letter

While content is certainly more important than format or style, all of us recognize that style and format are to a letter what dress and appearance are to an individual. None of us would appear for a job or interview sloppily dressed or covered with mud. Similarly, when we send a letter, it speaks for us in a business situation. A single error may nullify an otherwise well-written letter. The receiver of the letter can evaluate the message only by the letter's total impression. If even a minor aspect suggests carelessness, slovenliness, or inaccuracy, the reader loses confidence in the more fundamental worth of the message. Therefore, an examination of format and style is important.

Most organizations and companies use one of three format styles in their letters—the semiblock, the block, and the simplified style. Examples follow.

(Semiblock)

THE FORD MOTOR COMPANY

THE AMERICAN ROAD
DEARBORN, MICHIGAN

January 12, 1974

Dr. Herman Weisman
7807 Hamilton Spring Road
Bethesda, Md. 20034

Dear Dr. Weisman:

Thank you for the opportunity to help you prepare your students for what the business world will require in the field of report and letter writ-

75-1243

ing. The importance of this field cannot be emphasized too much, for the ability to write good reports and letters is a basic requirement for business success.

All of the Ford Motor Company training courses on writing stress three things:

1. Be brief — Business men do not read letters and reports for pleasure, so make them short and to the point.
2. Be specific — Do not make the reader interpret your letter.
3. Be conversational — Stilted, formal letters build a barrier between the writer and the reader.

The type format preferred by our company is described in the enclosed copies of our standards on internal and external communications. Also enclosed are copies of letters and reports written by our employees.

If there is any additional information you might desire, please feel free to contact us.

Very truly yours,

FORD MOTOR COMPANY

W. J. Gough, Jr., Supervisor
Offices Services Section
Administrative Services Department
Finance Staff

WJG:rn
Enclosures

(The Block Style)

HILL AND KNOWLTON, INC.

Public Relations Counsel
150 East Forty-second Street
New York, N. Y. 10017

February 1, 1974

Professor Herman M. Weisman
7807 Hamilton Spring Road
Bethesda, Maryland 20034

Dear Professor Weisman:

I am sorry that it has taken so long to reply to your letter of November 27, requesting information about forms used in business correspondence.

Recently Hill and Knowlton, Inc. issued a *Secretarial Guide to Style* as a guide to the forms preferred when writing letters, memoranda, and reports. As can be noted from the samples given at the back of the Guide, indentation is preferred, but block form is also considered acceptable. Single space for letters and memoranda is also preferred, but length will sometimes dictate style.

I am sorry that we are not able to provide specimens of letters, memos, and reports, but feel that the material contained in the Guide will be helpful. As indicated in the introduction, the booklet is a supplement to the *Complete Secretary's Handbook* and the *Correspondence Handbook*, but does not conform in all instances to these aids.

I hope that we have been able to supply some of the answers you are seeking. We shall be happy to assist you if you wish additional information.

Sincerely yours,

Thelma T. Scrivens
Administrative Assistant
Education Department

TTS:lt
enclosure

Some organizations may use all three styles or combinations of them. In recent years, some Federal agencies have decreed the use of the simplified style to increase the productivity of their typists. A good many companies have style manuals for letters and memoranda issued by personnel of their companies in order to achieve uniformity and excellence. If the organization or company for which you work has such a manual, follow its recommendations.

(The Simplified Style)

HANSON PRINTING MACHINERY

2222 22 Street S. E.   Washington, D. C. 20020

10 July 1974

Dr. Herman M. Weisman
7807 Hamilton Spring Road
Bethesda, Maryland 20034

Dear Dr. Weisman:

Enclosed is the Handy Type Index and Price List you recently requested from American Type Founders.

We are area distributors for American Type Founders and stock practically all of the currently popular faces in the Washington, D. C. area. You may be assured that prompt attention will be given your order when it is received.

We will be pleased to furnish any further information concerning printing equipment or supplies that you may need. Please give us the opportunity of serving you.

Yours very truly,

W. Wayne Gilbert

WWG:mje
Enclosure

The semiblock differs from the block format only in that the paragraphs in the body of the letter are indented. The block and semiblock formats are used in about 80 to 90 percent of all typed business letters. The simplified letter is very efficient in that it eliminates the need of indenting and tabulating by the stenographer, but many readers are disturbed by its unbalanced appearance. Management consultant firms have a tendency to use this form because it gives the appearance of efficiency. Many advertising firms also use it because of its breezy appearance. The simplified letter form will, on occasion, omit the salutation and use a subject line instead. It frequently omits the complimentary close and omits the dictator's and typist's initials.

Mechanical Details

Stationery

Stationery should be of a good quality of unruled bond paper, preferably white, 8-½ x 11 inches. United States government agencies use 8 x 10 stationery. Company or organization preprinted letterhead should be used when appropriate. For continuation pages, white bond of the same size and quality as the letterhead should be used. Some companies also use shorter letterhead—8-½ x 7 inches. This size is used for for letters of one or two very short paragraphs. Tissue sheets are used for carbon copies.

Framing a Letter on the Page

A typed letter should be so placed on the page that the white margins serve as the frame around it. Practice and convention call for more white space at the bottom than at the top unless the length of the letter and the size of the letterhead make this impossible. The side margins should be approximately equal and usually should be no wider than the bottom margin. Side margins are usually from one to two inches, depending on the length of the letter. The bottom margin should never be less than one inch below the last typed word. The right-hand margin should be as even as possible. The body of the letter proper should be placed partly above and partly below the center of the page. A short letter makes a better appearance if it is more than half above the center of the page. The placement of the letter on the page, frequently referred to as centering, can be achieved by starting the first line of the inside address at a chosen depth to give the most pleasing appearance of white space above and below the letter. The shorter the message, the lower the inside address is placed.

Heading

When preprinted letterhead paper is not used, the writer includes a heading to help identify the source of the letter. The heading includes the address, but not the name, of the writer. The street address is on one line and the city, postal zone, and state on another. The date follows the city and state line. The heading is placed at the right side of the page in block form. See examples below. The heading helps to frame the letter on the page.

    1303 Springfield Drive   87 Cherry Creek Lane
    Fort Collins, Colorado 80521   Minneapolis, Minnesota 55421
    November 8, 1974   March 29, 1974

                605 West 112th Street
                New York, N. Y. 10026
                December 21, 1974

One and one-half inches from the top of the paper is the usual top margin. If the letter is short, the top margin will, of course, be longer. The placement of the heading should be planned so that the last line ends at the right-hand margin of the letter.

The Date Line

The purpose of the date line is to record the date the letter is written. When preprinted letterhead stationery is used, it is the first item typed on the page. If ordinary bond paper is used, in practice, the date line has become the last element of the heading. (Note the examples above.) The date line can be centered or placed on the right or the left depending upon the style of layout. Sometimes the design of the letterhead will suggest one position or the other. The date line is placed so that it ends flush with the right-hand margin of the letter. Its usual placement is about four spaces below the last line of the letterhead. If the letter is very short, the date line should be dropped to give a better balance to the page. The month should be spelled out in full and no period is used after the year.

Inside Address

The purpose of the inside address is to identify the receiver of the letter by giving the complete name and address of the person or organization to whom the letter is being sent. It is usually placed four to six lines below the date line. No punctuation is used at the end of any line in the address. No abbreviations should be used in the address. Write out street or avenue, as well as the name of the city and the state. Use a courtesy title, such as *Mr., Mrs., Miss,* or *Ms.* when addressing an individual. When a title follows the name of an addressee, it is written on the same line, except where it might be unusually long; then the title is placed on the next line:

    Mr. Thomas B. Morse
    1620 Dakota Avenue
    Cincinnati, Ohio 45229

    Mr. LeMoyne Patterson, President
    General Speed Machine Co., Inc.
    43 Madison Avenue
    Milwaukee, Wisconsin 53204

Dr. Sylvester Tarkington
Assistant Director, Engineering
True Ohm Resistor Corporation
Suite 2400
Pennsylvania Building
Philadelphia, Pennsylvania 19102

Attention Line and Subject Line

The attention line is intended to direct the letter to the person or department especially concerned. An attention line provides a less personal way of addressing an individual than placing his name at the head of the inside address. It is losing favor, however, since modern practice is to address letters to individuals. If a subject line is to be used, it should be typed below or in the place of the attention line.

The J. B. McConnell Company
1812 Atlantic Avenue
Brooklyn, New York 11233
Attention: Mr. Joseph Rich, Comptroller
Subject: Payment for Crystal Diodes

Gentlemen:

The attention line is placed two spaces below the last line of the inside address; it should not be in capitals, nor is it usually indented. Practice varies on how it appears. Here is an example:

General Motors Corporation
General Motors Technical Center
P. O. Box 117, North Penn Station
Detroit, Michigan 41202
Attention: Mr. Raymond O. Darling
           Educational Relations Section

No end punctuation marks are used. The purpose of the subject line is immediate communication of the topic of the letter. Therefore, it should be placed conspicuously. Practice varies on exact placement or introduction of the subject line. Some company correspondence manuals use the word *Subject*, or *Reference*, or the abbreviation for reference, RE. Others omit these. The prevailing use of the subject line in Government and Defense Department correspondence has stimulated its wide use. Also, the National Office Management Association encourages its use. The subject line may be placed on the same line with the salutation or two line spaces below the last line of the inside address or so placed that it ends at the right-hand margin of the letter. Here are some examples:

Avco Manufacturing Corporation
420 Lexington Avenue
New York, New York 10017
Subject: Our Purchase Order No. T1052

Avco Manufacturing Corporation
420 Lexington Avenue
New York, New York 10017
Gentlemen: <u>Our Purchase Order No. T1052</u>

Avco Manufacturing Corporation
420 Lexington Avenue
New York, New York 10017
    <u>Our Purchase Order No. T1052</u>
Gentlemen:

The first letters of important words in the subject line should be capitalized.
Underscoring may or may not be used depending upon personal preference.

Salutation

The salutation originated as a form of greeting. The simplified letter form
omits it entirely, as old-fashioned and superfluous. It is placed two line
spaces below the inside address or six spaces if an attention or a subject
line is used. Use a colon after the salutation. Conventionally, the word *Dear*
precedes the person addressed unless a company or group is the recipient
of the letter:

Dear Mr. Smith: (Preferred) *or*
Dear Sir:

Dear Dr. Jones: (Preferred) *or*
Dear Sir:

Dear Mrs. Smith: (Preferred) *or*
Dear Madam:

Dear Ms. Smith: (but never Dear Madam when Ms. is used.)

Gentlemen: is the correct form in addressing a company or a group of men,
and *Ladies:* is correct when addressing two or more women. Abbreviations
should not be used in the salutations other than *Mr., Mrs., Miss, Ms.,*
Messrs., or *Dr.* Titles like *Professor, Lieutenant, Governor, Senator,* and
Secretary should be written out. The first letter of titles in the salutation,
including *Sir*, should be capitalized.

Body of the Letter

Begin the body of the letter two line spaces below the salutation. Whether
the body is indented, of course, depends upon the style used. Single-space
letters of average length or longer. Short letters of five lines or less may be
double-spaced. Always double-space between paragraphs in single-spaced
letters. Quoted matter of three or more lines is indented at least five spaces
from both margins. Very short letters which are double-spaced should use
the semiblock form to make paragraphs stand out as separate units. Double-
spaced material does not receive extra space between paragraphs.

Complimentary Close

By convention, the purpose of the complimentary close is to express fare-well at the end of the letter. In a simplified style, the complimentary close is frequently omitted, but conventionally, the complimentary close should be typed two lines below the last line of the text. It should start slightly to the right of the center of the page, but it should never extend beyond the right margin of the letter. The comma is used at its end. Only the first word of the complimentary close is capitalized.

Yours truly,
Very truly yours,
Sincerely,
Sincerely yours,

are the proper and conventional complimentary closes most frequently used. *Yours cordially* implies a special friendship; *Respectfully yours* implies that the person addressed is the writer's social or business superior.

Signature

When the sender's name and title are printed in the letterhead, they are frequently omitted after the signature. Otherwise, the name and the title are typed three to five spaces directly below the complimentary close. First letters of each word are capitalized; no end punctuation is used. The letter is not official or complete until it is signed. Only after the dictator or writer has signed the letter does he become responsible for its contents. In routine letters, a secretary may, at times, sign the name of the official sender. In such a case, the secretary or stenographer signing the sender's name adds her initials. Sometimes her initials are preceded by the word *per*.

Your truly,   Sincerely yours,
/S/ Richard C. Smith   /S/ Charlton Billings, per M. B.
Richard C. Smith   Charlton Billings
Manager   President

Respectfully,   Cordially,
/S/ Tom Swift   /S/ Marvin Troupe
Tom Swift   Marvin Troupe, Ph. D.
Assistant Engineer   Chairman, Sociology

If a company instead of the writer is to be legally responsible for the letter, the company name should appear above the signature. The use of company letterhead does not absolve the writer; so, if you want to protect yourself against legal involvement, type the company name in capitals a double space below the complimentary close; then leave space for your signature before your typed name:

Yours truly,   Sincerely yours,
STERLING PRODUCTS   AERO SPACE, INC.

James Joyce   Ted Blum
Design Engineer   Technical Editor

Reference Information

In order to identify the dictator or official sender and the typist, the dictator's initials followed by the stenographer's are typed two line spaces below
the signature, flush with the left-hand margin. The dictator's or sender's
initials are conventionally typed in caps, with a colon or a slant (/) and
the typist's initials following; for example: HMW: mhb or HMW/mhb.

Enclosures

If there are enclosures, type *Encl.* one line below the reference initials. If
there is more than one enclosure, type *2 Encls.* or *3 Encls.*, as the case may
be. If the enclosures are significant, list their identifications:

    2 Encls.   (1 Contract No. N2021)
                (2 Proposals No. 42C)

Carbon Copy Notations

When a letter requires copies for other than the addressee, the designation
cc: should appear in the lower left-hand corner of the carbon copies. Under
certain conditions, the dictator may wish to have the addressee informed
of the distribution of copies, in which case the notation should also appear
on the original. In other instances, the sender may not wish the addressee
to know that a carbon copy was sent to a certain individual; then the letters
bc (standing for "blind copy") are penned or typed under the *cc* notation.
The carbon copy notation appears two lines below the enclosure data.

Postscript

The postscript, years ago frowned upon, is now being used more and more
in business correspondence. It is used to add extra emphasis to some particular item or to give additional information. It is placed two line spaces
below the referenced information, flush with the left-hand margin. The
initials *P.S.* are sometimes used but are lately being omitted because they
are considered unnecessary.

Second and Succeeding Pages

When a letter carries over to succeeding pages, use plain white bond paper,
the same quality as the first page. Some firms have preprinted second-
page stationery. Always start at least one inch from the top of the page.
The name of the addressee should appear at the left margin. The page
number, preceded and followed by a hyphen, should be centered and the
date should be at the right—forming the right-hand margin. Sometimes

these elements are single-spaced, one below the other, on the left margin. The second page should carry at least three lines, although five lines are preferable. Never divide a word at the end of a page.

New Hampshire Electronics   -2-   July 6, 1974

or

New Hampshire Electronics
Durham, New Hampshire 03824
Page 2
July 6, 1968

The Envelope

The first thing the recipient notices about your letter is its envelope. The envelope deserves as much care as the letter it contains. Assure its accuracy by checking it with the address printed on the letterhead of the company to which you are writing. The elements of neatness and attractiveness are just as important in typing the envelope as in typing the letter. There are two standard sizes of envelopes in use today—the 9-½ x 4-⅛ and the 6-½ x 3-⅝. The larger size is in greater favor because of the greater ease in folding the letter sheet.

Addresses on envelopes should be written in block style, double-spaced if the address is in three lines and single-spaced if the address contains more than three lines. Postal authorities prefer that the state appear on a separate line and that the zip code be used. Foreign countries are typed in capitals. Use no abbreviations or punctuation for the end of the line. The person's name, title, and the name of the branch or department should precede the street address, city, and state. *Special delivery* or *registered* should be typed in capitals several spaces above the address. If an attention line is used in the letter, it should be typed on the envelope in the lower left-hand corner. If a *Personal* notation is desired on the envelope, it should appear in the same position as the attention line.

Mr. Frank J. Hill
Senior Vice President
Hill and Langer Company
93 Mill Pond Road
Dobbs Ferry, New York 10522

Personal

The Sawyer Corporation
62 Broadway
New York, New York 10004
Attention: Personnel Department

The Memorandum

The memorandum or memo is a form of communication that is receiving increased use in business and industry (its plural is either memorandums or memoranda). Until recent times its use has been restricted to inter-office, interdepartmental, or interorganization communications. However, it is now being circulated with greater frequency out of the originating organization. Its format is similar to that of the letter, but its tone is impersonal. Most organizations have printed memorandum forms for interoffice or interdepartmental communications. Its format has been highly conventionalized, although details will vary from one organization to another.

The purpose of a memo is to circulate information, to request others to take care of certain work, to keep members of an organization posted on new policies, to report on an activity or situation, etc. A memorandum should consider one subject only. The subject is stated in the subject line. The primary purpose of the memo is to save both reader and writer time. Amenities and courtesy are sacrificed for conciseness. Although the body of the memo is similar to the body of the letter, its other elements have a specialized format. There may be a preprinted heading identifying the company and department originating the memo. Instead of the inside address, the following three lines are used:

TO:

FROM:

SUBJECT:

A date line is placed, usually, in the upper right-hand corner. A project or file number may also be included. In place of the complimentary close, there may be a signature of the writer or the writer may either sign or put his initials following his name in the "From" line. The memorandum has been borrowed from the practice of military correspondence. If the memorandum is a long communication, it may be organized into a number of sections and subsections. Memorandums are frequently the most convenient mechanism to convey reporting information.

(Sample)

GENERAL DYNAMICS CORPORATION
INTRA-CORPORATION COMMUNICATION

TO: Mr. F. R. Crane   NEW YORK
FROM: John Doe   DATE: December 18, 1974
SUBJECT: Executive Orders Manual

Here is your copy of the General Dynamics Executive Orders Manual containing the formal instructions, policy statements, announcements, and other advices of the President.

The Orders in the Executive Orders Manual have superseded all previous Executive Orders and include all Orders currently in effect. The Manual will be kept current by the Office of the Vice President — Organization.

Your receiving a personal copy of the Executive Orders Manual carries with it responsibility for complying with the instructions and policies in the Manual and for insuring compliance by all personnel under your direction.

Enclosure   J. D.

OFFICE MEMORANDUM   DATE: November 27, 1974

To: PUBLIC RELATIONS STAFF

From: J. K. ABBOTT

We have arranged through Glen Cross to hear tapes of the first three talks in the special program on "Orientation for some of the Company's non-insurance personnel." These tapes will be run on next Monday, Tuesday, and Wednesday, December 2, 3, and 4 at 4 P.M. in our conference room. Anyone interested is invited to listen in.

Subjects of the first 3 talks are as follows:

1. Monday — "Organization and History of the Company" — Mr. Meares
2. Tuesday — "The Ordinary Insurance Operations Organization and Actuaries Place in the Company" — Mr. Phillips
3. Wednesday—"Our Product, Ordinary Insurance, and Annuity"— Mr. Ryan

Special Types of Correspondence

Employment Letters

For a period of time, I was associated with an electronics company. One of my duties was personnel administration and recruitment of technical and professional personnel. During that period, I had the opportunity to read letters of application from several hundred people seeking positions with our company. I also interviewed a substantial number of applicants. I suppose within this group was a good cross-section of young people looking for their first job, as well as those wishing to change from one position to another. The application letters fell into four general categories.

The first included those so poorly written and/or so carelessly and sloppily done that I sent them into the wastebasket at once on the assumption that the person was so careless that he could not possibly be a first-rate, professional employee.

The second group contained those letters which did not give enough information to enable me to form any judgment of the writer's prospects. Because our company was public relations-conscious, these were acknowledged in polite terms but not filed.

The third included unsolicited applications written by people with fairly good backgrounds, training, and experience, but because no vacancy existed or seemed likely to occur for their particular backgrounds and capabilities, there was no reason at present for our further correspondence. However, these letters were acknowledged individually; the reply would indicate that we were impressed by the applicant's background and that we would keep his inquiry on file. If we had information about suitable vacancies at other companies, this was mentioned.

Finally, a few of the letters presented a clear picture of an individual who seemed capable, personable, well-trained and experienced for the opening, and having the personal qualifications most organizations want in their staff. Such persons were invited for a personal interview. When the interview confirmed first impressions, the individual was further interviewed by department heads in whose section or division his talents and experience would be appropriate. If the interviews further confirmed earlier impressions, a job offer ensued.

The situation which I have just described is the typical one in the employment market. A problem many people have in applying for an opening is that they do not have a clear conception of the conditions their letters meet. Many job applicants, no matter how capable or experienced, are so obsessed by their own personal situation that their approach in applying for the job is unfortunate. In present-day business and industry, the job application letter plays a prominent role in placing individuals in appropriate openings. It might be well to picture the setting into which an application letter arrives. For purposes of illustration let us imagine that you are an employer. Let us say you have an opening for a very responsible position which pays a good salary. You are anxious to find a qualified person to fill this vacancy. You place an ad in the Sunday edition of the city newspaper because you want it to receive a wide reading. Since you are very busy, you make it a blind ad—that is, you ask respondents to send replies to a box number. In this way, by not identifying yourself and your company, you need not reply to applicants who are not qualified.

Response to the ad brought you more than one hundred envelopes bearing applications. Although you are delighted to get so many applications, you are also overwhelmed by the problem of how to select the right person for the position. As you look at the envelopes, you notice one that is lavender and perfumed! "Well, I can get rid of that one easily," you tell yourself, and you throw it promptly into the wastebasket. You notice another envelope which has misspelled your street address. Also, the envelope bears the remains of the writer's breakfast. "Well," you tell yourself cheerfully, "I wouldn't want anyone as careless as that filling this position," and you throw that one also unopened into the wastebasket. As you leaf through the pile, you find several that you can dispose of quickly, but about one hundred applications still remain. So you roll up your sleeves and start digging into the mail. You open the first letter; the page seems neat and well done. You begin to read: "Regarding your ad in last Sunday's paper, I'd like to be considered for the position." As you read the letters, each seems to be like the previous one. Most begin with a participial phrase like "Replying to your ad," or "Having seen your ad," or "This is in reply to your ad."

After about the twelfth letter, you find one which has an original beginning. You read that letter with eagerness. The first few sentences are quite interesting and appropriate. They lead immediately into qualifications which are appropriate. Moreover, the writer is able to describe himself, his experience, and background in such a way that he appears to be an individual apart from the others. This man has character; he has ability; he has background; and he has experience. "I want to see this man," you tell yourself, and you put his letter carefully aside. You return to the rest of the letters, and for most of them, you go no further than the first few sentences because they sound so alike. You are looking for persons like the one whose letter picked up your interest and made himself come alive. After almost two hours, you have four piles of letters: those to be thrown away; those to be acknowledged politely; those to be acknowledged with more care and then filed for later use; and those—about seven or eight—from persons whom you want to interview.

From this hypothetical but typical situation, we can see that the reception your application letter will receive depends on:

1. Its initial impression, its appearance. If the letter is neat and well-framed on the page, it colors favorably the mood in which it will be read.

2. The first few sentences. If they are interesting (not bizarre) and lead the reader to the applicant's qualifications, the prospective employer will continue to read. The employer is interested in the qualifications of the writer. The employer has a good idea of what he is looking for. The writer should emphasize at once his qualifications which he believes to be the most important to the opening. (An applicant's qualifications include, of course, experience, education, and personal matters.)

3. Establishing one's individuality. An employer becomes dizzy reading several scores of applications. One applicant, even though qualified, begins to sound like another. Hence the writer needs to set himself apart from other applicants. He needs to make himself remembered. This, of course, is not easy, but we shall soon see how it can be done.

Analysis

Many persons, even the very experienced and capable, approach an application for a position from the point of view of their own eagerness for it. They forget that a prospective employer has his own viewpoint and needs; so, unfortunately, the letters they write express only their eagerness for this position, how much they would like the position, and how they feel the position would be right for them. Despite any great show of sincerity on the applicant's part, the employer cannot get interested. He is looking for a person who knows the employer's requirements and wants and who can show how the applicant's experience, training, and personal qualifications can be of benefit.

The first step—one that begins before the actual letter is written—is to analyze the position which is available. This analysis must be based on the information describing the opening. This, of course, is found in ads or announcements revealing the availability of the opening. A good device is to itemize the position's qualifications listed in the announcement. Then, in a column alongside, list an inventory of your own qualifications, your background, your training, your experience, and your personal characteristics which would be appropriate for the opening. This sort of inventory helps you understand, through comparative analysis, the type of qualifications and their importance to the employer. This also allows you to look objectively at your own qualifications in the light of what the prospective employer is looking for. If you do not have what the employer wants, there is no point in applying. Save yourself the time, energy, and emotional investment. It is true, sometimes, that your qualifications may not be exactly what the employer is looking for, but you may have some of them, plus additional ones which would be attractive to the prospective employer. There is no formula to gauge just how many qualifications you need to make it appropriate for you to apply. This depends on your analysis and knowledge of the requirements of the type of position being advertised.

If the ad reads:

> Process engineer wanted
> with five years' experience
> in aircraft sheet metal, draw
> die work. Must know time
> estimation.

there is no point in applying if you have just been graduated from engineering school and lack experience in the aircraft industry and lack knowledge of estimating.

Yet, it is also true that employers are looking for certain personal qualities which no amount of schooling and actual work experience can provide. There have been many surveys to identify the reasons why employees lose their jobs. Incompetency is very low on the list. Very high on the list are personal factors—matters such as absenteeism, belligerency, hypochondria, alcoholism, drug addiction, inability to work with others, and so forth. Personnel managers, therefore, scrutinize applications very carefully for clues on prospective employees' instability, health and emotional problems, and personality difficulties. Employers are always on the lookout for prospective employees who will be responsible, who have stability, initiative, and characteristics of potential growth. Employers want not only individuals who have the personal knack of getting along well with others but also those who may have leadership and managerial abilities. Demonstration of such factors is eagerly searched for in application letters.

In your inventory, you should note rather carefully those qualities which would be attractive to employers. Granted that all of us have a sincere belief in our own abilities and that we are popular and capable of working productively with others, we must be aware that mere assertion is not enough. What is required is evidence in the form of proven accomplishments. If you have been elected to an office in your fraternity or a campus organization, you have demonstrated some leadership abilities. Recall any experience in extracurricular matters that offered challenges. Indicate instances of productive accomplishments.

The Form of the Application Letter

In modern practice, application letters are in two part. Part One consists of a one-page letter of about four paragraphs featuring the most significant qualifications pertinent to the position available. Part Two consists of a résumé, as it is sometimes called, or a personal record, or data sheet, which gives a detailed account of the applicant's background, education, experience, personal qualities, and references.

The Letter. The application letter might be considered a sales letter; its content is structured like a sales letter to attract attention and to create interest on the part of an employer. It speaks with conviction about its "merchandise," and it should stimulate the recipient to the desired action.

To attract attention, the application letter should have an opening which will intrigue the reader; to create desire, the letter should describe the major qualifications of the applicant in such a way that the reader will be attracted to him. The description, however, is not enough. The tone of the language must be sincere and there must be proof concerning the qualifications in order to clinch the reader's conviction that the applicant is qualified.

The final element—stimulation of action—is a request for an interview. Few people are hired on the basis of a letter of application alone. The purpose of an application letter is to create an interest in an applicant and enough conviction of his desirable qualifications that the reader is stimulated to ask the writer to appear for an interview.

The Application Letter's Beginning. We have seen that the very first sentence or two of an application letter is critical. Good opening sentences are unhackneyed, simple, and direct. There are a number of ways to begin a letter of application to interest the reader.

The Novel Opening. We have seen in our earlier hypothetical situation that a majority of the letters begin with the commonplace, "In reply to your Sunday's ad," or "Having read your ad in the morning *Post*," or "I have learned that you are looking for a man . . . in the ad in yesterday's paper," or "Regarding the ad you had in the recent issue of. . . ." Prospective employers, therefore, are attracted by a more novel but appropriate opening of a letter. Here are some examples of effective and novel beginnings:

> I've peeled logs, sweated behind a skidder and cussed a loader; but a forest means more to me now than just hard work and a chance to make a dollar. Now a forest means good logging practices, market procedure, and forest management.

> I am a young engineer eager to secure a position where I can earn my keep and, at the same time, learn how to be a better chemical engineer. I understand from Mr. Ben Holloway, of your Synthetics Department, that you are expecting to add a Junior Engineer to your staff. May I have a moment to acquaint you with my background and qualifications for this position?

> Although there is a striking difference between the lofty crag-bound mountains of Colorado and the low fog-bound mainland islands of Alaska, I sense uncommon satisfaction from the prospect of a possible job assignment in your country. I have used the ordinary methods of getting to jobs in timber cruising — foot, horseback and automobile — but never have I traveled by boat. The prospect of lightering from lodging to cruise and cruise to cruise intrigues me. And, believe me, I don't get seasick! Leastways, I never have.

In your effort to be original, be careful that you do not become bizarre. No one can deny that the following two openings are novel, but they defeat their intended purpose by their oddness:

> Stop! Go no further until you have read my letter and résumé from beginning to end. I feel sure you will find me to be the best qualified metallurgist for your position in the quality control department.

> Symbiosis is the biological term used to indicate cooperation between two or more parties toward a mutual goal. Your company has an outstanding training program for young executives. I feel I am qualified to meet your expectations.

Name Beginnings. If the applicant has learned of the availability of a position through a person whom the prospective employer knows, or whose name or title will command the respect of the prospective employer, then beginning the letter with the name of that individual can be very effective. Here are some examples of name beginnings:

> Professor J. K. Wagner, Department Head, Forest Recreation and Wildlife Conservation at Colorado State University, called me into his office this morning to see if I was interested in the permanent ranger position described in the notice you sent him.

> The Placement Office of our university has informed me that my qualifications might be of interest to you for the Management Trainee Position in your Overseas Service Department.

Dr. James C. Donovan, Vice President of Engineering Research in your Denver office, suggested this afternoon that I write you concerning an opening in your Chicago office. He said, "Ted, because I know you have an unusual five-year educational combination of engineering and management, I think you have just the right background, and, of course, the ability to make yourself quite useful to the Industrial Engineering Department in our Chicago office. Would you be interested in dropping them a note at once?" I was happy to hear of this opportunity and am acquainting you with my interest and background.

Summary Beginning. Another very effective way to begin is to present immediately a summary of the most significant qualifications for the opening. Here are some examples:

Since graduating from Wharton School of Business three years ago, I have served as Assistant Budget Officer in Keely and Company, with the principal duties of preparing budget requirements, undertaking statistical studies, making comparisons of methods, costs and results with those of other refining companies.

My four years' work in the field servicing of electronic testing equipment, as well as a degree in Electrical Engineering, make me confident that I can qualify for the position you have open in your Quality Control Department.

Question Beginning. Beginning a letter with a question can be very effective. Psychologically, when anyone is asked a question, he pauses to listen. This method may be quite effective when one is applying for an unadvertised position. Here are some examples:

Do your plans for the future include graduating Electrical Engineers with practical experience who are free of military obligations? If so, please consider my qualifications.

Wouldn't that Mechanical Engineer you advertised for in the Sunday *Post* be more valuable to your department if he had as much as three and a half years' experience as maintenance technician in the Navy?

Describing Your Qualifications

The letter's beginning has the purpose of capturing the reader's attention. The applicant must now create an interest in his qualifications within the reader. The applicant has four major qualifications: education, experience, personal qualities, and personal history. Just which of these four elements are emphasized in your letter will depend, of course, on your analysis of the requirements of the position. The principle is to write first and most about that qualification which is deemed most important to the position available. If the ad reads: "Wanted — experienced design engineer," obviously the employer will want to know first and most about your experience as a design engineer. While he will be interested to know that you were graduated *magna cum laude*, that fact may not be as important to him as four years' experience designing certain kinds of instrumentation. If you have experience which is appropriate to the requirements of the opening, that is the most important qualification. Therefore, discuss it first and most.

If your strongest point has been your education, then that is what you need to start with. If your personal qualities are your strongest point, then show that you can develop or are capable of gaining the kind of experience necessary to do the job and that you are capable of growing into more responsible positions within the company. After you have presented your strongest point, take up other qualifications you may have that are important to the position. Details should be specific. The applicant should give the name of the firm, the kind of position he held, and the duties connected with the position. If the applicant accomplished anything outstanding in the job he should tell about it.

General and ineffective: "For three years, I served as a technician."

Specific: "For three years, I served in the model shop of the research and development department of the Radio Corporation of America in Camden, N. J. My work consisted of making breadboards, fabricating prototype models from rough engineering sketches of development engineers. This required not only my expert use of a soldering iron, but also machine shop tools and equipment."

In discussing education, focus attention on those subjects directly or closely related to the position being applied for. For example, a student applying for a job with an agricultural machinery manufacturing company might indicate his educational qualifications as follows:

My major area of study was in the field of machine design, including such courses as kinematics of machines, dynamics of machinery, machine design, machine analysis, agricultural power, and agricultural machinery. Some other pertinent courses include statics, dynamics, strength of materials, fluid mechanics, machine drawing, physics, chemistry, and mathematics and descriptive geometry. In my Special Problems Course in Agricultural Machinery, I designed and built a corn conveyor feeding machine presently being utilized on our Agriculture Hog Farm.

By its substantiation of qualifications the central portion of the application letter creates an interest in, and a desire for, the applicant.

The final paragraph of the application letter has the duty of persuading the reader to grant the applicant an interview. Mention should also be made, either in the last portion of the letter or in the middle portion, that a detailed résumé or personal data sheet with complete information on the applicant is enclosed. Appropriate ways of closing the letter are as follows:

May I have an interview with you at your convenience? You can reach me on the telephone at AC 303-482-2620 between the hours of 6:00 and 9:00 P.M.

May I have fifteen minutes of your time to substantiate any statement in my letter and answer any further questions? You can call me at AC 301- 921-2228.

I shall be in Chicago from December 22 to January 2. Would it be convenient to talk to you about the possibility of employment?

The Personal Data Sheet or Résumé

The personal data sheet or résumé is an inventory of the applicant's background and educational experience, personal qualities, history, references, and other pertinent information. The purpose of the letter of application is to interest the prospective employer in studying your data sheet in detail to see how appropriate your qualifications are for the position available.

The data sheet organizes and outlines elements of information and data of interest to the employer. Effective data sheets are always attractively arranged for quick and easy readability. Centered at the top of the sheet are the name and address of the applicant. In those professions or pursuits in which appearance is important to the accomplishment of the job, an application size photo is attached to the personal data sheet. Jobs in teaching, sales, theatrical pursuits, public relations, and advertising conventionally require photographs. In most other professional employment and activities, a photo is not required. Information included in the résumé is usually arranged under the headings: (1) Personal data,
(2) Education,
(3) Experience, and
(4) References.
The applicant who has been out of college a number of years and whose experience has been fairly extensive, places the experience section before that on education. It is customary to list experience in a reverse order because prospective employers are interested chiefly in what an applicant has done most recently. Under "Experience," data should tell not merely the title of the job but should also specify as exactly as possible the specific duties which were involved. Accomplishments should be included. Recent graduates should include summer and part-time jobs. If an applicant has worked his way partially or fully through college, he should so indicate. Similarly, under "Education," list the latest degree first. Follow through in reverse chronological order education through high school. List college courses important to the opening. Indicate all accomplishments and extracurricular activities.

Personal references are important only if an applicant has no work experience. Be sure you have permission to list your references. No more than three or four are necessary.

The résumé or data sheet always bears a date to assure the prospective employer that the contained data are current.

Included below for study are several letters of application and their data sheets:

631 So. Grant Avenue
Fort Collins, Colorado 80521
January 26, 1975

Pabco Building Materials Division
Fibreboard Paper Products Corporation
Florence, Colorado 81226

Gentlemen:

Do your plans for the future include graduating electrical engineers with a varied background of experience? If so, please consider my qualifications.

My curiosity in the "magic" world of electronics prompted me to enlist in the navy where the opportunity to become an electronics technician was available. Not only was I granted this opportunity but also the chance to pass the knowledge I acquired on to others as an instructor and to operate and instruct airborne equipment. Too, I was able to make first class petty officer during my four year enlistment.

The experience in the navy was but a mere acquaintance with electronics and an invitation to learn more. With this invitation and the assist of Uncle Sam in the form of the GI Bill, I enrolled in the College of Engineering at Colorado State University where I took such courses as Electronic Circuits, Transmission Lines, AC and DC Machinery, Power Distribution, and Network Analysis.

To meet the financial requirements in acquiring my degree it was necessary to supplement the GI Bill benefits with part-time and summer employment. Besides financial assistance these jobs offered acquaintance with and some practical experience in the business, agriculture, and petroleum fields. All of these jobs helped to increase my sense of responsibility, confidence, reliability, appreciation toward fellow workers, and initiative. I was also able to meet with and mix with people of various characters.

One of the experiences I feel most challenging, enjoyable, and beneficial at Colorado State University was that of president of the Newman Club — a student religious, educational, and social organization. No classroom offered the leadership, organization, and confidence training this office did.

The knowledge, training, and experience I've acquired have been mainly to my own satisfaction to date. I now want to put it to use where it would most benefit others. Therefore, should I meet your qualifications, may I have an interview? I may be contacted at the address and phone number listed in the résumé until further notice.

Yours truly,

Victor P. Baniak

Enclosure

RESUME
Victor P. Baniak
631 So. Grant
Fort Collins, Colorado 80521
Phone: 484-0935

Personal Data ─7

Age:	27
Height:	5-7
Weight:	150
Health:	Excellent
Marital status:	Single
Special interests:	Music, drawing and lettering, photography, skiing, travel.
Military Record:	4 years in U.S. Navy as aviation electronics technician; honorable discharge with AT1 rating; recommended for re-enlistment.

Education

BSEE, Colorado State University, June 1975.
>Important courses: Electronic Circuits, Feedback Systems, Electronic Analog Computers, Network Analysis, Transmission Lines, Calculus, Mechanics of Materials, Thermodynamics, Engineering Economy, Technical Writing, Psychology, Public Speaking.

Certificate of completion and AT3 rating. AT (A) School, Naval Air Station, Memphis, Tenn., August 1972.

New York State Regent's Diploma (with honor). Catholic Central High School, Troy, New York, June 1968.

Diploma, Public School No. 12, Troy, New York, January 1965.

Experience

Part time, September 1972 to present: Driver, Cooper-Michael Motors, Ft. Collins, Colorado; transported new cars from Denver to Ft. Collins; required driving ability, responsibility, reliability.

Summer 1972: Roustabout, Pan American Petroleum Corp., Midwest, Wyoming; helped maintain and repair oil wells and flow lines when with connection gang, formed part of crew to pull rods and tubing when on pulling unit; jobs necessitated familiarity and use of large manual tools, constant safety precautions, alertness, physical stamina, coordination and balance on derricks; a better understanding and knowledge of oil production was gained, a new and interesting experience.

Summers 1970 & 1971: Summer Hand, Rutger's University, New Brunswick, N.J.; aided in any way possible with research projects which consisted mainly of small grain research; harvested, cleaned, weighed, and calculated yields of grain samples, accompanied extension specialists to demonstrations throughout the state, painted small signs, learned to use calculator.

June 1963—July 1965: Clerk-Manager, Popp's Variety Store, Troy, N. Y.; had complete charge of store during the day, ordered merchandise from salesmen; this was first and perhaps most important experience in coming in contact with people of various professions, gave me a sense of responsibility, importance, economy, confidence, first experience in the business world.

Military Service

July 1965—July 1969: U.S. Navy; boot training at Great Lakes Naval Training Center, Airman (P) School at Jacksonville, Fla., Nov. 1965—Feb. 1966, AT(A) School at Memphis Naval Air Station, Feb. 1966—Aug. 1966, AEW Radar School at Norfolk, Va.,

Sept. 1966, radar instructor at Norfolk Naval Air Station, Sept. 1966 to time of separation in July 1969; went from recruit to PO1 in minimum time intervals required for advancement in rate, flew as radar instructor in airborne exercises, was radar crewman in search missions over Atlantic.

References

V. W. Popp, 72 Menand Rd., Menands, N.Y.; Chief Accountant for Cluett Peabody Co. in Troy, N.Y.

Dr. Donald A. Schallock, Rutger's University Farm Crops Dept., New Brunswick, New Jersey; Extension Weed Specialist.

Prof. J. E. Dean, Colorado State University, Ft. Collins; Head of Electrical Engineering Dept.

Prof. C. H. Chinburg, Colorado State University; Asst. Prof., Electrical Engineering Dept.

> 728 West Laurel
> Fort Collins, Colorado
> April 9, 1975

Supervisor
Mesa Verde National Park
Colorado 81330

Dear Sir:

Professor J.V.K. Wagnar, department head of Forest Recreation and Wildlife Conservation at Colorado State University, called me into his office this morning to see if I were interested in the permanent ranger position described in the notice you sent to him.

The saying "opportunity knocks once" immediately flashed through my mind. To find this opportunity open at this point in my career was a pleasant surprise.

Why? I have two reasons: (1) I have long had a hobby and interest in Indian lore and archeology which would be of tremendous value in this type of position and (2) I have trained in a Forest Recreation major for National Park work hoping to eventually secure a position where a study and interpretation of Indian culture is carried on.

Professor Wagnar stressed the background needed for this position and the education necessary to qualify. My background has included various jobs where public relations were required. Helping manage a swimming pool one summer gave me invaluable practice in controlling groups of people and in gaining their confidence and interest. At Yellowstone National Park on busy weekends I helped guide tours to points of interest giving interpretive talks along the trail. From my hobby of Indian lore I have gained a knowledge of various Indian cultures which directly relate to Mesa Verde's Indian culture.

The Bachelor of Science degree in Forest Recreation and Wildlife Conservation from Colorado State University I will receive on June 6th of this year was based on a thorough training for National Park work. I have taken a number of electives which would help me in National Park Service employment as the enclosed résumé sheet points out. If any questions arise as you read this data, I'll be glad to answer them.

I will be in Mancos, Colorado from May 22–29 doing research for the University on increased tourist visits to this community resulting from your location near the park. I will be happy to call at your convenience for a personal interview. If this is not convenient for you, I can be reached at my Fort Collins address after May 31.

Sincerely yours,

Robert D. Wize

RESUME

Robert D. Wize • 728 West Laurel • Fort Collins, Colorado 80521
Hunter 2-3835

(Picture)

Major Interest

Forest Recreation

Basic Qualifications

University training in forest
recreation
Varied experience with the
National Park Service and the
United States Forest Service

PERSONAL INFORMATION

Physical Details:
Age, 22 years
Height, 5 feet 10 inches
Weight, 180 pounds
Health, excellent
Nationality:
American
Church:
Presbyterian

Marital Status:
Single
Professional Membership:
National Society of Parks and
Recreation
Special Interests:
Outdoor sports—especially
mountain climbing and skiing.

EDUCATION

Bachelor of Science Degree in Forest Recreation and Wildlife Conservation, Colorado State University, Fort Collins, Colorado, 1975

Major Courses in Forest Recreation

Principles of Wildland Recreation
Recreation Facilities
National Park Management
Recreation Policy and Management
Field Recreation Studies and Management
Problem in Forest Recreation
Ten week Forestry Summer Camp in the field — 1972
Spring Quarter 1974 — field recreation studies

Related Courses in Wildlife Management and Forestry

Principles of Wildlife Management
Cover Management
Wildlife Ecology
River Basin Wildlife and Recreation
Wildland Administration
Forest Ecology

Diploma, Aurora High School, Aurora, Colorado, 1971
College Preparatory Major

EXPERIENCE

Summer—1974—Grand Canyon National Park, Arizona
              Fire lookout and small fire suppression
Summer—1973—Yellowstone National Park, Wyoming
              Laborer on high altitude trails
              Assisted in guiding tours of tourists on heavy
              use weekends
Summer—1972—Forestry Field Summer Camp at Pingree Park
              Colorado State University
Summer—1971—Skyline Acres, Denver 7, Colorado
              Assistant swimming pool manager and lifeguard

UNIVERSITY ACTIVITIES AND ORGANIZATIONS

Member of Lancers-Sophomore men's honorary organization
Member of Forestry Club
    Offices held:
        Secretary
        Associated Western Forestry Clubs vice-president delegate to
            1971 and 1974 conventions
Member of Seigga Ski Club

REFERENCES

(with permission)

Mr. Stanley C. Jameson Mechanical Engineer Douglas Aircraft Inc. Denver, Colorado 80201	Mr. L. N. Leonard Yellowstone National Park Wyoming 83020
Mr. R. K. Parker 1875 Clarkson Denver, Colorado 80201	Mr. H. K. Rissler Grand Canyon National Park Arizona 86023

Follow-up Letters

The application letter is a sales letter. Good salesmen know a sale frequently depends upon repeated efforts and follow-ups. Follow-up letters, after an interview, create a favorable impression upon a prospective employer. People admire determination and persistence. After you have been granted an interview, express your appreciation and reemphasize those particular qualifications which you have determined during the course of the interview are important to the prospective employer. If you are able to present additional information or additional qualifications to enhance your application, it is well to do so in your follow-up letter. The follow-up letter should be neither too long nor too short and should be timed, in most cases, to reach your prospective employer the day following the interview:

Dear Mr. Fischer:

I appreciate the time you spent with me yesterday discussing the sales engineering position you are seeking to fill. Your description of some of the problems this position entails offers a challenge I would like to tackle. I knew that selling was in my blood from the moment I sold my first subscription to the *Minneapolis Morning Tribune* as a paper boy more than ten years ago. The thrill of selling was reinforced a number of years later during three summers as a vacuum cleaner salesman for Kirby Company. My work as a Junior Applications Engineer for Moore Hydraulics Company during the summer of my senior year at Engineering College clinched my resolve that Sales Engineering would be my life's pursuit. It also gave me the kind of invaluable experience needed for engineering sales by working with customers on problems which needed modification of off-the-shelf Moore engineering products.

The products of your organization, Mr. Fischer, were pointed out to me early in my laboratory courses by my professors as among the finest in the industry. Because a salesman must believe in the products he sells, I would be proud to have an association with your organization. The problems you mentioned in our conversation yesterday aroused my enthusiastic interest. I would like very much to help you meet them and look forward to doing so.

Sincerely yours,

Company Replies to Job Applications

COLLINS RADIO COMPANY
CEDAR RAPIDS, IOWA 52403

March 5, 1975

Mr. F. R. Billings
3030 North 10th Street
Milwaukee, Wisconsin 53206

Dear Mr. Billings:

It was a pleasure to meet you when we were on your campus recently interviewing prospective graduates concerning an association with Collins Radio Company.

Our Company has a number of openings which, we believe, offer an opportunity for the career you seek. We would like to have you come to Cedar Rapids at our expense to get better acquainted with Collins Radio Company by seeing our headquarter's facilities and to meet some of the members of our staff. It will fit our schedule better if you are able to plan your visit any Monday through Friday between March 18 and April 15. However, please arrange to come whenever it is most convenient for you.

You will need one full day here from 8:30 A.M. to 5:00 P.M. to cover the program we have planned. Select whatever mode of transportation is best for you. Cedar Rapids is serviced by United Air Lines and the Milwaukee and Northwestern Railroads, although you may prefer to drive. We sincerely hope your wife will be able to accompany you to Cedar Rapids as her considerations are most important in the selection of your future home.

When you have reviewed available transportation schedules and have selected a convenient date for your visit, please advise me. If timing is a problem, call me collect at Empire 3-2661, Extension 505. We will be happy to reserve hotel accommodations for you as soon as we know your plans. We will confirm completed arrangements. Since we will be having a number of candidates in during this period, it will be helpful if you will suggest an alternate date or two, as we are able to accommodate only a limited number each day.

An application blank is enclosed for you to complete and return to my attention, along with a copy of your college transcript. If your transcript does not include courses completed recently, please attach a list of these, as well as the courses you will complete before being graduated.

We are looking forward to your visit to Cedar Rapids and to discussing the part you and Collins Radio Company may play in the bright future of the electronics industry.

Very truly yours,

J. J. Field, Assistant Director
Treasurer's Division

JJF:DT

COLLINS RADIO COMPANY
CEDAR RAPIDS, IOWA 52403

May 14, 1975

Mr. James C. Mitchell
3520 East Jackson Boulevard
Lansing, Michigan 48906

Dear Mr. Mitchell:

Miss Williams has forwarded your completed application and your letter so that we may consider you for a possible association in our Controller's Division. We sincerely appreciate your interest in our Company.

We should like to congratulate you on your efforts to acquire a specialized Accounting education. You will find it an exceptionally fine background for any field of endeavor in which you engage.

The Accounting system of Collins Radio Company is necessarily somewhat complex because we are engaged in both Government and commercial enterprise. Your application indicated you have completed approximately nine semester hours of Accounting work with La Salle Extension and Drake University. At the present time, we are using, almost exclusively, Accounting majors in our Controller's Division, which has the responsibility for the Company's accounting program. Because of this fact, we think it unfair to you to consider your application at this time. Most of your competition in that division would have from two to three times as many hours of specialized accounting background as you have, in addition to considerable exposure in the area of economic theory. Most likely your progress here with that type of competition would be seriously limited.

May we suggest that you continue your Accounting studies and perhaps get in touch with us when you have completed that curriculum. Your application would seem to indicate you have the ambition and aggressiveness which we like to find in Collins' employees.

Thank you for your interest in Collins Radio Company. May we extend our best wishes for your continued success in your Accounting studies.

Very truly yours,

J. J. Field, Assistant Director
Treasurer's Division

JJF:DT

The Letter of Inquiry

Letters of inquiry are probably the most common type of correspondence in business and industry. This letter seeks information or advice on many diverse matters, technical and otherwise, from another person who is able to furnish it. Sometimes the inquiry may offer potential or direct profit to the person or company addressed. Frequently the person addressed has nothing to gain and much time and energy to lose. Most organizations will answer all letters of inquiry as a matter of good public relations. Whether they send the inquirer the information requested frequently depends on the writer's ability to formulate his inquiry. A properly formulated inquiry will make the receiver want to answer it and makes his job of answering easier. A poorly written request may go unanswered or may receive an answer of little value to the inquirer.

A well-formulated letter of inquiry will have the following organizational pattern:

1. The opening paragraph should be a clear statement of the purpose of the letter. It should define for the reader the information desired or the problem involved: what is wanted; who wants it; why it is wanted.

2. The second paragraph should lead into the inquiry details. Its wording should be specific and arranged in such a way as to make the answering of it as easy as possible. A good technique is to state specific questions in tabulated form. However, the request should be reasonable. The writer should not expect a busy person to take several hours to answer a long, involved questionnaire. Many concerns will not answer inquiries from people who are not customers, unless they know why the writer wants the information. This is especially true if information requested is of a proprietary nature and would be of benefit to a competitor.

3. The final paragraph should contain an expression of appreciation with a tactful suggestion of action. The letter might conclude with a statement that the writer would return a similar favor or service.

The letter of inquiry falls into two categories: the solicited letter and the unsolicited letter. The solicited letter is written in response to an advertisement inviting the reader to write for further information about a certain product or service. The unsolicited letter is one in which the writer has taken the initiative for making his request or asking for information or advice. Here are some examples of solicited letters:

Venard Ultrasonic Corporation
Herricks Road
Mineola, N. Y. 11501

Gentlemen:

Please send me some further information on the features and costs of the Venard line of Ultrasonic equipment, which you recently advertised in *Electronics* magazine.

Sincerely yours,

Bakelite Company
30 E. 42nd Street
Room 308
New York, N. Y. 10036

Dear Sirs:

I have noticed your advertisement in the September 10 issue of the *Wall Street Journal.* I would appreciate your sending me the booklet, "Products and Processes."

In addition, I am particularly interested in epoxies — and their application in the construction of dies and molds. I would like to know the physical characteristics of epoxies, such as bearing load limits, heat resistance, and wear resistance. Perhaps a representative of your company might call on us to discuss our possible uses of epoxies.

Yours truly,

John L. Ebersohn

Here are some examples of unsolicited inquiry letters:

PRD Electronics
202 Tillery Street
Brooklyn, New York 11201

Gentlemen:

At the recent IEEE convention, I noticed an interesting photograph in your exhibit of an engineer calibrating micrometer frequency meters.

Can you supply me with an 8 x 10 glossy print, along with some rough ideas and notes, about what is actually being done in that operation. We would like to use it in the "New Production Techniques" section of our *Electronics Production* magazine?

Sincerely yours,

John Sewell, Associate Editor

The Ford Motor Company
Dearborn, Michigan 48126

Gentlemen:

Will you give us the benefit of your experience? We, as a university, want to prepare our students going into business and industry to be able to write effective letters, memoranda and reports. To make our course work as practical as possible, we would like to know what business and industry require in these areas. Will you give us your ideas?

Do you have a preference for, or a prejudice against, any of the forms presently used in business correspondence — semiblock, block, complete block, the National Office Management's Simplified Letter Form, or the hanging indentation form? Why? Is the form used universally throughout your company?

We have another favor to ask. Examples are often the best teacher. May we ask for specimens of your letters, memos, and reports which may be helpful to train our students along the line you suggest?

We hope you can help us in preparing our students to meet the communication problems they will find in business and industry.

Sincerely yours,

Security Chemical Company
Lima, Ohio 45802

Gentlemen:

Would you have your Agricultural Division recommend appropriate fumigants for my gladiolus corms while they are in winter storage and a soil sterilant for the control of gladiolus thrips? Please let me know where these materials may be obtained in commercial quantities locally.

Won't you send me a copy of your garden catalogue?

Sincerely yours,

Answers to Inquiries

Replies should begin with a friendly statement indicating that the request has been granted, or granted to the extent possible. Then there should follow the complete and exact information that was desired by the requester, including whatever explanatory data might be helpful. If part of the information wanted cannot be provided, this fact should be indicated next and accompanied with an expression of regret and an explanation of the reasons why the complete information cannot be given. Additional material which might be of value to the requester is also included. Finally, the reply could end with the courteous offer to provide any further information that might be wanted that it is possible for the writer to provide. Here are some examples of answers to inquiries:

Mr. John Billings
Purchasing Agent
Johnston Laboratories
Stockertown, Pa. 18083

Dear Mr. Billings:

Your letter of September 8, requesting the minimum shipping quantity and price per pound on bulk shipment of Dioctyl Phthalete has been referred to this office for reply.

We are pleased to quote below our current price schedule for Flexol Plasticizer DOP, which is our product trade name for this material:

Plexol Plasticizer DOP in tank car or compartment tankcar lots. 28¢ per lb. in tank truck lots (4,000 gal. minimum). 28¢ per lb. in tank truck lots (1,000–4,000 gals.)............29¢ per lb.
Delivered to your plant at Stockertown, Pa.

We appreciate your interest in this chemical of our manufacture, and if we can be of any further assistance, please communicate with us. Our district sales office, located at 12 N. 6th St., Philadelphia, Pa., is convenient to you. Their telephone is MArket 8-5555.

Sincerely yours,

James Mason Chemical Company

THE CUMBERLAND COMPANY
PAPER MANUFACTURERS

CABLE ADDRESS   CUMBERLAND MILLS
CUMBERLAND BOSTON   CENTRAL MILL
                                                 COPSECOOK MILL

80 BROAD STREET, BOSTON, MASSACHUSETTS 02401

30 December 1974

Professor Herman M. Weisman
7807 Hamilton Spring Road
Bethesda, Maryland 20034

SUBJECT: Correspondence and letter-writing practices

As you see, Professor Weisman —
we rate as an ardent supporter of the — simplified form of letter-writing — so-called.

In fact, we're credited with being the originator of the practice. We've been using the simplified system of letter-writing, and successfully, for at least 25 or 30 years. Here's why:

There's no reason in the world for using the endearing term in the salutation part of a letter. In our opinion, the simplified form of correspondence automatically eliminates this decadent practice, and the communication becomes "alive" immediately. That's what we call the fast start in letter-writing. Of course, there are many other reasons, too numerous to mention here in favor of the practice. But for the benefit of your students, perhaps the easiest way for us to demonstrate our technique is this: We'll arrange to send to you, automatically, copies of typical correspondence covering a period of two weeks. Through this means, you can determine by actual experience, instead of theory, how we operate in progressing the practice of simplified letter-writing.

Incidentally, perhaps we should tip you off in advance that most of our correspondence is with people engaged in the advertising profession. And may we further add that the simplified form of letter-writing evidently appeals to them. We've never had even the semblance of a squawk from our customers to the effect that the technique is "flip" or undesirable.

Please believe, sir, we welcome this opportunity to be of some service to you. Here's wishing you much continued success in your worthy educational endeavors. With heartiest of season's greetings to you for the New Year, we are, as always

Promotionally optimistic,

s/Randy Raymond

RR/emb
Advertising Department

R. J. REYNOLDS TOBACCO COMPANY
QUALITY PRODUCTS SINCE 1875

Winston-Salem, North Carolina 27104
December 19, 1974

Dr. Herman M. Weisman
7807 Hamilton Spring Road
Bethesda, Maryland 20034

Dear Dr. Weisman:

We are sorry to have been delayed in responding to your friendly inquiry, and hope this reply will be of service for your purpose.

No single structural arrangement is used, to the exclusion of all other generally acceptable styles, in the preparation of outgoing letters of this Company. The same applies to internal correspondence, such as reports and memoranda.

Only necessary file copies are retained of the correspondence. We therefore do not have specimens available to furnish as examples. However, we are glad to cooperate by telling you of practices here in preparing internal correspondence. Under the date, a heading: "MEMORANDUM FOR:" is often typed, along with the name of the individual addressed — as example, Mr. John J. Doe. On the next line, "SUBJECT:" frequently is included, together with the subject matter — as example, Anniversary Arrangements.

If the subject matter is comprehensive, we usually set off the main topics in the body of the memo with Roman numerals. Smaller divisions are alternately lettered and numbered—with a capital letter identifying each major subtopic and an Arabic numeral indicating its individual subordinate parts. If further subdivision is needed, a small letter is used at the start of each.

Of course, in many of our memos and reports the subject matter allows for a more simplified form of presentation. For instance, in making a suggestion we sometimes include the recommendation in the initial paragraph and list the supporting reasons below it, with an Arabic numeral at the start of each reason given. There are also occasions when an entire memo can be confined to a single short paragraph.

While we realize that the procedures mentioned are not unusual, we are glad to pass them along as information.

We happen to receive letters from a large number of college students. Quite a few of those letters include sections prepared in a report style. With only a very few exceptions, we would have no criticism on the organization and structure of the ones we have received. We think that speaks well for the instruction those students receive in class.

Every good wish.

Sincerely,

W. S. Koenig
Personnel Department

WSB:bz

P.S. If you are ever in this vicinity and would like to see products of ours being made, you are cordially invited to stop by for a personalized tour in our plant. Enclosed is an information sheet used locally about these tour arrangements.

The Quotation Letter

The quotation letter is a reply to an inquiry about prices. Complete description of the product is usually given in an accompanying bulletin.

The Oakland Company
St. Paul, Minnesota 55102

Attention: Mr. Elmer Peterson, Purchasing Agent

Gentlemen:

Thank you for the opportunity of quoting on our Auto Rapid Calculator.

Model 12 PD (8 column, fully automatic)	$595.00	
Taxes: 6% Federal Excise	35.70	
2% State Sales	11.90	
	$642.60	
Less credit on your Model 42X		$642.60
(approximately 20 years old)		−100.00
		$542.60

The attached specification sheet describes this extremely popular Auto Rapid model offering what we believe is the greatest value in an automatic calculator on the market today. You will note that it incorporates the many exclusive features that have made the Auto Rapid famous for speed, accuracy, and simplicity. All our calculators carry a one-year guarantee. Terms are 30 days net, and we have a calculator ready for immediate delivery.

Be sure to call us if there are further questions.

Sincerely yours,

John Doaks, Sales Dept.

Mr. John Kane
Chicago Paint Company
1217 Marshall Street
Chicago, Illinois 60623

Dear Mr. Kane:

Thank you for your letter of September 12, requesting a sample and price quotation on our Triethanelamine. We are sending you, without charge, 8 ounces of Triethanelamine for your evaluation.

Our current price schedules on this material are as follows:

In less than carload or less than truckload lots
(in 55 gallon drums) .... 25.5¢ per lb.

On less than 55-gallon drum lots
In 1 gallon containers .... 45.5¢ per lb.
In 5 gallon containers .... 30.5¢ per lb.

Shipments less than 55 gallon drums are F. O. B., South Charleston, West Virginia; 55-gallon drums or less are F. O. B. delivery point of rail carrier (when shipped by rail carrier) or delivered to your plant (when shipped by truck) at Chicago. Shipments are made in non-returnable 1, 5, and 55 gallon containers holding 9, 45 and 500 lbs., net, respectively, for which no extra charge is made.

This opportunity to be of service is appreciated. If we can be of further assistance, please do not hesitate to write us again. Or, if more convenient, our Chicago District sales office is located at 930 South Michigan Avenue in Chicago.

Yours very truly,

George Smith, Sales Manager

Claim and Adjustment Letters

As long as the human equation operates in business and industry, mistakes will be made, claims will be filed, and adjustments will have to be made. No matter what the cause, the claim or complaint needs to be expresed calmly, courteously, and objectively. Facts must be stated positively and truthfully. No matter how tempted the writer is and how justified he may be, any impatience, sarcasm, or discourtesy must be avoided. Vituperative or sarcastic language throws obstacles into the proceedings and makes adjustments more difficult and almost certainly will cause further delay in solving the problem.

The following suggestions constitute a practical guide in writing a letter of claim:

1. Explain what the problem is or what has gone wrong. Give necessary details for identifying the faulty product or service. Include dates of order shipment and arrival or nonarrival. Specify breakage or the kind and extent of damage. Give model number, sizes, colors, and whatever information is necessary to enable the reader to check into the matter. It is proper to include a statement of inconvenience or loss which has resulted from the cause of complaint.

2. Motivate the reader for the desired action by appealing to his sense of fair play or pride.

3. Include a statement of what adjustment you would consider to be fair. Here are examples of claim letters:

Mr. A. H. Brown
Contracts Administrator
Homes Electronics
14 Lincoln Way
Lexington, Mass. 02173

Subject: Our Purchase Order No. BDLA83760

Dear Mr. Brown:

Our project engineer, Mr. T. H. Sherman, has informed me that despite an earlier agreement, your company has decided not to provide the feature of easy means and convenient interchangeability between crystal diode and bolometer in the instrumentation you are developing for us.

Please refer to our Purchase Order No. BDLA83760, which indicates that the instrument under development will conform to specifications.

I am sure this misunderstanding can be cleared up in time for delivery to be made in accordance with the required dates.

Sincerely yours,

R. H. Fanwell
Assistant Purchasing Agent

Knight Laboratories
2330 Eastern Avenue
Los Angeles, Calif. 90022

Gentlemen:

Our last order included 1 x 100 Deserol Ampules. The order was delivered, but the Deserol drug was not included within the packaged order, which arrived by Railway Express. Since your shipping document listed seven boxes of Deserol and this item was also included in your invoice attached with the shipment, we want to let you know of this discrepancy. What is more important, we are badly in need of this drug for filling our prescriptions.

Sincerely yours,

Barry Nelson

The Adjustment Letter

How adjustments are met is usually determined by company policy. Most companies grant adjustments whenever a claim seems justified. Claims are usually decided on their individual merits. However, every complaint or claim, no matter how trivial, is answered courteously and promptly. Where a claim is granted, the adjustment letter has the following structure:

1. The writer is thanked for calling attention to the difficulty or problem.

2. The problem may be reviewed and explained. Alibis are considered poor form; whenever possible and expedient, whatever caused the difficulty is dealt with frankly.

3. The writer grants the adjustment; he emphasizes his sincere desire to maintain good relations with the customer, and finally, appreciation for the customer's business is expressed.

When an adjustment is refused, the letter begins, similarly, on a positive note. The writer is thanked for calling attention to the difficulty. The situation is reviewed; particularly, the facts surrounding the claim are examined from the point of view of the decision.

The adjustment is refused with an explanation. The writer needs to show the reader that he understands the reader's problem. He must also, with friendly candor, make sure that the reader will understand the writer's situation. Here are some samples of adjustment letters:

Mr. R. H. Fanwell
Assistant Purchasing Agent
King Electronics
1231 Jay Street
Brooklyn, N.Y. 11222

Subject: Your Purchase Order No. BDLA31767

Dear Mr. Fanwell:

Your letter regarding the above purchase order has been reviewed with our Technical Personnel who have been working with your engineers on this instrumentation. Unknown to us, work requirements for this Purchase Order were left to informal verbal understandings between some of our technical personnel and your engineers. We (our Contracts Department) had interpreted your requirements in accordance with the written specifications associated with the purchase order and, therefore, had ruled that incorporation of a means for convenient and easy interchangeability between the crystal diode and bolometer was not called for in the purchase order. Our interpretation was based on the note in purchase BDLA31767, which reads: "Items 1a, 1b, and 1c to be in accordance with Proposal No. 1401 of May 24, 1960." Nowhere within Proposal No. 1401 is the requirement of interchangeability of crystal diode and bolometer mentioned.

Since we have discovered that there has been a verbal understanding between our technical people and your project engineer, we will, therefore, agree to accomplish, without any further cost, the interchangeability requested. Delivery of the low-powered unit will be accomplished within the next two-week period. We would appreciate notification of acceptability of that aspect of the work as soon as you have tested the delivered unit.

Sincerely yours,

Mr. Barry Nelson
City Drugs
12 University Avenue
Denver, Colorado 80206

Dear Mr. Nelson:

Thank you for calling our attention to the omission in your order of the Deserol drug. We did not intend to include it with your order. Listing it on the Bill of Lading and on the invoice was a mistake.

Deserol is packaged one ampule per box, seven boxes to a shipping container. Since it is kept under refrigeration, we normally ship it by air parcel post, special delivery, rather than with regular orders. Because of this, and because we were uncertain of the amount you desired, we did not include it in your shipment. If you will send us your order indicating how many Deserol ampules you would prefer, we would be glad to make immediate air shipment.

We appreciate your interest in our products. Won't you call on us if we can be of any further assistance?

Very truly yours,

The Letter of Instruction

Technical personnel, because of their specialized knowledge, are frequently called upon to instruct others who have a need for their specialized knowledge to perform a particular task. Letters of instruction are a convenient instrument toward this end. If the matter is a complex one, requiring diverse contributions by a number of people with a number of devices, tools, and equipment, the letter may be more conveniently structured as a report. Frequently, however, matters can be related most conveniently in the form of a letter or a memo.

Instructions are mastered more readily and are better remembered and carried out more intelligently if the person being instructed knows the reasons for the procedures indicated. The letter of instruction, therefore, usually begins with a covering explanatory statement which provides the background for the writing of the letter. The language of the letter is structured in accordance with the background, experience, and intelligence of the person for whom the instructions are intended.

The imperative mood is used for giving specific instructions. The imperative is the form of verb used in stating commands or strong requests. The subject of the verb is not expressed but understood. (The rest of this paragraph, for sake of illustration, is structured in the imperative mood.) Be careful in using the imperative that you do not appear unduly brusque or imperious. This is especially advisable in dealings with clients or persons outside your organization. The imperative mood is used for the specific directions required, so be sure the instructions are in parallel grammatical construction; be sure, also, that your instructions are complete and that all words and statements are specific and clear. Whenever the reader is to use his own judgment, so indicate. Be sure there is logical order in the presentation of the steps of your directions. When instructions can be arranged in successive or chronological order, organize them in accordance with the demands of the situation so that they ensue as a planned sequence of activity. Write instructions concisely and precisely; avoid roundabout expressions, vague and unnecessary words and generalities. Use such devices as numbering, underlining, and indentation of headings because these help clarify the indicated procedures. Include the time or date by which the action must be accomplished.

Mr. G. W. Schmidt
Field Engineer
Colorado Sugar Corp.
Berthoud, Colorado 80513

SUBJECT: Instructions for recording soil and drainage data

Dear Mr. Schmidt:

The data you collect on a soil and drainage investigation go to the Central Design office. The men who use your data have probably not seen the investigation site. In order that your data be complete and consistent, it is desirable that you follow a standardized procedure for recording such data.

Use Form D16, a copy of which is included. This form is available from the Engineering Office. Divide the recording of data into three main parts:

A. Generalized site description
B. Specific soil and material description
C. Symbolic profile representation

A. Generalized site description

Use the spaces at the top of the form to record:

1. Area — the particular land development project.
2. Pit Number — the assigned number.
3. Photo Number — if the area is covered by an aerial photograph survey, note the photograph number and the scale of the survey. Example: 17-30, 1:8100 (Photos are filed by number under the survey scale).
4. Ground water: record the depth of the ground water. If ground was not present at the time of the survey, write "none."
5. Date: Date is important in conjunction with water table fluctuations.
6. Surveyor: Write your last name.
7. Site Description: Note all conditions at the particular site. Land use, crop condition, if cultivated, natural vegetation, slope, salinity and alkalinity.

B. Specific soil and salinity description

1. Indicate any changes in the soil profile by drawing a horizontal line across the form using the depth marks as guides.
2. Write out the soil texture for each horizon. Abbreviations can cause confusion.
3. Indicate the structure of each horizon since structure is a guide to permeability.
4. Estimate the permeability of each horizon using standardized permeability classes.
5. Write out the moisture condition in each horizon.
6. Indicate any motling by standard nomenclature. Motling, when present, gives an indication of past ground water conditions, Example: Even when ground water was not encountered at the time of the survey, dark, rust-brown motling near the surface might indicate drainage problems sometime in the past.
7. If free water occurs in the pit, indicate by writing "Water Table" at the point of Free Water Surface. Note the nature of the water table, true or perched.
8. Indicate any samplings of soil materials by blocking out the depth marks, "marks on the form corresponding to the depth of the sample."
9. Indicate water samples, by writing "Water Sample" in the super-saturated zone.

C. Symbolic profile representation

1. Use the space on the left side of the form, D16, to show the complete soil profile in symbolic form.
2. Use standard soil and material symbols.

If you follow this method of recording field data, you can be sure your data will be complete and clearly understood by those using it.

Sincerely yours,

Mr. James Mitchell
19 Cherry Creek Lane
Jericho, New York 11753

Dear Mr. Mitchell:

We are sorry that our salesman did not explain that all Hollywood beds are delivered unassembled. I am sure you understand that it would be most awkward and inconvenient to ship in protected cartons a bed completely assembled. Such bulk would add to storage and transportation costs and would be reflected in the price we would have to charge you. Actually, assembling one of our Hollywood beds is a very simple process which can be done in four very easy steps. Won't you try them? We are sure you will find our instructions very easy to follow:

1. Place the two sections of the frame in a parallel position on the floor, so that both headboard plates are at the same end. Swing out the cross rails, until they are at right angles (90°) to the side rails.

2. Slip tension clamps on cross rails to your left, wing nuts facing out. *Do not tighten the clamps.* Overlap right and left cross rails so that left cross rail is on the *bottom* at the head end and on *top* at the foot end. Slide tension clamps to center of overlap of cross rails. *Do not tighten clamps.*

3. Measure the width of the boxspring across the bottom. Adjust the width of the bed frame to the same measurement *(tighten both tension clamps securely with the wing nut)*. Insert casters, glides, or rollers, whichever you have. Exert pressure downward on the frame until they click in.

4. Place bed spring and mattress on the frame which should hold the boxspring securely. If not secure, loosen clamps, adjusting frame for snug fit and tighten the clamps.

Thus you will have assembled the frame of your comfortable, but sturdy, Hollywood bed in not more than 15 minutes. Please do not hesitate to check back if you have further difficulty. I am attaching, also, for your information, drawings illustrating each step.

Sincerely yours,

The Letter Report

Frequently, in a reporting situation, the information to be transmitted requires no more than two, three, or four pages. Then it is most convenient to transmit it in letter form, if the information is being sent outside one's organization. If the information is being sent within one's organization, the memo form is used.

The letter report's appearance is exactly that of the business letter. It will have all the conventional elements of the letter. The printed letterhead of the originating organization may be used. The same rules governing the mechanical format of the letter govern the letter report. The tone and style of writing, however, do not have personal elements of a letter, but rather, the objective style of the report(see chapter 12).

A subject line is advisable, as well as internal headings within the subject-matter covered. Reports in letter form are frequently called informal reports because they deal with situations of lesser scope, with subjects of temporary interest, or with situations requiring immediate action which cannot receive profound or thorough investigation. Progress reports are frequently in letter form. Here is an example of a letter report:

SMALL BUSINESS ADMINISTRATION

Washington, D. C. 20416

July 14, 1974

Subscribers to
Management, Technical, Small
Marketers Aids

Last February we wrote you about our publishing program. We invited your comments on uses, good features, needed subjects and prospective authors. Your response was heart-warming. Here are some of the conclusions suggested by your replies taken as a group:

1. First of all, you told us that you do read the Aids and that you analyze what they say in terms of your operations. Many people spoke of routing each new issue to key associates for similar study. A number mentioned keeping the back-numbers together in a notebook for handy reference. Still another large group said they use the Aids in customer- or employee-relations work, passing on certain issues of special interest. Some use them in discussions and training sessions. Quite a few re-read old Aids as "refreshers" when related problems come up.

2. You liked our short format with a summary at the start and frequent subheadings in boldface type. You asked us to continue listing a few references for further reading at the end of each Aid. Some of you hoped for "more elementary material"; others requested just the opposite. (We'll try to strike a balance and issue some of both.) You urged us to keep the language and examples down to everyday businessmen's levels. (We'll work extra hard to do so.)

3. On suggested future subjects you came up with scores of interesting ideas. The major areas of concern are human relations and communications, techniques of control, selling, and money management.

4. More of you proposed potential authors than we had expected. This is all to the good. Many different fields of specialization are represented. You may be sure that all proposals will be carefully considered; some you will see in print later this year.

We are grateful for the thoughtful help of all who wrote in regarding these publications. We wish we could have replied to each one of you individually but the numbers were too great. Your interest and cooperation in the work we are trying to do on behalf of small businesses are very welcome.

To those subscribers who, for one reason or another, did not reply to our earlier letter, a word of encouragement: It's never too late. We are always eager to get good new ideas.

Sincerely,

Wilford White, Chief
Management Services Division

The Memo Report

The memo report resembles the letter report, except that it is intended primarily for members of the originating organization. There is a tendency, however, to circulate memo reports to persons outside the originating organization. Like the letter report, it is used in those report situations that are informal, of immediate importance, and of lesser scope. Because its material is intended for one's own organization, it is less formal than the letter report. It will be presented in the format of the memo, with "From," "To," and "Subject," lines. Internal heads will be used as appropriate. For study purposes, several examples of memo reports are included.

WESTINGHOUSE

FROM   Materials Engineering
        East Pittsburgh Works

DATE   February 15, 1975

SUBJECT Plaster Molding

CLEVELAND WKS., Lighting Div., Dev. Engrg., Sec. Mgr., J. W. Wagner

In your letter of February 5, you asked several questions regarding plaster molding as a method of producing one-of-a-kind castings to suit the needs of your Industrial Designer. Specifically you asked about the type of plaster, processing of molds, and types of alloys that could be handled. The castings to be made are required to have good surface finish and be as accurate as possible.

First of all, I want to point out that surface finish and accuracy (precise control of dimensions) are direct functions of pattern equipment regardless of the process used. Inherently, of course, different molding materials will have an effect but they cannot overcome defects in pattern equipment.

As far as molding processes are concerned, I believe there are several that can be considered for this application. Plaster molding is one as you mentioned, but you might also consider the CO_2 molding technique, good dry sand molding such as applicable to critical core work, and investment type molds. Depending on the degree of surface smoothness and precision you need any one of these processes may be applicable.

There are two basic plaster molding processes — one using a permeable, formed mixture, the other using a straight plaster mix which is of low permeability. Plaster molding techniques do require considerable control in order to produce good molds and castings. Most commonly used are gypsum plasters which work well on aluminum and copper base alloys except those copper alloys that are poured above about 1100°C. In this temperature range, the plaster tends to break down giving very poor surface finish. Also, plaster molds are usually poured hot to assure that all moisture is driven off the mold. This is less critical with the foamed, permeable plasters but drying cycles are important and cannot be short-cut without trouble resulting. I am not familiar enough with plaster molding techniques to be able to spell out specific procedures for you to follow, but I suggest you contact one or more of the gypsum plaster manufacturers who can give you the details you want.

You may also want to consider the investment type molds for this application but I suspect set-up and operation of this type of procedure for just occasional castings would not be justified. You would, however, get excellent finishes and dimensional control.

Sand molding procedures can also produce the type of quality you indicate if they are tailored to this need. Either the CO_2 process or dry sand procedures would be suitable. Surface finish would be directly related to sand composition in either case. The grain size and distribution of sizes of the sand would be important. I think Jim Drylie or one of his foundry sand experts could give you a good run down on this possibility; although, I don't know how familiar the foundry is with CO_2 molding. One big advantage to using a "precision" sand process lies in the greater size range that can be handled. If your interest in these special castings may range to large castings, i.e., larger that would fit in about a 12″ x 12″ x 12″ space, then plaster may have limitations. These larger items could be handled more readily in a sand process.

Sperry Gyroscope Company has used a carefully controlled sand process for producing a variety of "precision" microwave components. They claim very good dimensional control and surface finish. Unfortunately, I don't know the details of their sand mixes, but they use both dry and green sand practice.

I suggest you contact one or more of the following for detailed processing information:

Investment Molds "Duracast"	Kerr Manufacturing Company 6081 Twelfth Street Detroit, Michigan 48208
Plaster	U. S. Gypsum Company
Sand	Archer-Daniels-Midland Company Federal Foundry Supply Division 2191 West 110th Street Cleveland, Ohio 44102
	B. F. Goodrich Chemical Company 3135 Euclid Avenue Cleveland, Ohio 44115 ("Good-Rite CB–40" binder)

For the CO_2 process possibilities, I think DuPont or Linde Air Products or any good foundry supply house can give you detailed information.

I know what I have written doesn't exactly answer your question concerning details of the plaster molding process, but I think the other processes are worth considering, especially since sand procedures would not be foreign to your present foundry operations.

If you want me to pursue this in more detail for you, please let me know.

K-90, Matls. Eng., Nonferrous Application Group, Group Mgr., W. J. Robertson

October 21, 1975

MEMORANDUM

TO: H. A. Franklin

FROM: L. B. Jerrold

SUBJECT: Survey of Costs Involved in Revising our Precision Test
             Equipment Catalog

I. Purpose.

   In accordance with your verbal instructions, a survey was made to
obtain costs relative to revision of our Catalog. The problem, of course,
is to produce a convincing and effective sales tool within the limits of
a minimum budget. Therefore, a comparison was made of various approaches we might follow.

II. Survey Data.

   A. Reproducing present Catalog as is. (Total of 86 printed pages
      consisting of 46 pages on one side and 22 printed pages on both
      sides.)

Cost Factors				Amount	
1. Engineering				—	
2. Illustrations				—	
3. Composition				—	
(Typesetting)					
4. Offset Printing (Boro					
Offset Co.)	$5,000.00	(5M)	$ 9,200.00	(10M)	
5. Covers	1,250.00	(5M)	2,500.00	(10M)	
Total —	$6,250.00	(5M)	$11,700.00	(10M)	

   B. Correct Inaccuracies of present Catalog pages, same size and
      format.

Cost Factors		Amount	
1. Engineering (25 rewritten pages)	$ 600.00	1 man week
2. Editorial Supervision	600.00	1 man week
3. Illustrations (20 new photos:		
pasteups, etc.)	2,100.00	6 man weeks
4. Materials	300.00	
5. Composition (Varityping)	360.00	
6. Offset Printing (Boro		
Offset)	10M	9,200.00	
7. Covers	2,500.00	

      Total — $15,660.00

   C. Minimal Redoing of every page of Catalog; written and graphic
      data added; inaccuracies corrected; pages revised and/or rewrit-
      ten. (Same format.)

Cost Factors	Amount	
1. Engineering (50 rewritten pages)	$15,000.0	6 man months
2. Editing Supervision	5,000.00	2 man months
3. Illustrations (25 new photos;		
25 new curves; paste-		
ups, etc.)	11,200.00	8 man months

4. Materials   600.00

5. Composition
     Varityping (Anzel) $800.00
     Letterpress (Atlantic) $1,200.00

6. Offset Printing (Boro Offset)
     Uncoated paper $9,200.00
     Coated paper $13,120.00

7. Covers   1,687.50

     (Uncoated Total) $42,600.00   (10M)
     (Coated Total) $46,520.00   (10M)

D. Complete Revision and Restyling of Catalog to expand technical presentation; to consist of 80 sheets, perhaps printed on both sides, having a total of 120–150 printed pages: in one or two colors; Kromecoat cover and plastic binding.

Cost Factors	Amount	
1. Engineering	$30,000.00	12 man months
2. Editorial Supervision	11,250.00	4½ man months
3. Illustrations	22,500.00	15 man months
4. Materials	600.00	

Total — $64,350.00

Two methods of printing are open to us — letterpress or offset.

a. Letterpress (Estimate of Brooklyn Eagle Press)

1. Composition	$5,000
2. Lockup and Makeready	2,000
3. Printing	2,500
4. Paper (Coated)	5,600
5. Ink—2 colors	2,400
6. Cover and Binding	3,960
7. Engraving	4,230
	$25,690
Engineering	64,350

Total—$90,040.00   (10M)

b. Offset (Estimate of Times Litho. Co.)

1. Composition	$4,000
2. Printing	
1 color	10,560
2 colors	14,110
3. Cover and Binding	3,960

     (1 Color Total) $18,520   (10M)
     (2 Colors Total) 22,070   (10M)

Engineering $64,350.00

     (1 Color Total) $82,870   (10M)
     (2 Colors Total) 86,420   (10M)

III. Conclusion.

Dissatisfaction seems general with our present catalog which is a patchwork of insertions and deletions of the past several years. Much of the information it contains is no longer appropriate, is often inaccurate or misleading, and is sometimes inadequate. On top of a dull, colorless, haphazard format, it has no attractiveness, little freshness, less style, and nothing to make it distinctive. In an industry becoming more and more competitive, and within a framework of sales operation wherein customers must be reached, persuaded, and sold not by a sales force but through an indirect media, the catalog assumes a most important sales function. A complete overhaul of its technical and graphic data and a restyling of its format looms as an absolute necessity to make it meet the first requisites of any catalog, accurate descriptions which have appeal.

A preliminary survey gives indication that it is possible to have a complete revision and restyling of the catalog on the order of the professionalism of the Hewlett-Packard for an overall cost of approximately $65,000.

Discussion Problems and Exercises

1. Most technical personnel have secretaries type their letters. Do you think a "Federal case" has been made in this chapter on correct format, layout, and typing considerations? Defend your answer.

2. Should a letter dealing with technical matters be concerned with the "You Psychology"? Before you answer, recall that technical writing is objective and impersonal. Defend your answer.

3. For training purposes, build up a collection of letters. Obtain them from friends and relatives in business, industry, and from various types of organizations. Study and analyze these letters in accordance with the various principles propounded in this chapter. In this analysis, write down for each letter the writer's purpose. If there are several purposes, note if they are organized from an overall objective; if there is more than one purpose, do they create confusion? Rewrite such letters. Are there letters in which the writer does not seem to be successful in establishing his purpose? How would you revise those letters?

4. Write a letter of instruction to a technician in the field on how to compose a letter using the principles discussed in this chapter. Then evaluate your letter for the following factors:
   completeness of information
   clearness
   accuracy
   conciseness
   courtesy
   correctness of form
   tone
   Should a letter of instruction be courteous? To what extent?

5. Should courtesy be a primary consideration in a letter of complaint? Why?

6. Write a letter of application for a summer job for which you are qualified.

7. Study the want ads in the classified section of a major newspaper or the listing of positions available in a professional magazine of your field. Write a two-part application letter for a permanent position on your graduation from college. Try at least three different types of opening paragraphs—original, summary, name beginning. What effect does each type of opening have on the rest of the letter?

8. Now compose a letter for an unsolicited opening—that is, write to a company for whom you wish to work although you do not know that it has a position available. How much of the previous letter (problem 7) can you use? Should your résumé or data sheet be slanted for each job you apply for, or does it remain constant? Why?

9. An influential friend of your family, or a relative, who commands respect in the field in which you seek employment has recommended you for a position. Apply for it by letter.

10. Write follow-up letters following interviews for positions in problems 7, 8, and 9. Write job acceptance letters for problems 7 and 8.

11. You have not been offered a job in the situation mentioned in problem 9. Write a follow-up letter.

12. Write letters of acceptance for problems 6, 7, and 8.

13. One of the largest irrigation systems in Colorado has an opening for the position of assistant supervisor. The applicant must have a degree in irrigation, agricultural or civil engineering, and must be between the ages of 25 and 35. Business ability is necessary. Write the Fort Lyon Canal Company, Box 176, Las Animas, Colorado. The applicant has written the following letter:

> 704 S. College Ave.
> Fort Collins, Colorado 80521
> October 5, 1974

For Lyon Canal Co.
Post Office Box 176
Las Animas, Colorado

Gentlemen:

A coyote can live very well in a zoo, but he is not happy there. He longs to trade the security of his stuffy zoo quarters for his home on the prairie where he can battle the elements for himself and get away from dense population. I am like the coyote in many respects. I grew up in the Arkansas Valley and learned to love it. Cities, heavy traffic, and design rooms do not appeal to me. I am home in the Arkansas Valley working to try to increase its productivity. That is why I want to be assistant supervisor of your company.

I grew up on irrigated farms under the Fort Lyon and farmed there before entering the Air Force. As you know, my father farmed in the Arkansas Valley for 54 years, most of which was under the Fort Lyon. I am well acquainted with the problems of your company and of the farmers, a quality which would not only be useful in making decisions, but also in gaining the confidence of the farmers.

I will receive a Bachelor of Science degree from Colorado State University next month in Agricultural Engineering, irrigation option. My technical electives were taken in irrigation, and most of my nontechnical electives were taken in the fields of business and economics.

I would like to have an interview on any Saturday this month. If this would be impossible, I can appear at your convenience.

> Respectfully yours,

How do you think the prospective employer will react to the opening? If you deem an unfavorable reaction, how would you rewrite the opening? Does the applicant meet the specifications called for in the ad? What specific information is lacking? Can this be supplied in the résumé? Structure a résumé supplementing the letter. What merits do you find in the letter? What elements would you improve upon?

14. What is your reaction to the following letter?

> Mr. J. R. Barnes, CLU
> General Manager
> New York Life Insurance Company
> 1740 Broadway, Suite B-304
> Denver, Colorado 80202
>
> Dear Mr. Barnes:
>
> I want to hitch my wagon to a star . . . not any star, but the New York Life star of the insurance galaxy.
>
> I can help put your district even higher into the insurance sales atmosphere than it is now . . . if I can only be given the opportunity.
>
> I will graduate from Colorado State University in June, 1975, having completed my degree in the business administration field of study. I have no outstanding achievements that could be cited from my college records, but I have what I believe to be most essential to help you reach even higher sales records . . . the sincere desire to make a name for myself and your district by selling more insurance my first year than any previous salesman.
>
> I believe I am qualified to do this with your able training assistance in the insurance sales field. I have previously worked in the direct sales category for Curtis Publishing Company, Wearever Aluminum, Norge, Frigidaire. Maytag, Zenith and RCA on commission sales. Other sales work included advertising copy sales work for my hometown paper in high school and college paper at Kansas State College all in the period from age 16 until the present time.
>
> I like meeting people, making friends, doing things for these people, and most of all I like to make money. My motives are not entirely selfish because I don't have to say that to make this money in the insurance field one has to work hard . . . right? This I am ready to do.
>
> If you will glance up the left margin of this letter you can see that every paragraph starts with "I" . . . this is because I am thinking of what I can do for you and New York Life, if given the opportunity. May I have a personal interview with you at your convenience for furthering this discussion? My home telephone number of HUnter 2-5713, in case you are in town and wish to call.
>
> Sincerely yours,

What attitude is prevalent in it? What kind of person do you think would write such a letter? Do you think it would impress a sales manager? How would you revise this letter?

15. What is your reaction to the following letter? Why? How would you revise it?

Room 115 Green Hall
Ft. Collins, Colorado 80521
May 13, 1975

Personnel Manager
King's Clothing
Lakeside Shopping Center
Westminster, Colorado 80030

Dear Sir:

Beautiful women! What do these words bring to your mind? Are the images tall, slender, immaculate, and graceful models? Do you see them sitting in a pseudo-swing, posing in a horse and buggy, or simply stepping from a fashion runway?

Without a doubt they will be lovely and a pleasure to look upon. I would like to be able to make these visions come alive for you and your customers. Fashion shows are the media through which I would work. My experience in this field was acquired by modeling for an American dress chain store and in fashion shows of Job's Daughters and high school. I love to organize and carry through brainstorms using lovely models and select clothing such as you carry.

But even the best fashion show will not succeed without commentary that compliments and emphasizes hidden features that make an outfit complete but are not obvious at a first glance. Journalistic training on the Denver Post, my high school newspaper and *CSU Collegian* provide experience in composing proper comments while extensive college speech training make appearing before the public a pleasure.

Being 25, female and in excellent health further qualify me to work with you to advance interest and sales through fashion shows. May I come in and talk with you about your position available as a fashion show director? I shall be happy to call HA-2-0288 at 2:00 Monday for an appointment.

Enclosed are three letters you might be interested in referring to when considering me for employment. Also included is a recent photograph.

Thank you for your consideration.

Very truly yours,

Enclosures 2

16. Write two letters:
    a. A letter to a company, a government agency, or an educational institution in which you ask for information about a product, a piece of equipment, a bulletin, or a program of study.
    b. A letter giving the information requested in 16a.

17. a. Write a letter to a manufacturer ordering six separate catalog items; order a sixth item, but request certain changes from the catalog specifications. Ask for your order to be acknowledged and the acknowledgment to include a price list.
    b. Write the acknowledgment to the order in 17a; indicate two alternatives to the special revision of the catalog item. Explain the advantages and disadvantages of each.

18. Write two letters:
    a. A letter giving instructions to a group of workmen under your supervision.
    b. Write a letter report in response to the letter reporting the results after the instructions have been carried out.

19. Recall a product or service you have had that was unsatisfactory. Write a letter to the company concerned telling of your lack of satisfaction and asking for a refund or adjustment.

20. Write two replies:
    a. One granting the adjustment.
    b. The other refusing the adjustment.

21. Write a letter to a consulting engineer describing a technical problem; ask him whether he is interested in solving the problem for you and what his fee would be.

22. a. Write a letter to an eminent scientist inviting him to be the speaker at the annual banquet for your student scientific society.
    b. Write a letter the scientist might write accepting the engagement.
    c. Write a letter, after the banquet, thanking the scientist for helping to make your student affair a success.

23. Write a letter requesting information about price, efficiency, cost of operation, and maintenance of a piece of equipment for your college laboratory.

24. Reply to problem 23 above.

4

Report Writing—Reconstruction of an Investigation

Role of Reports

The society of man depends on cooperation. Man's range of knowledge within the last few generations—and most of all within our own life span—has increased faster than his ability to assimilate and understand it. No one person, even within the narrowest field, can grasp all there is to know about his field. Today's knowledge and experience represent a coordinated effort. Efficient communication has therefore become an important research tool and an important process in the standard operating procedure of every human enterprise. Today's scientific genius does not work in an ivory tower laboratory. What he does depends on the efforts of other fine workers. In turn, his efforts influence the efforts of others. Scientists of various countries are researching energetically for a cure for cancer, but every competent scientist knows that his efforts alone are not enough. He must acquire and use the knowledge, experience, and help of others. How does he get it? The answer is obvious. He must know what they are doing and the significance of what they are doing. The role of reports and technical papers is to relay this information.

Here is where you and I come in. Some of us may be engaged in exciting cancer research, and some of us may be engaged in reporting to the boss on how many nuts and bolts are left in the storage bin. But all of us are concerned with organizing the facts of an experience we are engaged in because the facts are important for someone to know. A decision or action may wait upon these facts. The decision may be as commonplace as ordering more nuts and bolts or as dramatic as finding the link between cancer and a virus.

What is a Report

One of my colleagues, a teacher of literature, tells a story which is appropriate. He was lecturing on Scott's poem *The Lady of the Lake,* which begins:
"The stag at eve had drunk his fill . . ."
Halfway through the poem he paused in his enthusiasm to notice that one student looked puzzled. He decided to take more time in his explanation, but no matter how carefully he pointed out the poet's technique, skill, and meaning in each successive line, the puzzled look on the student's face remained. He finally stopped and asked the student, "Is there anything in the last few lines that you don't understand?" "It's not the last few lines," said the perplexed student. "It's way back at the beginning. What's a stag?"

What's a report?

The origin of the word tells us much about its meaning and function. *Report* comes from the Latin *reportare,* meaning "to bring back." A working definition of a report might be: *organized, factual, and objective information brought by a person who has experienced or accumulated it to a person or persons who need it, want it, or are entitled to it.* Reports may contain opinions, but the opinions are considered judgments based on factual evidence uncovered and interpreted in the investigation.

Facts—The Basic Ingredients

The basic ingredients of a report are facts. A fact is a verifiable observation. Webster says: a fact is "that which has actual existence; an event." Facts are found through direct observation, survey techniques, experiment, inspection, experience, and through a combination of these processes as they occur in the research situation. A fact is different from an opinion. An opinion is a belief, a judgment, or an inference[1]—a generalization which is based on some factual knowledge. An opinion is not entirely verifiable at the

---

[1]An inference is a conclusion derived deductively or inductively from given data. However, an inference is considered a partial or indecisive conclusion. The term *conclusion* is reserved for the final logical result in a process of reasoning. A conclusion is full and decisive.

time of the statement. Sometimes opinions are little more than feelings or sentiments with a bit of rational basis. *Opinion* and *judgment* are synonyms and are frequently used interchangeably. Judgments imply opinions based on evaluations, as in the case of a legal decision. Neither opinions nor judgments are facts and must be distinguished from facts in technical writing and reports.

FACT: Jack Nicklaus has won more golf tournaments than any other golfer.

OPINION: Jack Nicklaus will win the Masters' this year.

FACT: A circle is a closed plane curve such that all its points are equidistant from a point within the center. A square is a parallelogram having four equal sides and four right angles.

OPINION: A circle is more pleasing to the eye than a square.

FACT: The water in the metal drum is 30 percent saturated with filterable solids.

OPINION: That water is unfit to drink.

Since facts are verifiable observations, a direct investigation is inherent in their disclosure.

Another point which should be emphasized is that facts, the chief ingredient of reports, are expressed in unambiguous language. The purpose of every technical communication is to convey information and ideas accurately and efficiently. That objective, therefore, demands that the communication be as clear as possible, as brief as possible, and as easy to understand as possible.

Forms Reports Take

Information and facts dealing with a simple problem or situation may be reported simply. Thus, if your supervisor tells you, "Joe, find out how much it will cost to replace the compressor on the number two pump," all that is necessary for the most efficient reporting of the information called for, after you have found the answer, is to pick up the telephone or pencil a note on a memo pad and say, "Boss, a new compressor costs $56.28."

But if your supervisor's supervisor wants to know the cost, you may have to "dress up" your penciled note into the form of a typewritten memo in several copies, with additional information to provide the background for the statement that a new compressor costs $56.28. If customers or clients get involved, the note may have to be enlarged to include comparative costs of several makes of compressors. And if that same pump has given you trouble for the last two years, you may have to dig up more facts and information—perhaps to find the reasons for the continued difficulty or whether another type of pump or a different system is called for. Then the matter begins to become more complex and the complications become sections of information and facts to be reported. Recommendations—what to do about it—become the most important part of your report. Also, when a customer is involved, instead of using a memo form you report in a letter. If the situation gets fairly complicated, requiring extensive investigation and reporting, the report becomes more formal and many of its elements receive formal structuring, format, and placement. You may end up having a letter of transmittal, an abstract, a table of contents, a list of illustrations, the report proper divided into formal sections, an appendix, bibliography, and perhaps an index, the whole bound in a hard cover. In other words, the report writer designs his report for the particular use and particular reader it will have. But we are getting a little ahead of ourselves.

Because of the many and complex circumstances which call for reports, it is difficult to classify reports rigidly. In some organizations there is little formality attached to report writing. Each writer puts down what he has to say, often haphazardly. In other organizations, particularly large ones, numerous and often elaborate forms are devised and given names; e.g., record report, examination report, operations report, performance report, test report, progress report, failure report, recommendation report, status report, accident report, sales report, etc. Given five minutes you could think of at least thirty different types of reports you have come across.

Actually, the facts and analyses marshalled to meet a certain reporting situation take many forms. There is no universal "right" form to clothe all reports. "Form," a learned colleague once said, "is the package in which you wrap your facts and analysis. Choose (or design) a package that is suitable for your material, your purpose, and your reader" (1:6).

Types of Reports

There are many bases for classifying reports: subject matter, function, frequency of issuance, type and formality of format, length, etc. A traditional classification divides reports into two descriptive categories:

1. Informational
2. Analytical

The Informational Report

The informational report, as the term implies, presents information without criticism, evaluation, or recommendations. It gives a detailed account of activities or conditions, making no attempt to give solutions to problems but confining itself to past and present information. Many informational reports, nevertheless, contain inferences which suggest to the reader the conclusions the writer would like him to reach; e.g., "There are fourteen nuts and bolts left in the storage bin." *Inference:* We should order another trainload. In this category belong the routine daily, weekly, or monthly reports on sales, inventory, production, or progress. Often such reports are mere tabulations and follow a definite pattern or have a preprinted form requiring only fill-ins. The best-known example of the informational report is the annual report of a corporation to its stockholders.

The Analytical Report

The analytical report goes beyond the informational report since it presents an analysis and interpretation of the facts in addition to the facts themselves.

The conclusions and recommendations are the most important and interesting parts of the report. The analytical report serves as a basis for the solution of an immediate problem or as a guide to future happenings. It is a valuable and frequently used instrument in all types of activity. Emphasis in this text is on the analytical report because the techniques applying to it apply equally well to the informational types of report.

Steps in Report Writing

Report writing is the reconstruction in written form of a purposeful investigation of a problem. While most technical writers do not engage in research, many are assigned to a research team or project to serve as interpreters or communicators of the investigation. The technical writer, therefore, must understand research methods and procedures. Particularly, he must be intimately conversant with the research he is reporting. Frequently the writer assigned to reconstruct the investigation by means of a report is in a position to help digest and find meaning in the mass of data the investigation reveals.

Within the writing process itself, the writer follows steps analogous to those followed by the experimental scientist. (It would be a good idea for you to review at this point chapter 2, "Scientific Method and Approach.") Just as a scientist observes and hypothesizes, so, by analogy, the technical writer examines his mass of data, reads it initially for meaning, tries to determine his readership, and tries to make sense of the data by way of establishing important ideas. In a second or succeeding phase, the scientist proceeds to experiment. Again by analogy, the technical writer organizes his material, sets up an initial order of continuity of data, may find upon closer scrutiny that such an organization is not effective, and proceeds to rearrange and reshuffle his data to make them more logical and understandable. In this way, the writer is *experimenting* with *his* data.

In the present textbook situation, however, it is being assumed that students taking a course in technical writing will be writing a report on an investigation that they themselves are conducting or are associated with in a capacity that permits them to serve as the interpreter or communicator of the investigation. The approach in this three-chapter sequence on report writing is based on the assumption that the student is following through on a combination of investigation and reporting, as perhaps was initiated by problem 12 of chapter 2, page 26, or problem 6 at the end of the present chapter.

The purposeful reconstruction of an investigation involves four major steps which follow very closely the experience of the investigation:

1. Preliminary analysis and planning
2. Gathering the data—investigating the problem or situation
3. Organizing the data
4. Writing the report

The report writer, whether he is the original investigator or serving as the interpreter or communicator of the original investigator, must follow through in his writing all the procedures the investigator followed in arriving at a solution to the problem.

Preliminary Analysis and Planning

You, the report writer, like the investigator, must ask yourself a number of questions: What is the purpose of the investigation? Who needs the answer to the problem? How will the answer be used? From the writer's standpoint, you must know the answers to such questions because the effectiveness of your communication depends on the answers. You may have to rephrase the questions to: What information is wanted? Who will read the report? For what purpose will it be used? What problem or problems am I expected to solve? These four questions resolve the larger question: What is the purpose of my proposed report? As the writer, you must be careful to distinguish between the purpose of the research and the problem studied in the research. The purpose of a research project is generally understood as the reason why the investigation has been undertaken; whereas the problem is what the researcher specifically hopes to solve. In considering the purpose of a research project, you should regard it as the explanation of the possible uses to which the results of a study may be put. Purpose concerns the probable value of the study. It offers the explanation of why the research was undertaken. In short, the problem concerns the *what* of the study, and the purpose concerns the *why*. When you, the writer, can formulate the purpose in your mind—actually write it down clearly and distinctly—you have started on the right road to answering your problem. Once the purpose has been determined, the next step is to define scope. It, too, is not to be confused with purpose. Purpose defines goals to be reached. Scope determines the boundaries of the ground to be covered. Scope answers the questions relating to what shall be put in and what shall be left out. (Scope is sometimes indicated by a client, if the investigation has been contracted for, but it is usually left to the judgment of the researcher.) As the report writer, you must check carefully with the reseacher your understanding of scope, purpose, and problem as well as other aspects of the investigation you are reconstructing in report form.

Having problem, scope, and purpose in mind, you are now ready to start on the next important step in your preliminary analysis and planning stage: blocking out a plan of procedure. Always keep in mind that the writing of a report is a reconstruction activity. You must go back to the equivalent investigative point where the problem has been defined in a clear-cut interrogative statement. With the problem defined, the purpose clarified, and the scope clearly set, you are ready to begin the work of solving the problem. This next step consists of breaking the problem or situation into its component parts. Again, ask yourself questions. What are the elements which make up the problem? Which elements are fundamental? Which are secondary? You, the writer, might set down on paper an outline or a procedural plan to follow. This outline might have the following scheme:

1. Statement of the problem in question form
2. Purpose or objective of the investigation
   a. Primary purpose
   b. Secondary purpose
3. Major elements
   a. First major element
      1. Component
      2. Component
      3. Component
   b. Second major element
      1. Component
      2. Component
      3. Component
   c. Third major element
      1. Component
      2. Component
      3. Component
   d. Additional major elements, etc.
4. Correlation of all elements into a final combination which constitutes the problem.

The purpose of this type of analysis is to: (1) help clarify the problem and

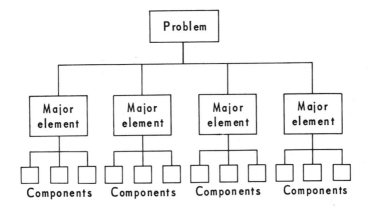

Figure 1
Block diagram showing analysis (breakdown of problem into its major elements and constituent components)

what composes it; (2) check and define the scope; and (3) help to formulate the task of the research. This analysis should lead to the provisional hypothesis which sends the research on its proper way.

If the problem is a new one, you may have to feel your way at first. As facts accumulate and you bring your reason to bear on the various elements of the problem, you begin to see a logic to them and can arrange them in accordance with the logic of the problem.

In the preliminary analysis stage, no part — purpose, scope, plan of procedure — is final. In the course of the investigation, something unexpected may call for readjusting your approach in any or all of the previous elements. The preliminary analysis has revealed the direction for the writer to follow. You, the writer, gather your information accordingly, or according to any necessary revision in recognition of unexpected turns in the investigation.

Investigative Procedures

There can be no real report without facts. Ideas, conclusions, evaluations, recommendations must grow out of facts. Before the writer begins to organize and construct his report, he must take the first step of gathering and assembling his facts — his data. Information may be gathered by one or all of several methods:

1. Searching all available printed material related to your problem or situation
2. Observing and examining the actual situation, condition, or factors of your problem
3. Experimentation (this is observation carried out under special conditions that are controlled and varied for specific reasons)
4. Interviews and discussions with experts or people qualified to give needed data
5. Questionnaires to be used when interviews are impractical

Searching the Literature

Searching the literature is the method used to learn new facts and principles through the study of documents, records, and the literature of the field. This type of research is used extensively in history, literature, linguistics, and humanities. Because it is almost the exclusive research technique used by historians, it is sometimes called the historical method of research. This research method is valuable in all fields whenever knowledge and insights into events of the past are required.

Human knowledge is a structure which grows by the addition of new material to the store which has been previously gained. An investigator has little chance of making a worthwhile new contribution if he is completely ignorant of what is already known about his problem. Before he begins an investigation, he must find out what has been written thus far about the problem. Six hours fruitfully spent in a library may save six months in a laboratory.

There are two goals in library searching:[2]

1. To find out if the information which is the object of the proposed research is already available, and
2. To acquire a broad general background in the area of research.

Efficient literature searching depends on knowledge of the structure of the source-finding devices and materials in the library. Therefore, a review of these devices is helpful.

The Card Catalog

The card catalog is the key to the library's collection of published material. The card catalog cross-indexes the books in the library by author, title, and by subject.

Literature Guides and Manuals[3]

As an aid to finding bibliographic and other reference materials in a field, guides to the literature of a subject area have been prepared for students and research workers. Examples of some are:

Bottle, R. T., *Use of Chemical Literature.* London: Butterworth, 1962.

Burman, C. R., *How to Find Out in Chemistry.* London: Pergamon Press, 1965.

Fowler, Maureen J., *Guides to Scientific Periodicals, Annotated Bibliography.* London: The Library Association, 1966.

Jenkins, Frances B., *Science Reference Sources.* Fifth ed. Cambridge, Mass.: M.I.T. Press, 1969.

Parke, N. G. III, *Guide to the Literature of Mathematics and Physics.* Second rev. ed. New York: McGraw-Hill Book Company, Inc., 1958.

Pearl, Richard M., *Guide to Geologic Literature.* New York: McGraw-Hill Book Company, Inc., 1951.

---

[2]This statement excludes the historical method of research, which employs library searching as its investigative technique. The historical researcher puts together in a logical way evidence derived from documents, records, letters, papers, and literature and from that evidence forms conclusions which either establish facts hitherto unkown or offer new insights and generalizations about past or present events, human motives, characteristics, and thoughts.

[3]Constance M. Winchell's *Guide to Reference Books* Eighth Edition (Chicago: American Library Association, 1967), is an excellent bibliographic source for obtaining information on literature guides, reference works, handbooks, encyclopedias, etc. for any given subject field.

Smith, Roger C., *Guide to the Literature of the Biological Sciences.*
Eighth ed. Minneapolis, Minn.: Burgess Publishing Company, 1971.

Bibliographies

Publication details of the literature for a subject area have been assembled
and published in bibliographies. Examples include:
Postell, W. D., *Applied Medical Bibliography for Students.* Springfield,
Ill.: C. C. Thomas, 1955.
Mellon, M. G., *Chemical Publications.* Fourth ed. New York: McGraw-
Hill Book Company, Inc., 1965.
Mavrodineanu, R., *Bibliography on Flame Spectroscopy Analytical
Applications 1800-1966.* National Bureau of Standards Miscellaneous
Publication 281, Washington, D. C.: U. S. Government Printing Office,
1967.

Handbooks

Handbooks are compact reference books on particular subjects. Examples
include:
Handbook of Chemistry and Physics. Fifty-fourth ed. Robert C. Weast,
Editor-in-chief. Cleveland, Ohio: Chemical Rubber Co., 1973.
Handbook of Experimental Psychology. S. S. Steven, Editor. New
York: John Wiley and Sons, Inc., 1951.
Handbook of Biochemistry Selected Data for Molecular Biology. Her-
bert A. Sober, Editor. Cleveland, Ohio: Chemical Rubber Co., 1968.

Encyclopedias

An encyclopedia is a book or set of books giving information on one field
or all branches of knowledge. Editors and contributors are specialists within
specific subject areas. Examples are:
International Encyclopedia of the Social Sciences. David L. Stills,
Editor. 17 Vols. New York: The Macmillan Company and The Free
Press, 1968.
Van Nostrand's Scientific Encyclopedia. Fourth ed. Princeton, N. J.:
Van Nostrand Company, Inc., 1968.
Kirk-Othmer Encyclopedia of Chemical Technology. Second ed. New
York: Interscience Publishers, 1964.

Books

Reference books and textbooks provide basic material. Monographs
(books, articles, or·papers written about a narrow, specialized technical
area) give much more detail. There is no simple way of finding all the books
on a given subject. The first place to look is in the subject index of a good

library or on the shelves themselves. There are a number of useful publications listing scientific books. Examples are:

Scientific, Medical and Technical Books Published in the United States of America. Washington, D. C.: National Research Council, 1958.

The Cumulative Book Index, a list of books in the English Language. New York: The H. W. Wilson Company. This is issued monthly and then cumulated in semiannual volumes.

The United States Catalog. New York: The H. W. Wilson Company. This is a comprehensive list of books printed in English arranged by title, author, and subject.

Abstracting and Indexing Journals

Journals which abstract and index publications within a particular field are the main reliance of investigators seeking papers on scientific topics. They provide scientists with an important way of keeping abreast of scientific progress. These journals furnish the investigator with an overview of published material on a given subject by a particular author under specific titles. Indexes usually provide no more than bibliographical references, but abstracts provide a digest or synopsis of the listed literature. Some better known examples are:

The Engineering Index, a monthly index of engineering subjects with brief annotations of the articles.

Applied Science and Technology Index; Industrial Arts Index prior to 1958, an index of industrial and trade subjects. The source also gives an annotated description of articles in many trade publications, monthly and cumulative.

Readers Guide to Periodicals Index, an index of general subject lists of articles from a great number of sources. It has a broad subject coverage of articles appearing in popular publications.

Biological & Agricultural Index, a cumulative subject index to periodicals in the fields of biology, agriculture, and related sciences, issued monthly except September. Prior to October 1964 issued as *Agricultural Index.*

Index to Legal Periodicals, a monthly cumulative author and subject index.

The New York Times Index, a monthly index to news stories in the *New York Times.*

Biological Abstracts, abstracts covering the complete spectrum of biological subjects, issued semimonthly.

Chemical Abstracts, abstracts covering more than 8000 publications listed by authors, subject, and chemical formula; includes patents in the chemical field.

Psychological Abstracts, monthly with annual indexes.

Sociological Abstracts, published nine times per year, presenting abstracts of significant articles and books in the sociological field.

Science Abstracts, monthly.

Section A: Physical Abstracts, covers mathematics, astronomy, astrophysics, geodesy, physics, physical chemistry, crystallography, geophysics, biology, techniques, and materials.

Section B: Electrical Engineering Abstracts, covers generation and supply of electricity, machines, applications, measurements, telecommunications, radar, television, and related subjects.

Science Citation Index, a quarterly international and interdisciplinary index to literature of science and technology.

Yearbooks

There are many yearbooks which provide excellent information on events and achievements during the specific year. Among them are:

The Yearbook of Agriculture

Facts on File, a weekly digest of world events

Yearbook of Education

New International Yearbook

World Almanac

Book of Facts

Yearbook of Labor Statistics

Suggestions for Library Searching

In the preliminary planning stage, you have broken your problem down into its component parts. Next, in order to find the state of knowledge of the problem and to secure a good general background for pursuing your investigation, you might begin by reading the most general treatments of the problem first, as, for example, in an encyclopedia. This can be followed by a more detailed but still quite broad discussion in a handbook. Next it would be desirable to search the library card catalogs for books on the subject. If there is a recent monograph on your problem, the library search may end at this point because such specialized books often contain bibliographies sufficient for most purposes.

If you cannot find a book which is complete or entirely up to date, you might look for a survey or review article in the professional periodical or trade publication which specializes in the area of your problem. Then appropriate abstract publications should be searched by working backwards in time until the necessary coverage has been obtained or until the year is reached which has been adequately dealt with in a book. Technical and professional papers will usually contain references to earlier works, and in this way the researcher can be carried backward to pick up references formerly missed.

Bibliography

One of the most troublesome problems the researcher has is the recording of information which he has acquired from his reading and setting it up in accessible form. This involves keeping accurate and complete bibliography cards and note cards. Bibliography cards provide a complete record of the sources of information used in library searching. The first step in making a literature search is to check with a librarian for aid in compiling a working bibliography — a list of possible sources of information. Most libraries have librarians trained in providing help in literature searching. Be sure to take full advantage of such available aid. While searching through indexes, abstracts, and card catalogs, you should keep accurate and complete records of the sources of information. In listing these sources, the following elements are recorded:

1. Library call number, usually in the upper lefthand corner (if the literature is obtained from a library).
2. The author's name, last name first, or the editor's name, followed by *ed.* If both the author and the editor are given, list the editor after the book title. The edition number, translator, and number of volumes are listed after the title, if applicable.
3. The title of the book underlined. The title of an article is in quotation marks, and the title of a magazine is underlined if the entry is an article.
4. Publication data:
   a. For a book, record the place of publication, the publisher, and the year of publication;
   b. For a magazine article, the name of the magazine underlined, the date, the volume, and the pages on which the article appears;
   c. For a newspaper article, the name of the newspaper underlined, the date, the page number.
5. Subject label is placed at the top of the card to allow filing of several bibliography cards under one topic.

Notes

Most professional research workers take notes on 3″ x 5″ or 5″ x 8″ cards because cards are flexible and easy to handle, sort, and keep in order. Three essential parts of the contents of a note card are:

1. The material, the facts, and the opinions to be recorded (include diagrams, charts, and tables if these are the data to be noted);
2. The exact source, title, and page from which they are taken;
3. The label for the card showing what it treats.

It is usually a waste of effort to try to take notes in a numbered outline form.

It is usually best to read the article or chapter through rapidly at first to see what it contains for your purposes and then go over it again and make necessary notes. Distinguish between the author's facts and opinions. Specific points, such as dates and names of persons, should be recorded with particular care. Notes may be recorded either in paraphrase, summary, or direct quotation from the source. The exact wording of the author should be recorded on the note card when the original wording is striking, graphic, or appropriate, when a statement is controversial or may be questioned, or when it is desirable to illustrate the style of the author.

In recording quoted matters, take care to copy accurately the words, capitalization, punctuation, and even the errors which may appear in the original. Use quotation marks to indicate exactly where the quoted passage begins and ends. Ellipsis may be indicated by the use of three dots (. . .) at the beginning or within the sentence, or four dots at the end of a sentence (. . . .). Before you return the reference source, be sure that

Author Card

QC20
.W54
Wiener, Norbert, 1894–1964.
 Differential space, quantum systems, and prediction [by]
Norbert Wiener [and others] Cambridge, Mass., M. I. T.
Press [1966]

 x, 176 p. 21 cm.

 Bibliography : p. 161–168.

 1. Mathematical physics. 2. Brownian movements. 3. Quantum
theory.   I. Title.

 QC20.W54   530.15   66–25165

 Library of Congress   [5]

Title Card

Library
Catalog
Cards

           Differential space, quantum systems, and
QC20   prediction
.W54
      Wiener, Norbert, 1894–1964.
 Differential space, quantum systems, and prediction [by]
Norbert Wiener [and others] Cambridge, Mass., M. I. T.
Press [1966]

 x, 176 p. 21 cm.

 Bibliography : p. 161–168.

 1. Mathematical physics. 2. Brownian movements. 3. Quantum
theory.   I. Title.

 QC20.W54   530.15   66–25165

 Library of Congress   [5]

Subject Card

QC20   Mathematical physics
.W54
      Wiener, Norbert, 1894–1964.
 Differential space, quantum systems, and prediction [by]
Norbert Wiener [and others] Cambridge, Mass., M. I. T.
Press [1966]

 x, 176 p. 21 cm.

 Bibliography : p. 161–168.

 1. Mathematical physics. 2. Brownian movements. 3. Quantum
theory.   I. Title.

 QC20.W54   530.15   66–25165

 Library of Congress   [5]

your bibliography and note cards include every item of information which you will need for use in your paper. Also note the library or source where the book or publication is obtainable.

In your library research, you should be concerned with the meaning, accuracy, and general trustworthiness of the material you are reading. In evaluating the reading matter, you should be asking yourself questions related to the author's competence and integrity. How good an observer is the writer? Did he have ample opportunity to observe and master the matters he writes about? Does the author have a reputation for authoritativeness in the field? How well does the author know the facts of his case? Was he personally interested in presenting a particular point of view? Is he prejudiced? Is he trying to deceive? Are his observations made at firsthand or did he receive information from others? Are there discrepancies in the writing which throw question on the reliability of the literature? Are the data sufficient to support the points the author is making? In short, are the data relevant, material, and competent?

As soon as the findings of other investigators are exhausted, it is time for you to do your own primary work. Procedure depends, of course, on the technical elements of your problem at hand.

Observation—Examining the Actual Situation, Condition, or Factors of the Problem

One of the most common and traditional means for conducting research is through direct observation. Observation is a cornerstone of the scientific method. Observation is careful examination of the actual situation, condition, or factors of the problem. If your problem requires you to use observation in its investigation,[4] it is important to decide whether you will use selective sampling or whether you will cover the entire field. Observation implies selection because our powers are limited. While we might want to make random observations, we do select the conditions of time and place. Frequently we find it necessary to examine and observe only a small portion of a problem. Inherent in observation is the necessity for recording what we examine and perceive. Human memory is not to be trusted. Observation must be followed by description. Observation should be recorded in precise, exact, and objective language.

Another essential point is that scientific observation tends to be quantitative. Numbers are used as part of the description where possible: how many, at what rate, with what value? The use of numerical measures permits a more precise description and makes possible the application of mathematics. Of course, not all matter and phenomena that are observed

[4]For academic purposes you will recall that we are assuming that the student technical writer is prosecuting the research of the problem he is to report. This assumption is deemed necessary in order that students know the processes in thought and activity that researchers experience.

are numerical. Qualitative statements have an important role. However, the observer must try to be objective and free of bias. Even though it is perhaps impossible for any observer to free himself completely from preconceptions and prejudices, it is important to arrange the conditions of the observation so the observer's bias will not result in distortion. Therefore, your observation should be given to others for checking. If possible, allow another observer or several observers to make independent observations of the same phenomena.

That this is necessary is illustrated by an experiment frequently repeated by psychologists. Very often psychologists will present an experiment during a meeting as follows: a man suddenly rushes into a room chased by another with a revolver. After a scuffle in the middle of the room, a shot is fired and both men rush out again, with barely thirty seconds elapsing for the incident. The chairman conducting this experiment will ask the group to write down an account of what they saw. These descriptions usually show that no two persons in a group will agree concerning the principal facts of the incident. A noteworthy feature is that over half of the accounts include details which never occurred during the staged incident. These experiments illustrate that not only do observers frequently miss seemingly obvious things, but, what is even more significant, they often invent false observations quite unwittingly. So in the research situation, the careful observer makes use of instruments wherever possible to aid his observations and promote greater accuracy and objectivity. Photography, the photoelectric cell, the thermocouple are frequently substituted for direct vision. Sound vibrations are converted into electrical oscillations and are analyzed and measured by instruments having a range far beyond that of the human ear. Thermometers, thermocouples, and pyrometers replace the sense of touch in estimating warmth and coldness. Where a phenomenon occurs rarely, or only for a short time, as in solar eclipses or earthquakes, optical and recording instruments are substituted for direct visual observation. Although many natural phenomena can be observed with the naked eye, many more become accessible with microscopes, binoculars, telescopes, and other optical and measuring instruments. Simple instruments such as meters and stethoscopes aid in direct human observation.

Here are some helpful suggestions for conducting observations under actual or controlled conditions:

1. Have a clear conception of the phenomena to be observed;
2. Secure a notebook or note cards upon which the data are to be recorded;
3. Set up entries or headings on the note cards or notebooks to indicate the form and units in which the results of the observation are to be recorded;
4. Define the scope of the observations;
5. Use care in selecting the basis of sampling, should sampling procedure be used.

Experimentation

Experimentation is observation carried out under special conditions that are controlled and varied for specific reasons. In an experiment, an event is made to occur under known conditions. Here, as many extraneous influences as possible are eliminated and close observation is made possible so that relationships between phenomena can be revealed. The sequence of the experimental process begins with the observation of the problem or difficulty. The experimenter formulates a hypothesis to explain the difficulty. Then he tests the hypothesis by the experimental techniques and draws his conclusion as to its validity.

Galileo is considered the father of the experimental method. I have previously mentioned his experiments with falling weights. I might call attention to another of his experiments to examine the nature of physical force. According to the physics of his time, which was still the physics of Aristotle, force was defined as that which, acting upon any object, causes it to move or to produce velocity or motion. If, for example, one pushed or applied force to a chair, the chair moved. When one stopped moving it, unless the force of gravity were operating or some other forces were applied to keep it moving, the object or the chair stopped moving.

Even by Galileo's day, the Aristotelian definition of force had become unsatisfactory because it could not explain certain commonplace phenomena. For example, it did not apply to the case of a projectile fired from a cannon. When a cannon fired its shot, all the force was applied at the moment of the explosion and then ceased. Nevertheless, the cannonball continued to move in a great parabola over a considerable distance. Here was an incident which contradicted the Aristotelian explanation of what force was and did.

Galileo set out to find a more appropriate explanation. In experimenting with falling objects, he realized that the motion of a body being drawn to the earth by the attraction or force of gravity was not affected by the weight of the object. So he reasoned that force could not be measured as merely proportional to the weight of the body on which the force acted. He thereupon came up with two hypotheses. One was that force might be proportional to the distance through which the object being acted upon travels; he was able to disprove this on mathematical grounds. The other hypothesis was that force could be proportional to the length of time during which it caused an object to move. In other words, the greater the force, the longer the time during which it operates to produce motion.

Up to this point, Galileo worked on this problem using mathematics and logic. He undertook to test his second hypothesis by experimental means. He constructed an inclined board along which he rolled a metal ball. With this equipment he was able to measure the distances covered by the rolling ball during various units of time fairly accurately; that is, he calibrated the inclined board to measure the time necessary for the ball to travel half the distance, two-thirds of the distance, three-fourths of the distance, and so on so that he could determine quantitatively the effect of force exerted by gravity. His experiments led him to the conclusion that force was to be defined as not merely that which produces motion or velocity but that which changes velocity as well.

According to Galileo's experimental finding, an object does not necessarily cease to move when the force causing its motion is removed. It ceases only to change its velocity. This now explains the action of the cannonball shot from a cannon. The cannonball's velocity remains constant as it rushes from the cannon's mouth (except for the effect of air resistance) until the force of gravity draws it earthward. Galileo's hypothesis and its experimental proof changed the whole concept of the physical theory of his day. That experiment has been called the foundation of modern mechanics.

Experimentation is not a technique used only by the scientist-researcher developing new instrumentations or discovering new laws in physics or in other natural sciences. Experimentation may be applied effectively also to business and ordinary pursuits. For example, the owner of a hardware store may use an experimental procedure to test the effectiveness of his show window in displaying a new power lawnmower. He might stand outside his store and count the number of persons who look at the show window and note how long they stop. He is merely observing, if he does just that. If he designs three different displays for his lawnmower, each installed for the same period of time on three successive days, and furthermore, if he records not only the number of people who stop to look at the display but also notes the number who entered his store on each occasion to inquire about the product, he is performing an experiment under varied and controlled conditions.

Before planning the actual experiment, the investigator should have a basic understanding of the nature of his problem and any relevant theory associated with it. The theory is the explanation of the problem. Of course, the theory or explanation is to be proved, but it serves as a guide to formulating his hypothesis for testing the answer to his problem. The experimenter should analyze his problem and get it down into words which express it in the simplest form. It is frequently possible to break most problems into parts which are more easily answered separately than together.

An experiment should not be conducted without a clear-cut idea in advance of just what is to be tested. The researcher should ask himself: Why am I doing this particular thing? Will it tell me what I need to know about my problem? The first thing in planning an experiment is to decide the kind of event to be studied and the nature of the variables which previous information suggested might be the controlling ones of the matter being tested. These variables may be divided into those which can be controlled and those which cannot. The ideal experiment is one in which the relevant variables are held constant except the one under study. The effects are then observed.

If observations are to be useful, the matter immediately under consideration must be separated from any which may confuse the issue. In scientific experimentation a device known as *control* is used to avoid such error. Controls are similar test specimens which, as nearly as possible, are subjected to the same treatment as the objects of the experiment, except for the change in the variable under study. Control groups correspond to the experimental groups at every point except the point in question. Controls are frequently used in medical research, as in the test of the polio vaccine where two groups of subjects were used. One group, the experimental one, was given the vaccine, and the other, the control group, was given a placebo — a pharmaceutical preparation containing no medication but given for its psychological effect.

The use of controls is not always sufficient to insure correct results. A story is told of a test of a seasickness remedy in which samples of the drug were given to a sea captain to test on a voyage. The idea of control was carefully explained to him. When the ship returned, the captain was highly enthusiastic about the results of the experiment. Practically everyone of the controls was ill and not one of the subjects had any trouble. "It is really wonderful stuff" exclaimed the captain. The medical research was skeptical enough to ask how the sea captain chose the controls and how he chose the subjects. "Oh, I gave the seasickness pills to my seamen and used the passengers as the controls!"

Rules of Experimental Research

John Stuart Mill, the great English philosopher, set up five rules for evaluating data in experimental research, or to use the philosophical term, "rules for search for causes of phenomena under inquiry." These canons are useful guides in the design and evaluation of experiments and serve as general principles to aid interpretation of data.

The first of these rules is known as *Method of Agreement*. It states that if the circumstances leading to a given result have even one factor in common, that factor may be the cause sought. This is especially so if it is the only factor in common. This principle is universally used but in and of itself is seldom considered as constituting valid proof of cause because it is difficult to be sure that a given factor is really the only one common to a group of circumstances.

The need for such caution is illustrated by the story of a young scientist who imbibed liberal quantities of Scotch and soda at a party. The next morning he felt rather miserable. So that night he tried rye and soda, again in very liberal amounts. The following day he was again visited by a distressing hangover. The third night he switched to bourbon and soda, but the morning after was no more pleasant than were the previous mornings. Being a scientist, he analyzed the evidence. Making good use of the method of agreement, he concluded that thereafter he would omit soda from his drinks since it was the common ingredient in his three distressing instances.

A second rule is known as *Method of Difference*. This canon states that if two sets of circumstances differ in only one factor and the one containing the factor leads to the event and the other does not, this factor can be considered the cause of the event. For example, if two groups of experimental guinea pigs are fed identical diets under identical conditions except that one group also receives a certain vitamin, the fact that the guinea pigs in the group receiving the vitamin outgrow the other group is evidence supporting the hypothesis that the vitamin is beneficial to the growth of the guinea pigs. It is evidence but does not conclusively prove the hypothesis. The result may have been due to other factors and circumstances. For example, the second group of guinea pigs might have been sick or been of a different strain and heredity.

A third rule is that of *Joint Method of Agreement and Difference*. This rule states that if in two or more instances in which a phenomenon occurs, one circumstance is common to all of those instances, while in another two or more instances in which a phenomenon does not occur that same circumstance is absent, then it might be concluded that the circumstance or factor present in all positive cases and absent in all negative cases is the factor responsible for the phenomenon.

The fourth rule, the *Principle of Concomitance*, states that when two things consistently change or vary together, either the variations in one are caused by the variations in the other or both are being affected by some common cause. As an illustration of this principle, Mill cited the influence of the moon's attraction upon the earth's tides. By comparing the variations in the tides with the variations in the moon's position relative to the earth, it can be observed that all the changes in the position of the moon are followed by corresponding variations in the times and places of the high and low tides throughout the world, with the high tides always occurring on that side of the earth nearest the moon. These observations have led to the conclusion that it is the influence of the moon which causes the movement of the tides.

The fifth rule is known as the *Method of Residues*. This rule recognizes the fact that some problems cannot be solved by the technique of experiment called for in the other four rules. The Method of Residues arrives at causes through the process of elimination. When a specific factor causing certain parts of a given phenomenon is known, this principle suggests that the remaining parts of the phenomenon must be caused by the remaining factor or factors (by the residue).

The concept of cause and effect is widely accepted in science. John S. Mill's rules for evaluating causal connection are useful if discriminatingly applied. That they should not be considered a foolproof formula has been demonstrated in the humorous story concerning the distressing effects of soda on the young scientist.

Keeping Records

Data should be entered in the laboratory notebook at the time of observation of the experiment. Among the most unforgivable practices are dishonesty or carelessness in recording the elements or factors in the experiment, neglect in keeping a record of everything done, or failure to take into account every small part of the components and every action of the apparatus. Without records, successes cannot be repeated and failures will not have taught any lessons.

The experimenter sometimes has difficulty in finding or knowing what to record. Each record of an experiment should include the purpose of the experiment, the equipment used and how it was set up, procedures, data, results, and conclusions. Sketches, drawings, and diagrams are frequently helpful. It is important to record what is actually seen, including things not fully understood at the time. Poor or unpromising experiments should be fully recorded, even those felt to be failures. They represent an investment of effort which should not be thrown away because often something can be salvaged, even if it is only a knowledge of what not to do the next time. The data always should be entered in their raw form.

Complete records are beneficial because the mere act of putting down on paper the what, where, when, why, and how of any piece of work will of itself generate ideas of how the work can be done in an improved manner. Where patent questions might be involved, it is desirable to witness and even notarize notebook pages at intervals. The witness should be someone who understands the material but is not a coinvestigator.

Interviews and Discussions with Experts or People Qualified to Give Needed Data

While the discussion technique is fourth on the list, the writer may actually turn to this means first. Once he has ascertained the problem, the writer may first need to determine what he already knows about the problem. After he has clarified his own thoughts, it may be profitable for him to turn to other people who are working in the field or on the problem he is investigating. Frequently, valuable information and shortcuts may be discovered by discussion with experienced and knowledgeable people. If the problem has been one which the researcher's organization has previously done work in, the researcher would do well to consult the files of his organization or company. He might talk to colleagues and superiors who have had experience in this area, but the researcher must prepare himself before beginning any interviews or discussions, especially if he must go outside his own organization to obtain information.

The Interview

The interview may be conducted by letter and by telephone, as well as in person. Letter and telephone interviews are less satisfactory. Direct contact with an individual and a face-to-face relationship often provide a stimulating situation for both interviewer and interviewee. Personal reaction and interaction aid not only in rapport but also in obtaining nuances and additional information by the reactions which are more fully observed in this face-to-face relationship.

Adequate preparation for the interview is a "must." Careful planning saves not only time but also energy of both parties concerned. The interview is used to obtain facts or subjective data such as individual opinions, attitudes, and preferences. Interviews are used to check on questionnaires which may have been used to obtain data, or when a problem being investigated is complex, or when the information needed to solve it cannot be secured easily in any other way. People will often give information verbally but will not put it in writing.

Here are points to consider to promote the success of your interview.

1. Make a definite appointment by telephone or letter, explaining carefully the purpose for the interview, the type of information required, and — this is very important — the significance or importance of the contribution the interviewee will make.
2. Select the right person who is in a position to know and has the authority to give the information you need. Find out as much as possible about the person to be interviewed before the interview.
3. Prepare for the interview. Know the subject-matter of the interview so that the interviewee does not feel he is wasting valuable time explaining basic things which you should have known beforehand. Whenever possible, outline the points for which data are required.
4. Have prepared questions and lead the interview. Do not expect the person interviewed to do your job for you by volunteering all information you may need or pointing out other matters which you did not know about or forgot.
5. Do only as much talking as is necessary to keep the interviewee talking. Do not contradict or argue with the interviewee, even if you know he is wrong. Sometimes erroneous information can be significant data. Be courteous. If there is a point requiring further explanation, ask a question which might show up the other side to the problem.
6. If the interviewee gets off the subject, be ready with a question which will get him back on the track.
7. If the person interviewed says something especially significant or expresses himself particularly well on some point, or if he offers statistics, figures, or mathematical formulas, or other matters which might require careful checking, ask his permission to write such statements down.
8. Do not prolong your interview. Should the originally allotted time have passed, remind the interviewee of this fact. It is up to him to ask you to stay longer in order to finish the conversation.
9. Immediately after the interview, record the answers which you have received from the interviewee. At a later time it is a matter of courtesy to send the interviewee the record of the interview to give him the opportunity to correct any errors you may have made or

permit him to change his mind about anything he has said. If you are planning to quote him, be sure that the interviewee is aware of this and gives you permission to do so.

The interview has the advantage that the interviewer can partially control the situation and can interpret questions, clear up misunderstandings, and get firsthand impressions which might throw light on data. The interview is a valuable instrument in those situations where the only data available are opinions.

Questionnaires

As a research tool, questionnaires require judicious handling. However, there are certain situations where a questionnaire must be used and can be used effectively. A much wider geographic distribution may be obtained more economically from a questionnaire than from an interview. Certain groups within the population are frequently more easily approached through a questionnaire; for example, executives, high income groups, and professional people. Questionnaires often allow for opinions from the entire family. A questionnaire can be filled out at the respondent's leisure and more thought can therefore be given to the response. Finally, by careful sampling methods a representative sample of a study universe may be made.

Like the interview, the questionnaire must have a reason for being used. It should be designed to require as little time as possible for completion. A covering letter should be included with a questionnaire that will motivate the reader to answer the questions. The letter should tell the use to be made of the answers. If possible, the writer should offer a copy of the summary of the results of the investigation and assure the reader that the task of answering will not require too much of his time.

The questions should be phrased in the form of a list, not in lengthy sentences or paragraphs. The wording must be structured with care to insure clarity and completeness. Where possible, questions must be grouped into logical arrangements. The first few questions should be easy to answer and should be interesting in order to secure cooperation. One question should stimulate interest in the next. Questions should be arranged in an order that aids the respondent's memory. The most effective questionnaires are frequently arranged for easy "yes" or "no" answers. Each question should contain one idea only. If the questionnaire becomes too long or involved the reader may get discouraged and neither finish nor mail it. Questionnaires should include a self-addressed and stamped envelope.

Much of the success of a mail questionnaire depends on its physical appearance, the arrangements of questions, the ease with which answers may be filled in, and the amount of space left for comments. A questionnaire of one page will bring more returns than one of two pages, and a post card is frequently the most effective questionnaire instrument. The sending of questionnaires should be accurately timed so that they reach their destination at a time when favorable replies may be expected. As far as possible, select periods when data or information may be fresh in the mind or when interest in a subject is ripe. Avoid vacation periods, periods when reports are made, and the close or the beginning of the year. (Refer to the sample questionnaire in the appendix, page 364).

Systematizing, Analyzing, and Interpreting the Data

The data you have obtained from your investigative techniques must be systematized, analyzed, and interpreted in order to answer your problem. The data that you have accumulated must be edited in order to ascertain the degree of their homogeneity, accuracy, and completeness. If gaps appear in the mass of data acquired, further steps may be needed to obtain the missing information.

Systematizing the Data

After you have decided that the raw data are complete and satisfactory, the next logical step is clear, methodical arrangement. This means that your raw or basic data must be so organized that they will lead you to the answer to your problem. This type of arrangement is best done by listing the elements of your problem on note cards and then arranging the raw data alongside the listed components of the problem. This systematization is chiefly a matter of linking the data or bases of evidence to the specific elements to which they relate. Make use of the block diagram, Figure 1, this chapter.

Analyzing and Interpreting the Data

The previous step consisted of editing, sifting, and arranging the data in an orderly relation to the elements in the research problem. You may now find it desirable or necessary to recast your data in a more refined way in order to derive inferences or conclusions. This is done by tabulation and retabulation of the raw data. Averages in various forms — modes, medians, quartiles, deciles[5] — may be used to obtain quantitative criteria of mass data. Frequently, only by averages will the complete meaning of mass data be comprehended, and they may then reveal qualitative differences. Data may have to be refined in a statistical form to eliminate the influences of such factors as cyclical or seasonal fluctuations. Intricate relationships may be revealed by tabulating, graphing, and diagraming. This

---

[5]*Mode* is the value or number that occurs most frequently in a given series; *median* is the middle number in a series containing an odd number of items (e.g., 7 in the series, 1, 3, 5, 7, 15, 34); *quartile* is the designating of a point so chosen that three-fourths of the items of a frequency distribution are on one side of it and one-fourth on the other; *decile* is any of the values of an attribute which separate the entire frequency distribution into ten groups of equal frequency.

refinement often results in greater insight and suggestions of inferences which will answer the components of the problem. By use of inductive and deductive reasoning, conclusions or answers to the problem can be arrived at.

Interpretation, as will be seen in chapter 10, is the process of arriving at inferences, explanations, or conclusions by examination of data and applying to them the process of logical analysis and thought. Interpretation is accomplished when phenomena are made intelligible by relating them to the requirements of the matter or problem being investigated. Individual facts and/or opinions are placed within a deductive or inductive system for inferential generalization. In short, interpretation, by use of observation, logical analysis, synthesis, experience, knowledge, and insight, arrives at explanations and answers to the questions the investigation must resolve.

Applying and organizing the interpreted data constitute the next step, which will be discussed in chapter 5.

Discussion Problems and Exercises

1. What is a report? What role do reports play in present-day industry and scientific endeavor?

2. The use of the scientific method in problem solving receives much emphasis throughout the pages of this text. Its purpose is to force the technical writer to think somewhat as the researcher thinks. Justify this emphasis. In what report writing situations would this emphasis be detrimental? Explain.

3. What sources of information are employed in the historical method of research? Its aspects of library searching are useful in what ways to other methods of research?

4. What is the role of the problem in the technical paper or report?

5. Locate and examine a report each from business, from industry, from a private, government, or educational research laboratory, and from a university experiment station. What are their differences and similarities? What is their purpose? intended use? readership? Can you classify these four reports? How was the research carried out in each? What methods were employed? Note their format, organization, style of writing, use of illustrations, and appendixes, if any.

6. Examine and analyze your experience and training for subject situations and problems you might investigate for writing a report along the lines indicated in this text. The following list may offer suggestions.
   a. How to design a unique anti-icing device for the home concrete walks and driveways.
   b. How to correct the problem of dampness in X building.
   c. How to prevent erosion of bentonite lining in irrigation canals.
   d. What are the comparative merits and faults of teaching machines and the classroom teacher?
   e. When and how should X township increase its water supply?
   f. How can X factory improve its production efficiency?

    g. What should the power level calibration be for the X-2 thermal nuclear reactor to insure safety from radiation?

    h. What formal training is needed by the trade and industrial instructor to enable him to give vocational guidance?

    i. Is high pressure hot water heating more economical than steam to heat X complex of buildings?

    j. What is the origin of the Ingleside Formation in Colorado?

    k. What is the effect of antihistamine on the antibody titer induced by influenza virus in mice?

    l. Can hormone implants in heifers produce faster maturity, improved breeding efficiency, and greater udder development?

    m. What causes the diurnal migrations of zooplankton in X lake?

    n. Can the X genus of sagebrush be permanently controlled?

7. Set up a work schedule listing the tasks and time table for accomplishing the tasks for the identification of a research problem, its investigation, and the writing of a report on the research.

8. Assemble a complete bibliography for the problem you will be investigating and reporting. Use 3″ x 5″ cards for your entries. Turn in note cards for two sources you have read.

9. a. Prepare a set of questions for an interview with an "expert" on an aspect of the problem you are investigating and reporting.

    b. Design a questionnaire for your use in obtaining information on your problem or an aspect of it.

10. If you are conducting an experiment or are doing field work and observation to obtain a solution to your problem, keep a laboratory notebook or journal to record your data.

11. After you have identified the problem you are to investigate for your report in your preliminary analysis phase, build a block diagram like the one pictured on page 85 of this chapter. Resolve or factor out the major elements of the problem and their components. Place these elements and components within their proper boxes. This type of block diagram analysis will be invaluable in prosecution of your research and its reporting in the next stage.

12. Discuss the four major steps in report writing. What inference is to be drawn from the fact that the writing process is the last step?

References

1. Anderson, Chester R., Alta Gwinn Saunders and Frances W. Weeks. *Business Reports*. Third ed. New York: McGraw-Hill Book Company, 1957.

Report Writing—
Organizing the Report Data

When the data for the report are in, we face the problem of organizing them into useful, functional form, interpreting and analyzing them to see what they mean. If your preliminary planning and analysis have been carefully done, the work at this stage is almost half accomplished. You ask yourself questions again: What facts have I found that are significant to my problem? What facts are most significant? What is their significance? How do these facts answer my problem? What is the answer to my problem?

The answers to these questions become the basis for the organization of the main body of the report.

Developing the Thesis Sentence

Having begun with a problem, the investigator has accumulated data for answering it. Raw data of themselves do not make a report. Raw data properly examined, analyzed, and interpreted through the methods of logical reasoning become the material out of which the report is designed. The first step in structuring a report is to develop a clear and concise statement of the answer. The answer may be positive or negative. The statement, specifically and clearly stated, is the payload which you are bringing to your reader. This payload often is called the *core idea* or *thesis sentence*. The thesis sentence is the answer to the problem or situation which you have investigated and are reporting. It is what you want your reader to know. The thesis sentence, or core idea, is a single comprehensive statement in complete sentence form which synthesizes the complete report and presents the problem's answer to the reader succinctly and clearly. A properly formulated thesis sentence is concrete evidence of how clearly and simply you, the writer, see your subject. It is the backbone for unifying all of the elements of the report.

Adequate research requires the clarified statement of where the research has gone, what it has meant, and to where it has led. Since clear thinking is the basis of clear writing, the writer must state the answer to the problem clearly and succinctly. So the first step in designing your report is to state clearly and concisely the core idea, or thesis sentence.

For the sake of illustration, we might examine a few hypothetical problems and see how a thesis sentence might be structured from the data.

1)	PROBLEM:	What causes the diurnal migrations of zooplankton?
	THESIS SENTENCE:	Migration of zooplankton is caused by an interplay of factors such as light, temperature, chemicals, and quest for food.
2)	PROBLEM:	Could a light, single-engine, pontoon-type aircraft be used effectively for fire suppression in the lake states?
	THESIS SENTENCE:	The Beaver, a light, single-engine aircraft, could be used effectively for fire suppression in the lake states.
3)	PROBLEM:	Are existing environmental conditions in Decatur County, Kansas, conducive to the propagation of bobwhites in sufficent numbers to support an annual hunting season?
	THESIS SENTENCE:	The environmental conditions in Decatur County, Kansas, are unfavorable to increased population of bobwhites.
4)	PROBLEM:	Can implanted sex hormones in dairy heifer calves be used to promote maturity, better breeding efficiency, and greater udder development?
	THESIS SENTENCE:	Sex hormone implantation in dairy heifer calves is useless in promoting maturity and is detrimental to breeding efficiency and udder development.
5)	PROBLEM:	Which of three wiring processes presently used in X Manufacturing Company is most efficient and best suited to the conditions of the factory?

THESIS SENTENCE: The printed circuit process is the most efficient and best suited for the special factory conditions of X Manufacturing Company.

The thesis sentence serves to keep the report writer from exceeding his limits. It serves to help him follow consistently his objective, and in certain situations it suggests to him points to be included and the order of their sequence. The thesis sentence is the test of the writer's mastery of his subject as a whole. Inability to formulate the thesis sentence is an indication that either the writer has not completed his investigation or he has not been able to analyze logically and interpret the data accumulated in the investigation.

In purely informational types of reports, it might be difficult to synthesize the material of the investigation into the thesis sentence; here, a summary statement is called for. But in all analytical types of reports, the thesis sentence is practical and necessary.

Although the thesis sentence is necessary in the construction of the report, it may not be found verbatim anywhere in the report itself. It might be compared to a working sketch which must be drawn (written) in order to point to the directions where the model or composition is going. The thesis sentence obviously cannot be written until all the research has been completed and all of the raw data analyzed in relation to the problem. How else is the answer derived? The answer to your problem becomes the core idea of your report. It is the dominant idea you want your reader to reach after he completes reading your report.

Formula for Organizing the Report Structure

Write the thesis sentence on a 3" x 5" card, then set it on a table before you. A reconstruction process then follows. Use other 3" x 5" cards and set down the major facts which have led you to the thesis sentence. Write each major fact or idea on a separate card. Place the card with your thesis sentence at the left of the table. Next, you make a mental equal sign, and then arrange the major facts or ideas of the data of your investigation which equal your thesis sentence. What you have arranged before you is your thesis sentence = one major idea + another major idea + another major idea + . . . as many major ideas as are necessary to lead you to the answer to your problem.

As stated above, the synthesizing process has been turned around. In actuality, the formula for the thesis sentence was derived from:

$$\text{major idea}_1 + \text{major idea}_2 \ldots + \text{major idea}_n \xrightarrow{\text{(yields)}} \text{thesis}$$
sentence.

In arriving at this formula, you may find it advisable to check on yourself by having each major fact on a separate card, followed by cards containing substantiating proof or data below it. Spread the cards out on a table and regroup the cards into the best sequence for an outline, beginning with the most logical or valid fact or datum relevant to the answer. Follow sequentially in the order of the logic. The logic for the structuring will be discussed later in the chapter. The major facts become headings in your report. To identify their place in the outline, number each of these and number their substantiating matter.

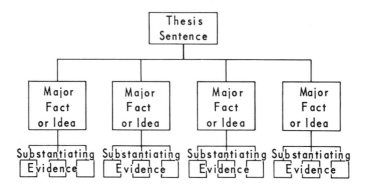

Figure 1.

Each block of major fact or idea relates to and answers the block of the major element of the problem diagramed in chapter 4. Substituted for the block of the Statement of the Problem is that of the Thesis Sentence. Replacing the block of the major elements of the problem are the blocks of the major facts or ideas. The blocks of substantiating data or evidence replace the blocks of components of the earlier diagram.

Patterns for Organizing the Data

I have just noted that the major facts and ideas and their substantiating data are to be organized in a logical sequence. Let us now see how the various kinds of content matter might lend themselves to effective arrangement for bringing the reader to the answer the writer has reached. In organizing the material of the investigation, the report writer must keep in mind the purpose of his report and the purpose for which the answer to the problem will be used by the reader. A report's organization must be functional. The writer arranges his data so that they meet the reader's specific needs.

A number of approaches in the organization of the report are possible. The one you choose depends on such factors as:

    Purpose

    Needs of the reader·

    Nature of the material

Temporarily we are going to skip the elements of introduction and conclusion in the report and examine the part which contains the main data, the body of the report. There are two major methods of organizing the data contained in the body of the report — the logical and the psychological.

Logical Method of Organization

The logical method builds its case step by step within the logic of its pattern, leading to conclusions and possibly recommendations. Since problems and their investigations vary in nature, the logic to the organization of the data is going to be different in each report. Accordingly, in the logical method, there are a number of patterns to the organization of the main body of data. Logical patterns include:

1. *Chronological Pattern.* The chronological pattern of organization is the simplest and often the most obvious way of arranging the data of an investigation. It follows a time sequence. It is the pattern most frequently used in informational types of reports, particularly those of a historical nature. Chronological order may also be used in reports using other organizational patterns where the presentation of a sequence of doing a thing is important to the exposition. In a technical experiment, for example, conditions or results within a sequence must be presented as they occur; otherwise the reader will be unable to duplicate the work. In an investigation of a historical nature, the chronology of events is of vital importance to the understanding of events, so the presentation of data must follow the proper time sequence. This type of organizational approach does not lend itself well to either complex situations or those situations where nontemporal relationships and special emphasis are important. The chronological approach serves in simple, uninvolved situations where the time order is important and special emphasis on particular matters is not required.

2. *Geographical or Spatial.* Data may be presented or arranged according to geographical or spatial relationships. For example, a report to the president of a corporation on how to allocate space to the various departments of the corporation in a new building might very logically begin with an assignment of the various wings moving east or west, as the case may be, or moving from the basement up to the various floor levels of the building. Or, in his State of the Union address at the start of a new Congress, the President of the United States might review the economic state of the country geographically, beginning with the northeastern states, working south and west across the country.

3. *Functional.* If you were reporting on the design of a new mechanism or instrument, a logical approach would be to examine each component or assembly by the function it performs in the working of the whole.

4. *Order of Importance.* It is sometimes appropriate to begin with the data most significant to the problem or situation, then move on to the next most significant, etc. This approach would be highly suitable in a situation where the researcher must find the best plan or procedure for accomplishing a certain end. Related to this pattern is the approach of

putting first those phases of a project in which the greatest success was achieved and following sequentially to the least successful phase.

5. *Elimination of Possible Solutions.* Within this pattern you would examine all possible solutions to a problem, beginning with the least likely and working toward the best possible solution. (This approach is opposite to that of No. 4 above.) This approach borrows from the narrative technique of building toward a climax.

6. *General to Particular (Deductive) Method.* The sixth structural pattern starts with the discussion of general principles and then deduces the specific applications from these. In reporting on the efficacy of inaugurating a company training program, for example, you might discuss first the general benefits to be derived from a training program and then arrive at a specific benefit from a specific training, as for example, report writing. You would spell out the details of how the class in report writing would result in the benefits previously said to derive generally from a training program.

7. *Particular to General (Inductive) Method.* This pattern is the reverse of No. 6. You might start with a specific training program — again, say, report writing — and then induce general benefits derived from such a specific program.

8. *Simple to Complex (Known to Unknown).* This pattern of organization is appropriate to situations where it is advisable to begin with the simple case or known situation and then move to more and more complex or unfamiliar grounds.

9. *Pro and Con.* In the report which investigates whether to do or not to do something, the material on which an answer is based might appropriately be grouped into data for and data against, or advantages and disadvantages.

10. *Cause and Effect.* This approach is particularly useful in the exposition of problems which deal with questions like: What is it? What caused it? or What are the effects of . . . ? The writer may begin with a fact or a set of facts (cause or sets of causes) and proceed to the results or effects arising therefrom. Conversely, the writer may examine and report the causes from which effects arise.

These ten patterns are not mutually exclusive and are frequently used in combinations.

The Psychological Pattern of Organization

With the psychological pattern of organization the most important data of the report are arranged in the most strategic place — the beginning. The psychological method follows no order of time or sequence in which the data are collected. Usually the conclusions are placed first and the discussion second. This report pattern — sometimes known as the *double report* — originated in business and industry, where the busy executive prefers to conserve his time and get at once to the matters he needs to know. When he is reading a report, the executive wants to know quickly what the problem was, what the answers to the problem are, and how the answers were derived. The psychological pattern permits him to see the conclusion first so that he can grasp the major issues. The study of supporting data is usually left to subordinates. The organizational sequence of the psychological pattern is as follows:

A short introductory section states the problem. This is followed immediately by a section of conclusions and recommendations. Because the executive reading the report may be interested in knowing how the conclusions were reached, the section on procedure follows immediately. Should he want to check points more carefully, he will continue to read the results based on the procedures. Then, if he wants all the details, he will turn to the remaining section, the discussion.

In the psychological organizational pattern the main body of the report does not contain "arithmetic" or other raw data. This material is put into the Appendix. The busy executive wants only the significant factors which have resulted in the generalizations making up the conclusions and recommendations. Technical subordinates will check the computations and the raw data in the Appendix if the executive wants full confirmation of the recommendations on which he is to act.

Just which of the two major methods of organization to use is determined by the writer after considering all aspects of the purpose and nature of his report, and of the needs of his readers.

The Outline

The synthesizing approach we have been discussing in this chapter can conveniently take form first as an outline. The outline is a schematic road map of the report and shows the order of topics and their relationships. The outline makes the writing of your paper easier and more effective and enables you to write with confidence and to focus on one stage of the paper at a time. You can see how the whole will take shape, and you will not be distracted while writing by the question of whether to put a particular piece of information in here or reserve it for later. The outline, then, permits the investigator to test the adequacy of his data. Subsequently, it offers the reader guideposts or road signs in the form of heads and subheads. The outline is the last step in the organization of the data for the writing of the report.

Outline Formats

Several types of outlines are used in practice. The simplest is the topical outline. Subjects or topics are noted in brief phrases or single words and numbered sequentially. The topical outline is a useful device in thinking through the organization of the paper. To begin, you may want to jot down key ideas and data related to the material of your paper. For example, let us take the case of a geology student whose research problem is tracing the origins of the intrusive body of rocks in Big Thompson Canyon, Colorado. Initially, he would write down the series of major facts and their components as follows:

1. Introduction
2. Historical background
3. The problem
4. Location and description of intrusive
5. Location and description of Silver Plume Granite outcrop
6. Specimen selection and field trimming
7. Microscopic analysis of specimens
8. The intrusive specimen
9. The Silver Plume specimen
10. Comparison of field and laboratory data
11. Conclusion
12. Recommendations

As you can see, topics are listed in a sequence without any indication of specificity, importance, or relationships, or that some are topics subordinate to major subject heads. In early planning, this type of outline is useful; it sets down on paper matters to be considered for more detailed structuring.

The second phase of outlining would focus on greater specificity, subordination, and relationships. The analysis entailed would reveal a breakdown of greater detail. This phase of the outline might be structured as follows:

Introduction
    Significance of problem
    Scope
    Historical background
    Definitions
Analyzing the Problem
The Intrusive
    Location of intrusive
    Description of intrusive
The Silver Plume Granite Outcrop
    Location of the outcrop
    Description of the outcrop
Specimen Selection
Field Trimming
Microscopic Analysis of Specimens
    The intrusive specimen
    The Silver Plume specimen

Field and Laboratory Data
  Similarity of occurrence in field
  Comparison of laboratory data
Conclusion
Recommendations

This second outline is an improvement over the first. It is more specific and it identifies relationships, allows some emphasis, and shows subordination of lesser items to more important ones. Nevertheless, it does little more than identify these matters. The writer still faces a painful task of thinking through the ideas and the status of the evidence to be presented. At some point in the composition process, the writer will have to reason through all points and details of the report. Many writers accomplish this by a third phase of outlining — the *sentence outline*. This form picks up the schematizing where the second phase left off. It places each topic within a complete sentence. Completing the thought of each topic enables the writer to test the context of the topic's environment; such deeper thought not only reveals flaws in the data as heretofore seen by the writer but also permits deeper probings, examinations, analysis, and interpretation of major points and minor details. The result is clarification of aspects of the problem and derivation of insights not previously attained. A concomitant and important benefit is an improved, orderly outline in complete detail, with properly designated relationships, emphasis, and subordination.

We might now examine a sentence outline based on the previous two phases:

Sentence Outline for

A PRELIMINARY REPORT ON THE POSSIBLE
ORIGIN OF THE INTRUSIVE BODIES IN THE
BIG THOMPSON CANYON[1]

Thesis Sentence: The intrusive body in Big Thompson Canyon, Colorado, is a recently exposed portion of a larger mass known as Silver Plume Granite which was intruded into the base of the Rocky Mountains in Pre-Cambrian times.

  I. This investigation undertakes to trace the geologic origin of the intrusive bodies in Big Thompson Canyon, Colorado.
     A. The purpose of this investigation is to serve as a field analysis problem for beginning geology students in properly analyzing and identifying geologic features.
     B. The meaning of certain geological terms, though frequently used by students, are not always fully understood and should be defined.
        1. An intrusive body was once a molten lava which cut across or was injected into overlying sedimentary or metamorphic rocks.

---

[1]Check this outline with the student report based on it in the Appendix, pages 393-404.

2. An acidic rock, such as granite, is derived from a molten source which was high in silica content.
3. A batholith is an igneous body more than forty square miles in area.
4. A stock is an igneous body less than forty square miles in area.
5. A dike is an igneous body that has cut across layers of sedimentary or metamorphic rock.

C. Field work was conducted by the author and other students at the intrusive in Big Thompson Canyon and at an outcrop of Silver Plume Granite west of Horsetooth Reservoir.
1. The intrusive in Big Thompson Canyon is located at the base of Palisade Mountain on State Highway 34.
2. The location of the Silver Plume outcrop is midway between the west edge of Horsetooth Reservoir and the town of Masonville.

D. Early preliminary observation and interpretation by geology students brought forth two varying opinions:
1. That the intrusive was a granite stock:
2. That the intrusive was a granite dike.

E. Extensive fieldwork and a library search have shown validity to both opinions.
1. Analyzed samples taken from the intrusive show that orthoclase feldspar, plagioclase feldspar, quartz, and biotite mica are present.
  a. Comparison of these constituents with constituents from a known outcrop of Silver Plume Granite shows that they are one and the same.
  b. Variations in color of the outcrops are due to magmatic differentiation.
2. The Silver Plume formation is composed of small batholiths and stocks.
  a. As the intrusive in question is relatively small in size, it can be assumed that it is an offshoot from a larger, underground body.
  b. Similar small intrusive bodies at other locations bear this out.

F. An analytical table of comparative mineral constituents and a map of the region validating this conclusion will accompany the report.

II. Field work and laboratory experiments combined with material gathered from a library search answered the question about the intrusive body.
A. Structurally, the intrusive and the outcrop of Silver Plume Granite are dissimilar, but of the same geologic period.
1. The intrusive is of a gray homogeneous granite with very little fracturing.

    2. The Silver Plume Granite outcrop is flesh-colored with large, individual crystals.

    3. Both of the bodies investigated were intruded into a Pre-Cambrian, metamorphic, country rock.

  B. Hand specimens were taken from the site of the intrusive and the known Silver Plume outcrop.

    1. A geology pick or prospector's hammer was used to break off good-sized rocks.

    2. The rocks were held in the hand and chipped to approximately 1″ x 3″ x 5″.

    3. The rocks were chipped in such a manner as to eliminate hammer bruises on the faces and also to expose only un-weathered rock material.

  C. The hand specimens, as observed under a microscope, were found to differ slightly in physical properties.

    1. The specimen from the intrusive was found to be a granite.

      a. The specimen from the intrusive contained orthoclase feldspar, 60 percent; plagioclase feldspar, 5 percent; quartz, 20 percent; and biotite mica, 15 percent.

      b. The color of the intrusive specimen was gray.

      c. The component grains of the specimen were one to five millimeters in diameter.

    2. The specimen from the known outcrop of Silver Plume Granite was flesh-colored with grain sizes from 1 to 30 millimeters in diameter.

  D. Comparison of field observations and laboratory data with information from T. S. Lovering's professional paper showed that the intrusive is a portion of Silver Plume Granite.

III. This preliminary research of the problem has led to a theory that may be fully validated following a more extensive investigation.

  A. The intrusive in Big Thompson Canyon is a portion of Silver Plume Granite which was intruded into the roots of the Rocky Mountains during Pre-Cambrian times.

  B. It is recommended that a more complete investigation be undertaken to determine definitively the relationship between the intrusive bodies in Big Thompson Canyon and that of the Silver Plume Granite.

Ideas and information in a report should be presented in a steady progression so that the reader feels he is getting somewhere as he reads. The reader must have important data clearly pointed out to him. The topical outline does not provide any easy mechanics for this. The outline arranged to show relationships and subordination is an improvement. The sentence outline, because it has put the writer through the chore of clarifying his thoughts about the main facts and their substantiating details, promotes a more logical composition and a more readable and comprehensible report.

The sentence outline just shown reveals the conventional structuring of this form. It uses Roman numerals to designate the main divisions and capital letters to indicate major subdivisions under each main division. Further subdivision makes use of Arabic numerals, and still further division makes use of lower case letters. Each entry in the sentence outline is composed in a complete grammatical sentence. This is demanded not only for the Roman numeral headings but also for divisions and subdivisions within them. Since a topic is not divided unless there are at least two parts, logic demands that there be at least two subheads under any division. For any Roman number I, there should be a Roman numeral II; for every capital "A," there should be a "B" in sequence. If only one point follows a main heading, make that point part of the main heading:

   I.
     A.
       1.
         a.
         b.
         c.
           1)
           2)
       2.
     B.
   II.

The sentence outline enables the writer to control the content and structure of his report in order to secure unity and coherence. Though more difficult to construct than the topical outline, the sentence outline is recommended because this method forces the writer to think his points out more fully and to express them as whole units. Each sentence in the outline can become the topic sentence of a paragraph in the final writing.

The Decimal Outline

The decimal system of outlining, though of more recent origin, is being used with greater frequency. It employs decimals to show the rank of heads and subheads.

   1. First main heading
      1.1 First subdivision of main heading
      1.2 Second subdivision of main heading
   2. Second main heading
      2.1 First subdivision of second main heading
      2.2 Second subdivision of second main heading
         2.2.1 First subdivision of 2.2
         2.2.2 Second subdivision of 2.2
            2.2.2.1 First subdivision of 2.2.2
            2.2.2.2 Second subdivision of 2.2.2

The decimal system enables easier revisions and additions. It can be used with a sentence outline or with a topical outline.

Format Considerations

The outline may be single-spaced or double-spaced, or single-spaced within groups of headings and double-spaced between such groups. The writer should be consistent, however. A very short outline may be arbitrarily double-spaced in order to fill the page. All symbols, letters, and numbers should appear in a straight, vertical line throughout the whole. All capital letters should be indented the same distance, and all Arabic numerals the same distance, as should all lower case letters. Three to five spaces for each level is conventional. Consistency in the amount of spacing will add neatness of layout and appearance. Excessive indentation will look as unattractive as too little or none.

Introduction, Body, Conclusion

The final draft of the outline can become the basis of the table of contents of your report. During the actual writing of the paper, the outline is used as the plan for the composition. Structurally, the outline, like the report, has a beginning, middle, and end, or to use the classical structural terms — introduction, body, and conclusion. The introduction and conclusion are fixed points at the beginning and end of the outline. The experienced writer spends much time on the outline because it saves him problems in the later writing of the paper.

Frequently, the body, which will begin with Roman numeral II, or if the decimal system is used, with 2., may have two or three major divisions within it. The conclusion section usually may be encompassed in one Roman numeral, although it is sometimes advantageous to separate the recommendations from the conclusions by creating another major heading preceded by a Roman numeral. This is appropriate if the recommendations are significant. Too many major heads could be symptomatic of inadequate analysis of the data.

In Conclusion

After you complete the outline, check it for sequence of major divisions. Are they in the right order? Look over your note material to be sure nothing important has been omitted. Check for duplication of headings. All main headings should add up to the thesis sentence. Are all main headings really main headings or should some of them be subheads? Check your subheads for proper subordination and sequence. Do all subheads within a major heading add up to the major heading? Should some of the subheads be promoted to main heads? Finally, check the wording of all divisions for clarity, correctness, and parallel construction. Usually, there are no more than four or five major heads. If a paper is to be long, the increased length should result from the use of more subheads, rather than from an increase in the number of major headings. The outline should have no less than three and no more than five or six major heads, which are, in the conventional outline numbering system, preceded by Roman numerals.

The sentence outline forces the writer to test the adequacy of his data. Not all data included are of equal significance. Evaluating facts correctly means giving them only as much weight as they have value in accomplishing the objectives of the report.

In summary, this type of planning gives the report coherence, continuity, and unity. By focusing on a central idea — the thesis sentence — the writer succeeds in presenting ideas and information in a steady progression. As a result, the reader feels he is getting somewhere as he reads; everything in the report is pertinent to the problem. The outline has enabled the writer to achieve an overview of the report and to see relationships that are decisive in contributing to the objective. These factors can only help the reader.

Discussion Problems and Exercises

1. Interview several prolific researcher-writers on your campus and off campus. Find out how each organizes his research data for report or professional paper writing. Find out what systems of outlining each person uses. Find out the researcher's individual reaction to the sentence outline. If he does not use it, would he recommend it to a beginning research writer? Does the researcher utilize the device of the "thesis sentence" or "core idea"? Report your findings to the class.

2. a. Explain how the thesis sentence is developed.
   b. Discuss fully the role the thesis sentence plays in organizing the material of a report.
   c. Develop the thesis sentence for your own report.

3. After you have derived the thesis sentence for the report you are writing, build a block diagram by using the major facts or ideas and their substantiating evidence within the blocks of the diagram. Use the formula recommended in this chapter.

4. Analyze the four reports you have located in problem 5 of chapter 4 as to the types of organizational patterns employed by the report writers.

5. Develop first a topical outline of one of the above reports, then a sentence outline.

6. Develop a sentence outline for your report by going through the three phases discussed in this chapter. The work you do in problem 3 should simplify your task.

7. Your text claims that the process of outlining promotes coherence, continuity, and unity of your report. Defend or attack this claim.

8. Evaluate, criticize and rewrite the student sentence outline on sagebrush control. While knowledge of the subject-matter would be helpful, it is not essential to correct the faults contained:

SAGEBRUSH CONTROL

Thesis sentence: The most effective method of sagebrush eradication in Western Colorado is either grubbing, railing, harrowing, blading, burning, plowing, or spraying.

 I. The geographical area studied, species of sagebrush studied, and purpose of this research need to be explained.
   A. The area included in western Colorado is largely covered by big sagebrush (Artemisia tridentata).
     1. The area of western Colorado includes thousands of acres.
     2. This area of Colorado is typical sagebrush land in physical characteristics.
     3. Big sagebrush (Artemisia tridentata) is the dominant species in western Colorado.
   B. The purpose and object of this research are economic in nature.
     1. Valuable land is presently occupied by big sagebrush.
     2. With only moderate costs sagebrush can be eradicated by various methods.
   C. Past and present research in the field is inconclusive for western Colorado.
     1. Past research has been general.
     2. Present research is still inconclusive as applied to small areas.
   D. A restricted approach is taken toward the problem.
     1. The most effective method of eradication only is sought.
     2. Library research, interviews and field work are the resources of information.
       a. Extensive library research is the basic source.
       b. Interviews with authorities in the field of range proves valuable.
       c. Verification of data through field investigation is necessary.
 II. There are several different methods of sagebrush eradication.
   A. Grubbing is one method of sagebrush eradication.
     1. Advantages of grubbing are few.
     2. Disadvantages of grubbing are many.
   B. Railing is a method of sagebrush eradication.
     1. Advantages of railing are many.
     2. Disadvantages of railing are few.
   C. Harrowing is a method of sagebrush eradication.
     1. Advantages of harrowing are moderate in number.
     2. Disadvantages of harrowing are few.
   D. Blading is a method of sagebrush eradication.
     1. Advantages of blading are few.
     2. Disadvantages of blading are many.
   E. Burning is a method of sagebrush control.
     1. Advantages of burning are many.
     2. Disadvantages of burning are few.
   F. Plowing is a method of sagebrush eradication.
     1. Advantages of plowing are many.
     2. Disadvantages of plowing are few.
   G. Spraying is another method of sagebrush eradication.
     1. There are many advantages to spraying.
     2. The disadvantages of spraying are few.
III. The most effective method of sagebrush eradication in western Colorado is either grubbing, railing, harrowing, blading, burning, plowing, or spraying.

6

Report Writing—Writing the Elements of the Report

Structural Elements of the Report

So far we have been examining the main discussion or body of the formal report. The formal report has other elements, usually arranged in a prescribed form. If you will recall, chapter 4 stated that there is no universal "right" form governing the arrangement of a report's elements. Elements should be arranged in an order which promotes greatest convenience of use to, and understanding by, the reader. Reader requirements vary as reporting situations and problems vary. Many research and industrial organizations and government agencies have developed style manuals, guides, and specifications to insure standards and promote effectiveness of reports. These guides prescribe forms and arrangements of report elements in keeping with the organization's requirements.

Specific details of form and content may vary considerably, but generally, elements of a report find arrangement in either the logical pattern of organization or the psychological pattern (the double report), as described in chapter 5. The present chapter follows the logical method of organizing the report's elements because that approach is universal. Once you, the student, have mastered the ability to structure and arrange the data of the investigation and the elements of the report within the framework of the logical pattern, you will be able to meet the requirements of any type of report pattern and form.

The elements of the formal report are:

Letter of Transmittal
Cover
Title Page
Abstract or Summary
Table of Contents
    List of Tables and Illustrations
The Report Proper
    Introduction
    Body
    Terminal Section
Bibliography
Appendix
Index
Distribution List

Let us examine each of these elements.

Letter of Transmittal

The major purpose of the letter or memo of transmittal is formally to present the report to the reader as a matter of record. The letter of transmittal indicates exactly how and when the report was requested, the subject-matter of the report, and how the report is being transmitted—as an enclosure of the letter or under separate cover. Its length varies. Frequently, it is necessary to say only, "Here is the report on . . . (the problem is indicated) which you asked me to investigate in your letter of December 21, 19———." If the report is sent to someone within one's own organization, the form used is a memo from the author to the recipient.

Letters of transmittal have been used occasionally to refer to specific parts of the report and to call attention to important points, conclusions, and recommendations. Some letters of transmittal may even go into certain elements which appear in the introduction, such as statements of purpose, scope, and limitation. The letter of transmittal may mention specific problems encountered in making the investigation. For example, a delay due to a strike or shortage of certain materials may have influenced the performance time, but not the data and their results; this situation may be mentioned in the letter of transmittal. Some letters may even offer acknowledgments.

The placement of a letter of transmittal varies in practice. It is sometimes bound within a report, or it may be placed within an envelope and attached to the outside of the package; in the latter case the address on the letter envelope may serve as the address for the report packet itself.

When the report is sent to the intended reader within one's own organization by hand, or through organization communication means, the memo of transmittal is frequently clipped to the outside cover of the report or placed within the cover for protection.

Examples:

805 Remington Street
Fort Collins, Colorado 80521
December 5, 1974

Mr. E. S. Murphy
Dept. of Physics
Colorado State University
Fort Collins, Colorado 80521

Dear Sir:

In compliance with your request for a complete report on the Power Calibration Curve for the CSU AGN-201 Reactor, this report is submitted.

This report defines the complete problem, develops the method, the theory for determining the power calibration curve, and discusses the results and conclusions of the experimental data.

The power calibration curve was found to be linear with the maximum power level of 101.5 milliwatts at 5×10^9 amps on channel 3 of the reactor control console. This compares favorably with 100 milliwatts determined by the Aerojet General Nucleonics Corporation.

It is recommended that in future experiments the data and data calculations be rounded off to less significant figures than used in this report. This can be done with sufficient accuracy still maintained.

From the indications of the data in this report, an interesting experiment for determining the effect of foil thickness upon activity may possibly be incorporated in the Ph-124 Experiments in Nuclear Physics course.

Yours truly,

Arthur R. Swanson

900 South College
Fort Collins, Colorado 80521
November 28, 1974

Dr. Herman Weisman
Room B-120, Engineering Building
Colorado State University
Fort Collins, Colorado 80521

Dear Dr. Weisman:

This report on an automatic parking lot attendant is hereby submitted in accordance with your assignment for November 28, 1974.

The report includes an analysis of the problems involved in such a system, the methods used in solving these problems, and conclusions and recommendations of the author.

Sincerely,

James B. Donnelly

Cover and Title Page

The formal report usually is bound within a cover. Identifying information is listed on the cover page; this consists of the title of the report and its author. Occasionally, the recipient's name is also included in the cover information.

The purpose of the title page is to state briefly but completely the subject of the report, the name of the organization or person for whom the report is written, the name of the person and/or the firm submitting the report (frequently the address of the report's maker is included), and date of the report. The title page may also have a number assigned to the report by the authorizing organization, as well as the number of the project concerning which the report was prepared. The title page gives the reader his first contact with the report. The various elements, therefore, should be arranged neatly and attractively on the page.

Examples:

Geologic Report
on
The Physiographical Development
of the
Colorado Piedmont Area

Submitted to
Dr. Herman M. Weisman,
Professor of Technical Journalism
Colorado State Univ.
Fort Collins, Colo.

by
Maurice DeValliere,
Geology Student
November 30, 1974

Report on the
Origin and Cell Movements
of the
Hypophysis
in the
Fifteen Hour Chick

Submitted to
Dr. Herman Weisman
Professor of Technical Journalism
Colorado State University
Fort Collins, Colorado

by
Annelee Johnson
Biological Science Student
Fort Collins, Colo.
February 27, 1975

Abstract (Or Summary)

The abstract is a very important element of the formal report since it may be the only section read by a busy reader or client. It is a brief, factual synopsis of the complete report. It is placed first on a separate page following the title page to save the reader wading through masses of technical data in search for answers to his questions. The gist of the essentials of the investigation is given: included are an explanation of the nature of the problem and an account of the course pursued in studying it; the findings or results, conclusions, and recommendations of the investigation are summarized in the order of their importance. Ideally, the abstract is no more than one typewritten page of about 200 to 250 words. The term *summary* is sometimes used interchangeably for this type of synopsis. Some writers prefer *summary* because the term abstract has been increasingly reserved for the synopsis published by abstracting journals, which summarizes the contents more thoroughly and which devotes less emphasis to purpose and conclusion than is usually found in that element of the formal report. The term *summary* might well be used in a purely informational type of report because the synopsis is no more than a summarization of the informational data within the report. In practice, the term *abstract* appears more frequently than *summary*.

Reports are frequently placed in the files of an organization's information system. To facilitate retrieval of the information of a document in a file, the document is first indexed; that is, the key topics discussed are identified. Many organizations require authors to provide *key words* which index or list the most important topics in their reports and papers. The key words, usually limited to five to twelve in number, are arranged alphabetically and placed below the abstract.

Examples:

Geologic Report on Physiographical Development
of the Colorado Piedmont area[1]

---

[1]From a student report by Maurice DeValliere.

The Colorado Piedmont is an area which lies topographically lower than the Colorado Front Range mountains to the west and the Great Plains Section of the High Plains Province on the east.

During the fall of 1959, the writer, under the direction of professors D. V. Harris and W. N. Laval of the Colorado State University Geology Department, made a detailed investigation of the Colorado Piedmont area for the purpose of writing a technical report, regarding the origin and possibilities for future expansion of the area. Library research was also employed.

The conclusions of this investigation are: (1) the Colorado Piedmont is an erosional feature which has developed upon the Tertiary sedimentary beds since the end of the late Pliocene uplift, (2) the major erosional process was stream activity which resulted in extensive stream capture, and (3) future expansion of the area is likely to occur along the north-western border, near the Wyoming state line, and along the eastern border, toward the Kansas state line.

Key words: Colorado Front Range; Colorado Piedmont; erosion; High Plains Province; physiographical development; Tertiary sedimentary beds.

AN EXAMINATION OF SOME METHODS USED BY HUNTERS FOR CARE OF DEER MEAT[2]

This paper examines some of the methods used by hunters for caring for deer meat. The objective was to determine just how a deer carcass was field-cared for. The data for the study were obtained by direct observations of deer carcasses and supplemented with questions asked hunters. An examination of 371 carcasses, over a two-week period, showed the methods of care given to deer carcasses to vary greatly. Search of the literature showed no similar study to date. The extremes of care varied from animals processed to those which were not field dressed. An average condition of care was determined. Many of the hunters live very close to the area sampled and do not take proper care of their deer, as they assume they will arrive home soon enough to care for it. Time, always a limiting factor, hurries a hunter in all of his procedures, including proper care of deer meat. Suggestions for proper care and a step-by-step procedure for field dressing a deer are presented. Ten recommendations to aid hunters in caring for deer are suggested. Nine bar graphs that are located in the Appendix diagram the data. A map of the area sampled and a form used for recording the data are also included within the Appendix. The most important recommendation for hunters is: Game should be treated like any good meat.

Key words: Deer; deer meat; field-care, carcasses; game; hunter.

POWER LEVEL CALIBRATION OF COLORADO STATE UNIVERSITY AGN 201 REACTOR[3]

This report provides the power calibration curve for the CSU AGN-201 Thermal Nuclear Reactor. The power calibration curve is a graph of reactor

---

[2] From a student report by Jack D. Cameron.
[3] From a student report by Arthur R. Swanson.

thermal neutron flux in the glory hole of the reactor as a function of channel 3 reading on the control console. In future experiments where materials are being irradiated in the reactor, it may be necessary to use the power calibration curve to find the thermal neutron flux to make the data meaningful.

The first phase of the report defines the problem. Channel 3 has its neutron detector located in the water outside the lead shielding of the core; hence it cannot measure flux directly. The second phase of the report develops the method for determining the reactor flux by using a standard neutron source to irradiate indium foils in a water tank. The third phase is the results and conclusions of the experiment data. The power calibration curve is the result. It is concluded that the power calibration curve is linear. The maximum power level was determined to be 101.5 milliwatts at 5×10^{-9} amps, on channel 3, as compared to 100 milliwatts found by the Aerojet engineers.

From the accuracy required of this report, it is recommended that future data calculations be rounded off to less significant figures than used in this report to avoid unnecessary calculations. Because of the accuracy requirements imposed, it was found that the thickness of indium foils does affect their activity sufficiently enough to consider the difference in thickness of the foils that are used in the nuclear physics laboratory at CSU. From all indications of this report, an interesting experiment on perturbation (foil thickness) effect could be incorporated in the Ph-124 nuclear physics course. Key words: Colorado State University AGN 201 Reactor; power calibration curve; thermal neutron flux; thermal nuclear reactor.

Table of Contents

The table of contents lists the several divisions and subdivisions of the report with their related page references in the order in which they appear. It is set on the page to make clear the relation of the main and subordinate units of the report. These units should be phrased exactly as are their corresponding headings in the text. A list of tables and/or illustrations follows the table of contents.

TABLE OF CONTENTS

LIST OF ILLUSTRATIONS

Introduction

The elements discussed so far have been preliminary. The report itself begins with the introduction. The purpose of the introduction is to answer the immediate questions which come to the reader's mind: Why should I read the report? Who asked for the report? When? Where did the information come from? Why was this report written? What significance has this report for me?

So the introduction gives the reader his first contact with the subject of the report. The introduction states the object of the report. It may be a restatement of the terms of the original commission either verbatim or as understood by the writer. It should give the background information necessary to understand the discussion which follows. The nature and amount of this information depend, of course, on the intended reader and his knowledge of the subject. In order to follow the report, the reader must know:

1. *The purpose of the investigation*
2. *The nature of the problem*
   Here will be found the Who, How, What, Why, and When of the investigation. Some writers distinguish between *purpose* and *object*, although the terms are often used interchangeably. If there are both immediate and ultimate goals, these must be identified. Either or both of these factors should be explained to the reader. The purpose and the object concern the *why* of the study, and the problem concerns the *what* of the study. The nature of the problem and the purpose are often grouped together but they are not the same, and this confusion should be carefully avoided.
3. *Scope, or the degree of comprehensiveness*
   Scope determines boundaries—what considerations are included and what are excluded. The delimitations of the investigation are carefully stated.
4. *Significance of the problem*
   The reader of a report consciously or subconsciously asks himself, "Why should I read this report?" The writer must answer this question in the introduction. He must tell the reader why the report is of significance to him. Needless to say, if the reader does not feel that the report holds any significance for him, he will not read it.
5. *Historical background* (review of the literature)
   The previous state of knowledge and the history of prior investigations on this subject are included. A review of the literature may be desirable to:
   a. Give the reader confidence in the investigator's awareness of the state of the art of knowledge of the problem.
   b. Enable the reader to get the necessary background quickly. Only enough of the past should be included to make the present understandable. Some reports continue previous investigations; the report must then provide the historical background needed to orient the reader properly. In beginning a report, the reader may have the following questions in mind: What has given rise to the present situation? What occasions have given rise to a similar situations? What significance has been attached to them in the past? Are there any conflicting views on this problem? A review of present conditions may also be necessary.
6. *Definitions are given of new or unusual terms or those having a specialized or stipulative meaning*
   Terms used only once are better defined in the context; those appearing constantly throughout the report should be defined in the introduction. If many such terms are to be used, you may include a glossary in the Appendix.

7. *A listing of the personnel engaged in the investigation, together with a brief sketch of their background and duties*

This will often give the reader more confidence in the data and conclusions obtained. The reader wants to know who the people are who obtained the data and assurance that they have requisite qualifications for working on the problem.

8. *The plan of treatment or organization of the report*

This element gives the reader a "quick map" of the major points, the order in which they will be discussed, and perhaps the reason for the arrangement. It serves also as a bridge between the introductory section and the body and discussion sections.

9. *Methods and materials for the investigation*

This item is included in the introduction only in reporting those investigations where procedures and materials are simple and *inconsequential* to the data obtained, as is illustrated in the sample introduction below from the student report on how hunters field-care their deer meat. In this instance, the method of investigation is a simple interview and questionnaire, supplemented by observation. The materials were forms for recording the data, a clipboard, and a pencil. However, in those investigations in which the data obtained are directly related to laboratory or experimental methods, or are dependent on elaborate procedures and materials, this element is placed in the body section.

These eight or nine points are general to most formal reports, although not absolutely necessary to all reports. There is no special significance in the order of their listing. Introductions may begin with the significance of the problem or with the historical background. In certain situations where the problem may be new and complex, definitions of terms may begin the introductory section. These eight or nine elements (or those which are used) are frequently organized as subheadings within the introduction. The purpose of the introduction is to orient the reader. Placed within it is the information the reader needs to follow the main body of the report easily and confidently.

Example:[4]

I. INTRODUCTION[5]

Objective and Scope

This paper examines some of the methods used by hunters for the care of deer meat. The objective was to determine just how a deer was field-cared for. The data for the study were obtained by direct observations of deer carcasses and supplemented with questions asked hunters at Ted's Place in the fall of 1957. Ted's Place is located at the junction of Highways 14 and 287

---

[4]Please refer to introductions in student reports appearing in the Appendix, pp. 366-404.

[5]From a student report by Jack D. Cameron.

in North Central Colorado. A Colorado Game and Fish Department's permanent big game check station is located here. Since all hunters, successful or not, are required to stop, it is an ideal location for a study of this kind.

An examination of 371 carcasses, over a two-week period, showed the methods of care given deer to vary greatly. The procedure used in determination of the methods proved to be satisfactory; however, many limitations were encountered. The inability of one man to check all of the hunters and their deer during a rush period at the check station was the greatest difficulty encountered.

Also the great majority of the hunters shot their deer in the northeastern mountains of Colorado, game management units eight, nine, and nineteen. Therefore, this limited sample may be biased as only one location within the state was observed. However, the data will hold true for this specific northeastern area.

Time, as is the case in much research, was limited. However, 16 percent of the total deer checked through Ted's Place were observed. Since only 10 percent of a population is usually required for a statistical analysis, the data obtained were considered sufficient.

This paper was prepared as a class assignment in technical writing. All data presented are original.

Review of the Literature

A thorough search of the literature uncovered many techniques and methods concerning the care of deer meat. However, this search showed no similar work of an actual examination of carcasses to determine the care given deer meat by hunters and the condition in which it was brought through a check station.

Much of the literature emphasized the great loss that occurs annually by improper care of game meat, but no specific study was found concerning examination of the field-dressed carcass.

At this date the author is unaware of any similar work in this specific field.

Methods and Materials

A direct observation of the deer carcasses and verbal questions asked the hunter, make up the data presented. By the use of a prearranged form, as shown in the Appendix, the information desired could be readily recorded. All information, observations, and questions were recorded on the form. Columns on the form were arranged to facilitate recording maximum information in a minimum of time.

As a hunter's car approached the station the form was readied. While the Colorado Game and Fish employee was checking the license, most of the direct observations could be made and recorded. When the employee was finished, the hunter was asked the following questions.

1. Did you save the heart, liver or kidneys?
2. What method was used to bring your deer from the field to your vehicle?

Only when the carcass was obscured by camping equipment and tarps were additional questions needed. These would include questions concerning the following:

1. Sex.
2. Appendage removal.
3. Splitting of the pelvic and/or brisket bone.
4. Tarsal and metatarsal gland removal.

The majority of the carcasses were located so the flesh could be felt to determine coolness. Although this is a relative factor, 18 percent of the carcasses were not cooled as the flesh was still warm to the touch. Many of these instances were very recent kills.

Co-operation from the hunters was very satisfactory. Less than ten hunters were curious about the questioning and asked for an explanation. The other 360 hunters evidently took the questions for granted as they assumed the author to be a Colorado Game and Fish Department employee.

The only materials used in this study were forms for recording the data, a clipboard, and a pencil.

Body (The Report Proper)

Although the term *Body* as a heading is seldom used, it is the major section of the report which presents the data of the investigation. In its stead, words descriptive of the material contained in the section are used. The body section or sections include the theory behind the approach taken, the apparatus or methods used in compiling the data with necessary discussion, the results obtained from the procedure, and an analysis of the results. In a scientific or technical investigation, the theory determining the procedure is explained and the method of procedure is recounted in chronological order; apparatus, materials, setups, etc., are described. The reader will place greater confidence in a report's conclusions if he knows how they were determined. Duties of the personnel are presented, and evidence is organized so that the reader can follow the thinking in orderly fashion.

The body is usually the largest section of the report. The various organizational approaches for the body, which are determined when the outline is developed, (see chapter 5) were discussed earlier. The pattern of organization chosen depends, of course, on the nature of the problem. The reader is guided through the body of the report, which may be structured into more than one section, by the major and secondary headings as set down in the outline. These headings will indicate the relationship of the main and subordinate units of the organizational plan. They should be in agreement with the table of contents. Paragraphs which represent units of the organizational plan and serve to bridge, introduce, and conclude topics help to provide a coherent text leading the reader through the discussion of the investigation.

Terminal Section

The terminal section may be either a summary of major points of the body, if the report is purely informational, or conclusions based on analysis of the data following logically from the evidence given in the body. The conclusions should, whenever appropriate, be listed numerically in the order of their presumed importance to the reader. Recommendations follow next. Conclusions and recommendations may be placed in one section or in separate sections. Conclusions deal with evaluations as to the past and present of a situation; whereas, recommendations offer suggestions as to future courses of action. Recommendations are often the most important part of the report, and their adoption or rejection depends on how they are presented. Like the conclusions, they should be positive statements. They should suggest specific things to be done or a course of action to be followed, and should, if appropriate, include an estimate of the cost involved. Some types of reports do not require recommendations, but usually the section on conclusions and recommendations is the most important part of the report. It is what the client is paying for.

Examples:

CONCLUSIONS[6]

1. One conclusion from this preliminary study of the construction of nomographs for the solution of two-phase flow equations is that although no workable nomographs were constructed, it is still possible to draft a correct solution to the problem. This will occur only after more study has been carried on by the author.

2. It is also concluded that the scale equations, the distances between scales, or the direction in which the scale equation is defined to be positive is in error in the solutions of this paper. When all these are done correctly, the nomograph will solve the equation for which it is constructed.

3. The nomographs, when constructed, will have the same limitations imposed on them as the equations on which they are based.

4. The nomographs will have, at best, only the accuracy which is possible on a ten-inch slide rule, since such a slide rule was used in all computations. Additional errors may be introduced by slight errors in the drafting of the scales, and errors in reading a small scale accurately. The total of these errors will be about two percent.

5. Although no correct nomographs were constructed, this paper may still serve a useful purpose to a designer by collecting the equations necessary and setting up a procedure to be followed in his design of a two-phase fluid system.

RECOMMENDATIONS

1. It is recommended that study of the art of nomography be continued until it is possible for the author to construct suitable nomographs.

2. The two-phase fluid designer may use this paper as a simplified approach to the paper by Lockhart and Martinelli as a basis for his design.

---

[6]From a student report by Paul C. Wergin.

3. The nomographs, when constructed, should be made to a larger scale than those shown in the appendix of this report.

4. The nomographs, when constructed, should not be used for computation with a fluid meter since such an apparatus depends upon a change of momentum.

CONCLUSIONS AND RECOMMENDATIONS[7]

Through the design and construction of the prototype model, it has been shown that an automatic parking lot attendant system is feasible. The prototype has been tested and operates as described in this report. The same circuits that operate the read-out can be used to supply control information to the coin-operated gate. The finished model is a compact unit weighing approximately 30 pounds and measuring 18 x 11½ inches around the base. It is 10 inches high. For the proposed system to become operational, several difficulties present in the prototype must be eliminated. The prototype has no provision for altering the fee rate structure which is built into it. This could be accomplished by redesign of the printed circuit boards using removable jumper leads, whereby various combinations of digital read-outs could be arranged. There are instances of price inequities in the present model such as in the example on page 11 of this report. In the prototype, a carry-over between rate change periods sometimes results in such a discontinuity of the accumulated fee. In the example given, the accumulated fee at 6:00 a.m. was 0.75. At the change over to the day rates, this was picked up at 0.80 due to the fact that no 0.75 combination is available on the day rate photocell disk. Ideally, the card would have been picked up at 0.75 and continued from that point. These small inequities may or may not be considered serious. If the inequities are considered undesirable, it may be necessary to attack the method of decoding the card from a different viewpoint.

The handmade rotary stepping mechanism which moves the photocell disk is not completely positive in its action. There are commercial rotary stepping solenoids available on the market which should eliminate this problem.

The author is of the opinion that the elimination of these difficulties may be accomplished in the production engineering phase and the system will be acceptable.

Bibliography

The bibliography is a list of references identifying literature which has contributed to the report. It is rarely included in the appendix. It includes references to both published and unpublished material, such as notebooks, reports, correspondence, and drawings. The items listed need not necessarily have been referred to in the text of the report. The conventional listing of the bibliography is alphabetical. The items should be numbered sequentially. (See pages 143-144 for examples.)

---

[7]From a student report by James B. Donnelly.

Appendix

The chief purpose of the appendix is to gather, in one place, all data which cannot be worked into the body without interrupting the report. If the body is self-contained, the detailed data in the appendix provide points of reference when questions arise. An appendix may not be necessary in a very short report. However, the appendix is indispensable to the type of report which uses considerable statistical information. Sometimes, all 8½" x 11", or larger, charts, tables, diagrams, photographs, and other illustrations are placed in the appendix of a typed or mimeographed report. The following items are usually saved for the appendix:

Charts
Tables
Computations and data sheets
Diagrams and drawings
Exhibits, graphs, maps, photographs, letters, and questionnaires
Records of interviews and other similar matters serving as data which are not found in the literature or are not revealed through the methods and procedures of the investigation are also placed in the appendix.

Index

An index is necessary only in voluminous reports where the alphabetical listing with page references of all topics, names, objects, etc., is useful for ready reference. Normally, the table of contents performs this function adequately.

Distribution List

The distribution list not only provides the names of persons and offices to receive the report but also serves as a control device to ascertain that the confidence of the sponsor or client is not violated. Distribution lists may be tacked to the inside of the front cover or placed at the very end of the report.

Writing the Report

The actual writing process begins after you have completed your research, accumulated your data, and organized your work into an outline. Before beginning to write, you might profit from reviewing your outline. You will find that your concept of the report has been growing and ripening in the recesses of your mind, subconsciously, for the most part. A final review will increase your awareness and freshen your viewpoint. It will also test your outline. Your review may stimulate the check of a note here or datum there and perhaps the revision of a point here or there. You are likely to experience a certain stimulation and excitement as you prepare to do the actual writing. As you review, thoughts about various aspects of the report will come rapidly into your mind. You may want to jot these thoughts down. After you have reviewed the outline, data, and notes, you will be ready for the actual process of the writing. Keep your outline before you and begin the writing. (Professional writers are the first to admit that writing is never easy, but writing can become easier with proper preparation and experience.)

The Writing of the First Draft

The first draft should be written in one sitting, if possible; so allow yourself at least several hours for this phase. Your report will have more life and will represent the sense of your material more closely if you write rapidly than it will if you pause to perfect each sentence before going on to the next. Save problems of spelling and grammar for the revision. To allow for revision, leave plenty of space in your rough draft between lines, between paragraphs, and in margins. If you follow a good outline that breaks your report up into logical stages, you will not be burdened by the strain of trying to keep a great deal of material in mind at one time. The outline permits you to concentrate on core aspects of your report. Some people find it is easier to write the body first and the introduction and the conclusion last. Others have a vivid interest in a certain aspect of the research and find they can "lick" an entire report by writing those parts first which come easiest to them. Thus, writing in stages is like hacking off small bits of a whole. Because of the outline, the writer can do this without losing continuity. It is probably better to make the first draft full and complete, even though you may feel it is wordy, since it is always easier to scratch out material and explanations than it is to expand. Your rough draft is a production of ideas rather than a critical evaluation of those ideas. The logical flow of ideas will be interrupted if you try to evaluate your material critically as you write; these interruptions slow down the writing process. Concern over spelling, punctuation, sentence form, and grammar also slows down the flow of writing. You should not worry about such mechanical details in the first draft. You will have ample time at a later stage to verify, check, correct, and revise. It is important that at first you write and write fully on all the aspects of the investigation that you have previously outlined.

Revision

Professional writers know that papers are rewritten, not written. After the rough draft has been completed, you ought to plan for a "cooling off" period. Let at least two or three days elapse, if possible, before you review your first draft. An interval of time allows you to approach what you have written, not from the closeness of the first heat of the writing, but from a distance. In the revision, take the point of view of your reader as best you can and read the report to see whether its objectives are being met and whether the problem is being answered. You should read your rough draft through from beginning to end without pausing for revisions. This first reading is necessary for an overview; it should not be used for the checking of details. Points may be noted by marks on the draft for attention at a later time. At this reading, you should ask yourself such questions as: Do the various parts of the report fit together smoothly? Does the thesis of the report come through clearly? Have I covered all points essential to the objectives or purpose of the report? Do they come through clearly? Is the information adequate and arranged effectively for content and organization? If the conclusion does not have the point you want your reader to get, or if the conclusion is not based upon adequate and clear data, you may have to revise until your material provides the answers to the preceding questions. Following the revision for content will come the reading and revision for mechanics and style. Read and be familiar with the material in chapter 12. Check for the following:

1. *Paragraphs.* Do the paragraphs hang together? Do all paragraphs have a topic sentence? Do the paragraphs have unity? Should any paragraphs be combined? Do any paragraphs need further development?
2. *Sentence structure.* Are all sentences complete grammatically? Do any sentences have to be reread for meaning? If so, break up or rewrite. Are sentences punctuated correctly?
3. *Style.* Is the style consistent and appropriate? Is the writing objective, concise, and clear?
4. *Word choice.* Can any deadwood be removed? Are the words as exact and meaningful as possible? Are there wrong, inexact, or vague words? Are there clichés, jargon, shop talk? Are there any nonstandard abbreviations? Are there any inconsistencies in names, titles, symbols? Are words spelled correctly?

The writer's best tools in these matters are a dictionary and a good handbook of grammar.

Format Mechanics

Your report is ready for typing after you have thoroughly checked and revised it. Use white, unruled bond paper, 8½″ x 11″, from 15 to 20 pound stock. Be sure to have sufficient carbon copies for filing and reference. If your report is to be printed or duplicated, you will need to consult with your printer or typist on the format considerations for reproduction copy. At any rate, your report should be typewritten with a black ink ribbon on one side of the page only.

Leave ample margins on all sides of the page. Margins on the left should be no less than 1¼ inches, preferably 1½ inches to facilitate binding. Margins on the right should be no less than ¾ inch. The margin should be no less than 1 inch at the top, no less than 1 inch, preferably 1¼ inches at the bottom. The bottom margin, including footnote space, if there are footnotes, should be no less than 1 inch, preferably $1\frac{1}{4}$ inches.

Manuscripts for publication are always double-spaced. Reports may be either double-spaced or single-spaced, depending on the conventions of your organization or specifications of your client. Any material that contains equations, superscripts, or subscripts is easier to read if it is double-spaced. Reports written as classroom assignments should be double-spaced to facilitate marking and correction. If the typescript is single-spaced, use double spaces between paragraphs. Two spaces separate paragraphs of double-spaced copy. Paragraphs, whether single-spaced or double-spaced, are indented five spaces.

The appearance of your report page will be improved if you follow these rules:

1. Do not start the first sentence of a paragraph on the last line of a page.
2. Do not place a heading at the bottom of the page with less than two lines of text to follow.
3. Avoid placing the last line of a paragraph at the beginning of a page.

Text pages of the report are numbered with arabic numerals. The placing of numbers on the page should be consistent. They may be centered at the top or placed in the upper right-hand corner. Numbering at the bottom of the page, either in the center or at the lower left, should be avoided because these numbers may be confused with footnote material. Prefatory or preliminary pages of the report (Title Page, Abstract, Table of Contents, List of Illustrations, Foreword) are numbered with small Roman numerals. Should blank pages be used for the sake of appearance, they are counted but not numbered. The Title Page is counted as the prefatory page (i) but is not numbered. Final assignment of numbers to pages might be delayed until all the pages are typed, unless the exact numbers of illustrations and their placement are known and planned beforehand. However, tentative numbers might be written in the upper right-hand corners very lightly in pencil.

Equations

Mathematical equations are generally centered on a page. Lengthy equations should be typed completely on one line rather than broken into two lines:

Poor:

$$\ldots\text{. the cut-off frequency of a rectangular guide is } f_c = \frac{c}{\lambda c} = \frac{c\sqrt{\left(\frac{m}{a}\right)^2 + \left(\frac{u}{b}\right)^2}}{2}$$

Preferred:

. . . . the cut-off frequency of a rectangular guide is

$$f_c = \frac{c}{\lambda c} = \frac{c\sqrt{\left(\frac{m}{a}\right)^2 + \left(\frac{u}{b}\right)^2}}{2} \tag{13}$$

Each new line of an equation should be positioned so that the equal signs are aligned with the equal signs of the preceding line:

$$\frac{\Delta Z_0}{Z_0} = \frac{1}{4\pi^2} \frac{w^2}{D^2 d^2}$$

$$K = \frac{\Delta Z_0}{2Z_0}$$

Many mathematical symbols must be penned by hand. The typist should leave the necessary space for hand lettering by the author. All complete equations are conventionally numbered consecutively within each chapter for easy reference purposes; the numbers appear within parentheses flush with the right margin of the text.

Example:

$$VSWR = \left(\frac{P_{s\ max}}{P_{s\ min}}\right)^{1/2} \tag{1}$$

$$P = c_1 s_{in}^2 \left(\frac{2\pi x}{g}\right) \tag{2}$$

A short formula within the text is set off by commas and is identified with an equation number, as for example:

It can readily be seen that when $\lambda = 2a$, $\alpha = 90°$, or the waves are not propagated down the guide. . . .

Equations — no matter how short — containing fractions, square root signs, sub or super-numerals, or letters require extra space above and below the text line and therefore should not be included in a text line but centered and placed on a line by themselves, as illustrated in preceding examples.

In summary:
1. Line up equal signs in a series of equations.
2. Keep all division lines on the some level with equal signs.
3. Divide equations only after plus or minus signs but before equal signs in second line.
4. In dividing equations, line up the second line with the equal sign or the first plus or minus sign in the first line.

5. Parentheses, braces, brackets, and integral signs should be the same heights as the expressions they enclose.

Tables[8]

You will help achieve clarity in tables if you observe the following rules:

1. Center all numbers within the column.
2. Align numbers by the decimal points or commas, if any. Otherwise, align numbers by the right-hand digits.
3. Use headings whenever possible in order to avoid repetition in the body of the table.
4. To indicate subdivisions within the tables, use single, horizontal rules below the heading at the bottom. Where there are more than two columns, use a single, vertical rule. Omit vertical rules when there is sufficient white space between columns.
5. Number tables consecutively with Roman numerals.
6. Center table numbers at the top and type the title in all caps.
7. Type the first word of a column head in initial caps, but type the other words in the column head in lower case. (See examples in chapter 7.)
8. Insert short tables within the text. Place longer tables on separate pages. Where there are a number of large or oversize tables following sequentially within a paragraph or page of text, place such tables in the appendix to avoid confusion.

Headings

Headings are used as mapping devices. The features of the report are marked by sectional and subsectional titles or heads. Your headings correspond to your major divisions and subdivisions in your outline and in your table of contents. As mapping devices, headings serve to show relationships and subordinations; they should clearly indicate the logic of relationship and subordination throughout the report. A single system is recommended here, although in actual practice a number of conventions of showing this relationship and subordination are used.

First-Order Headings

First-order headings are written in all caps and centered two inches from the top of the page. Such a head constitutes a major text division and corresponds to the Roman numeral of your outline. First-order headings begin on a new sheet of paper, and the text follows four typewriter spaces below.

---

[8]A fuller treatment of tables is given in chapter 7.

Second-Order Headings

Second-order headings are placed flush to the left margin, two spaces below the last line of the preceding paragraph and are typed in all caps. They correspond to the capital letters in your outline and indicate major subdivisions of a section. The text follows two spaces below.

Third-Order Headings

Third-order headings are written with initial caps and are placed flush to the left margin and are underlined.

Fourth-Order Headings

These headings are indented five spaces and are written with initial caps. They are the same as the third-order head except they are part of the paragraph of the text. The text follows on the same line.

Some reports follow through on the numbering system of the outline to conform with the various heads and their subheads of various rank. Certain organizations and many government agencies call for numbering of heads and subheads. The decimal system of numbering is frequently used.

FIRST-ORDER HEADING

SECOND-ORDER HEADING

Third-Order Heading

    <u>Fourth-Order Heading</u>. Text of paragraph follows on the same line as the Fourth-Order Heading.

Documentation

Bibliographic References

Documentation is required in reports, professional papers, and other serious types of writing for three reasons:

1. To establish the validity of evidence. All important statements of fact not generally accepted as true as well as other significant data are supported by the presentation of evidence for validity if the exposition within the text itself does not offer the proof of the data or the facts. Direct reference to the source is provided so that the writer's statements may be verified by the reader if he so chooses, or if the reader wishes to extend his inquiry into the borrowed matter beyond the scope of this particular writing.
2. To acknowledge indebtedness. Each important statement of fact, data, or information, each conclusion or inference borrowed by the

writer from someone else should be acknowledged. Also, a citation is desirable when a conclusion or idea is paraphrased or its substance is borrowed and presented.

3. To provide the reader with information he might need or want about the subject matter a writer has borrowed or obtained from another source.

When any report, professional paper, book, or other type of serious writing contains information obtained from other publications, books, articles, and reports, such sources should be indicated under a list of references or a bibliography. Previously, I indicated that your report should have a bibliography if it utilized information obtained from other sources. The bibliography follows the last page of the report and precedes the appendixes. The bibliography, as you will recall, is a list of the sources arranged alphabetically and numbered sequentially.

Scholarship has a long tradition of providing documentation for borrowed sources. The cited source material is documented in a footnote at the bottom of the page of the text upon which the borrowed material appears. However, this system of documentation is not as efficient as it might be. In this book, I have used what I consider a very efficient and very simple system of documentation. This system, instead of using a footnote at the bottom of the page, integrates the documentation reference within the text line following immediately the matter or source to be documented. The documentation reference begins with a parenthesis, then lists the sequential number of the bibliographic reference source being used, followed by a colon, the page numbers of that bibliographic reference, and then is closed by the parenthesis; for example, (9:113-119).

Let me illustrate again how this works. In chapter 8, I borrowed material liberally from scholars in semantics, communications, linguistics, and anthropology. On pages 183-184 of chapter 8, I quote from W. A. Sinclair's book, *An Introduction to Philosophy*. In this system which I am recommending, instead of a superscript following the words "is very close" I have a parenthesis, then the number 9, which is the sequential appearance of W. A. Sinclair within the list of references at the end of the chapter, followed by a colon and pp. 113-119 at the end of the phrase, then close parenthesis (9:113-119). This would tell the reader that this material was borrowed or quoted from the bibliographic source appearing in sequence in the *references* and that the matter quoted appears on pages 113 to 119 of that source.

Now let us take another example. Again in chapter 8, beginning on page 193 you will discover a quotation taken from Wilbur M. Urban's book *Language and Reality*. Following that, I paraphrase additional material. On the same page after the quoted material, there is a parenthesis, followed by 10, indicating the sequential listing in the references for Wilbur M. Urban's book *Language and Reality*, and following the 10, there is a colon with 107-115 following it, with a parenthesis and then a period. After the word *latent*, there follows (10:115-116). Now for the sake of further example and illustration, suppose that the paraphrase was of material which Dr. Urban had in several other chapters. You would thus have (10:115-116; 231-235; 412-418). This reference documentation indicates that the paragraph is a paraphrase of ideas expressed by Dr. Urban in the pages listed.

Now suppose you are going to paraphrase some ideas you have obtained from several sources — how would you do this in this system? Again, this is a very simple matter. For the sake of example, let us suppose that I am paraphrasing material from several sources within one paragraph. I will offer documentation at the end of the paragraph for the information I have presented. Following the information I have borrowed within that paragraph, I would begin with a parenthesis, then with the most important source — let us say it is source number 3 in my bibliography or references. I would have (3:8-12; 23:420-433; 13:56-57; 8:16-17, 43-50). This documentation will tell the reader that the material of the paragraph preceding was borrowed from the sources indicated and from the pages listed of those sources.

This system actually is simple, efficient, and provides the reader with all the needed information.

Footnotes

While the above system has eliminated the need for footnotes to document borrowed reference material, footnotes may still be used and required in the following instances:

1. To amplify the discussion beyond the point permissible in the text, and
2. To provide cross references to various parts of the report.

For example, on page 87 of chapter 4, I have a footnote of the first category. This footnote amplifies the information which I have presented in the text about the two goals in library searching. The footnote itself presents the additional information that these two goals are exclusive of the historical method of research, where this form of research utilizes the investigative techniques of library searching. If this footnote of about 75 words were placed in the text, it would interrupt the continuity of the text matter for the reader. Although interesting, related, and amplifying, this information is not essential to the understanding of the text. Therefore, it is placed in the footnote.[9]

In the second type of footnote, where cross references to various parts of the report are given, the writer is permitted to refer to material appearing in other parts of the report, such as the appendix or matters appearing in earlier or later portions of the report. This type of footnote aids the

---

[9]Footnotes are, as the term implies, notes at the foot of the page which the reader can read if he wants to or, after he reads the first line, can see whether he wants to continue or not. The flow of the main text is not impeded by this device.

flow of the text, helps in clarity, and sends the reader, if he so chooses, to related information he may want to examine at that particular point.

Bibliography

The Bibliography, as we have previously stated, is a list of references used by the writer in his report. When the list of references is large, it is sometimes convenient to classify items according to types of references, listing in separate categories all books, periodical literature, publications of learned societies and organizations, government publications, encyclopedia articles, and manuscripts and unpublished materials. In most instances, the alphabetically and sequentially numbered bibliography is the most convenient.

Each bibliographic reference should list the complete elements which will help to identify the source. These bibliographical elements are: author, surname first; title; place of publication; publisher; date; page numbers. In the case of a report or government publication that is classified, the security classification will be indicated. Listed below are examples of various bibliographic references and the conventional way of indicating them in a bibliography.

Models for Bibliographic Entries

1. *AF Regulation 80-41, Research and Development Management of Information Analysis Centers*, Department of the Air Force, Headquarters U.S. Air Force, Washington, D.C., 2 March 1970, 5 pp:

2. Albertson, Maurice L. and Herman M. Weisman, "Water Abstracts, A Proposed Abstracting Journal for Water Resources and Related Subjects," Unpublished paper read before the meeting of the American Society of Civil Engineers, Phoenix, Arizona, April 14, 1961, 14 pp.

3. *Anglo-American Cataloging Rules*, prepared by the American Library Association, the Library of Congress, the Library Association and the Canadian Library Association, Chicago: American Library Association, 1967, 400 pp.

4. Basso, Keith H. and Ned Anderson, "A Western Apache Writing System: The Symbols of Silas John," *Science*, 180:1013-1022, June 8, 1973.

5. Butler, J. A. V., *Inside the Living Cell*, New York: Basic Books, Inc., 1959, 174 pp.

6. Brady, Edward L., "The Information Analysis Center and Its Role in the Processing and Transfer of Technical Information," in *Miniconfrontation on Information Analyses Centres*, Directorate for Scientific Affairs, Scientific and Technical Information Policy Group, Organization for Economic Cooperation and Development, DA5/STINFO/70.11, Paris, 8 April 1970.

7. *Calloway Workshop, 1974*, Kansas City, Mo.: Calloway Productions, Inc., 1974, 105 pp (Mimeographed).

8. Eckschlager, K. (R. A. Chalmers, Translation Editor), *Errors, Measurement and Results in Chemical Analysis*, London: Van Nostrand Reinhold Company, 1969. 155 pp.

9. Evans, Robley D., "Gamma Rays," in *American Institute of Physics Handbook, Third Edition*, Dwight E. Gray, Coordinating Editor, New York: McGraw-Hill Book Company, 1972, pp. 8-190-8-218.

10. Finocchiaro, Maurice A., *History of Science as Explanation*, Detriot, Michigan: Wayne State University Press, 1973, 286 pp.

11. Luhn, H. P., "Selective Dissemination of New Information with the Aid of Electronic Processing Equipment," *H. P. Luhn: Pioneer of Information Science Selected Works*, (Clair K. Schultz, Editor) New York: Spartan Books, 1968, pp. 246-254. (Originally published by IBM Corporation, Advance Systems Development Division, Yorktown Heights, New York, November 30, 1959.)

12. *Medical Tribune*, p. 1, June 20, 1973.

13. *The New York Times*, pp. 1, 26, December 21, 1973.

14. Physics Survey Committee, National Research Council, *Physics in Perspective, Volume I*, Washington, D. C.: National Academy of Sciences, 1972, 1065 pp.

15. Rossmassler, Stephan A., (Editor), *Critical Evaluation of Data in the Physical Sciences — A Status Report on the National Standard Reference Data System*, June 1972, NBS Techinical Note 747, Washington, D. C.: U. S. Government Printing Office, November 1972, 73 pp.

16. Wallace, Robin A., et al, *Mercury in the Environment — The Human Element*, ORNL NSF-ED-1, Oak Ridge National Laboratory, Oak Ridge, Tennessee, Reprinted March 1971, 61 pp.

17. Weisman, Herman M., "Reaching Other People's Minds Through Written Language," *Technical Communication*, 20:16-19, First Quarter 1973.

18. Weisman, Herman M., *Information Systems, Services, and Centers*, New York: Becker and Hayes, Inc., a subsidiary of John Wiley and Sons, Inc., 1972, 265 pp.

19. Wolfe, W. M. (Moderator), "How Can Effectiveness of Analysis Centers Be Measured" Addendum to the *Proceedings of the Ad Hoc Forum for Information Analysis Center Managers, Directors, and Professional Analysts*, Battelle Memorial Institute, November 9-11, 1965, (unpublished manuscript n.d.) 53 pp.

A Note on Reports Prepared to Government Specifications

Contractors doing research for the government are invariably required to furnish reports covering such investigations. Procedures governing the preparation of the reports are determined by specifications. While the Department of Defense and other branches of the government have their own specifications guiding the preparations of reports, the report itself

varies little from the principles covered heretofore. The purpose of specifications is to establish procedures and standards by which the government
may insure the quality of the preparation and be assured that all the information required is covered competently by the report. The contract
will indicate in detail the type, frequency, and number of copies of the
report to be furnished. The specification defines the basic procedures, the
content, editorial standards, preparation of completed (reproduction) copy,
preparation of illustrative materials, security standards, and distribution.
Reports to the government may vary in size from a one-page form, monthly
progress report to final reports of many hundreds of pages. Government
specifications will indicate all the elements the report should contain. Such
definition insures uniformity and quality for the many thousands of projects
being carried out for the government. While the specifications establish
standards, patterns, and criteria, their requirements actually are not formidable to the writer. The specifications determine the what, how, and
when of a contract. They will define what goes into a progress report, a
quarterly report, or a final report; how it is prepared; and how and in
what quantities it will be forwarded or distributed.

Discussion Problems and Exercises

1. Prepare a letter of transmittal for your report; include reference to the
   commission of the investigation and other pertinent details.

2. Your major assignment for the term is to write a report on a research problem based on the outlines you prepared for problem 6, chapter 5. Your
   instructor may wish to break this assignment up into stages:

   a. First, write and turn in the Introduction to your report. Your instructor
      will return it with evaluative and suggestive comments.
   b. Write the "Body" sections of your report. Your instructor will offer his
      comments on this portion of the report assignment.
   c. Write the Terminal section. Your instructor will provide critical com-
      ments on this aspect also.
   d. Take your instructor's critical suggestions into consideration and rewrite
      the complete report. The final product you turn in should contain all the
      elements called for in this chapter.

3. What is the purpose of the Introduction? What should it include?

4. What is an abstract? What is an epitome? What is a synopsis? What is a
   summary? Answering these questions might require some library searching.
   Why does a report have an abstract? Should the abstract of a report be composed primarily with the requirements in mind for suitability for publication,
   as within the abstract journals of the field of knowledge covering the subject
   of the report? Or should the abstract of a report be composed to meet the
   requirements of its readers?

5. Conclusions and recommendations are frequently grouped together within
   the terminal section of the report. Are they the same? What is their relationship? How are they derived?

6. Write the conclusions and recommendations for a hypothetical investigation you have conducted as to whether a new stadium should be built before a new library on the campus of your university or college.

7. How would you revise the following introduction? Consider its completeness, effectiveness of organization, and language.

INTRODUCTION

The Problem

The Experiment Station of Colorado State University is, at present, making a study to determine the effects of different seasons of use on a mixed, grass-alfalfa pasture. The grass being used is Russian wildrye. Application of the results will probably be limited to northeastern Colorado, as the study is being conducted exclusively in this area. The experiment is now in its third year. Since there hasn't been enough time to show any definite results, or make any definite conclusions, this will be more in the form of a progress report.

Importance of the Problem

Use of Pasture Mixtures
Alfalfa, when grown in pure stands, causes livestock to bloat. One reason for the use of a grass-legume mixture is to prevent this. The legume also serves to fix nitrogen which is used by the grass. Such mixtures also improve productiveness and fertility of the soil.

Other Work
There have been other studies made on grass-legume mixtures. Most of these, however, have been studies made of their effectiveness or to find better mixtures. To my knowledge, this is the first time anything has been done with the Alfalfa-Russian wildrye mixture.

Application of Results
This study, when completed, should provide a basis on which to plan for season of use, and for intensity of use on this type of mixture.

Definitions

The following are some definitions and descriptions which will help to clarify the report for the layman.
Alfalfa (Medicago sativa—An herbaceous perennial legume, varies in height from 2–3 feet; flowers are purple; fruit, a spirally twisted pod, 1–8 seeds per pod.
Animal Unit Month—Number of animals grazed for one month (30 days)
Line Transect—An elongated sample plot, which has no width.
Russian wildrye (Elymus junceus)—a large bunchgrass; about 3 feet tall; leaves 6–18 inches, light to dark green; dense erect spike; a cool season grass. Introduced from Russia in 1927.

Acknowledgments

Thanks are to go to Roy Miller, of the Experiment Station, for the help which he has given on this report. The field work was all done by him and much assistance was received in compiling the data into usable form.

8. Criticize and rewrite the following abstract:

ABSTRACT

What are the effects of different grazing patterns on a pasture—mixture of alfalfa and Russian wildrye? This is a problem which is currently under study by the Experiment Station here at Colorado State University.

To find the answer to this problem, a 13.75 acre field was seeded with the mixture. It was then divided into six separate pasture units. For the past three years it has been grazed on a spring-only, summer-only, and spring-fall basis. In 1974 and 1975 a series of forty-eight transects were read on the units; eight transects per pasture. The study at present is considered only partially finished.

From what has been done so far, the spring-only treatment seems to be the best suited for this mixture. The spring-fall treatment seems to be doing poorest, although it has been grazed heavier than the others. The study should be continued for a few more years before any conclusions may be made.

9. Criticize and rewrite the following Terminal Section:

RESULTS AND CONCLUSIONS

Results

General

The data shows an increase of Russian wildrye on all pastures, with the exception of A, which had a slight decrease. There was a decrease of alfalfa in all pastures except D. (See Appendix B.)

By Treatment

The two plots which were given a spring-fall treatment (A and E) gave the worst results. However, it must be remembered that A was almost twice as heavily grazed as any other pasture. The spring only plots (C and D) showed the most improvement. This was the only treatment where the alfalfa increased. The Russian wildrye here showed nearly as much increase as it did on any other pasture. (See Appendix B.)

Conclusions

After such a short time, no real conclusions may be arrived at. The trend, as summarized under the results, could completely reverse itself by the time this study is completed. The best course of action, at this time, would be to continue the study for a few more years, and establish a more definite pattern. Statistical methods could also be applied after a few more samples are available.

7

Graphic Presentation in Technical Writing

Pictorial representation was man's earliest means of nonverbal communication. It served the limited needs of primitive and early civilized man well. Leonardo da Vinci was one of the first writers of science and technology who recognized the importance of integrating pictorial representation with verbal description as a way to communicate factual information clearly and efficiently. His notebooks established a useful tradition in technical writing because his graphic materials supplement his words and sometimes supplant them. The axiom that "one picture is worth a thousand words" is not the propaganda of the technical illustrator but a generalization which has much experience to substantiate it. It is also true that an improperly conceived, poorly integrated, or badly executed pictorial element may create irreparable confusion for a reader.

We will be concerned in this chapter with the types of visual aids which can help the technical writer to describe, define, explain, analyze, compare, classify, and narrate. This chapter is not meant to be a text on graphic aids. Its purpose is to identify and outline uses of graphic presentation of information in technical writing. Authoritative texts on techniques of graphic aids will be found in the bibliography. We will examine the uses of lists and tables, photographs, drawings, diagrams, graphs, charts, and flow sheets.

In technical literature, illustrations are seldom used for embellishment. Even though illustrations do perk up a manuscript, add to its attractiveness, and enhance its readability, their sole purpose in technical literature is functional. Graphic presentations are functional because they can reveal more clearly, concretely, and accurately than words matters of a statistical or complex nature. They are functional and efficient because they promote easier reader comprehension by their ability to organize, confirm, and underscore data and interrelationships which might otherwise be obscure.

Graphic presentation helps the mind of the reader to better digest, ruminate, and analyze facts and ideas. Illustrations not only help the reader but help the writer to organize and understand his subject. We shall see later in chapter 10 how illustrations of devices and processes serve as a point of departure in presenting descriptions and explanations. An illustration, like an image reflected in a mirror, presents a vivid, organized structure for the communication of required information.

Planning the Graphic Aid

After all data are compiled, their organization, analysis, and interpretation frequently can best be accomplished through graphic means. Certain relationships between quantities can be shown most clearly by graphs. The first step is to examine those data which are statistical, involved and/or technically complex to determine which might more clearly and concretely be communicated by graphic presentation. The nature of the material and the reader's knowledge level and requirements determine which type of graphic aid best meets the necessities of efficient transmission of the information. For example, statistical data in a report to a technical person might most accurately be presented in a table, while the same data for presentation in a semitechnical article might best be presented in a bar graph. The table offers more precise graphical data; the bar diagram offers only rounded numerical data. The type of graphic aid selected depends on the reader's ability to understand and use it.

Location of Graphic Aids

Visual-aid material should be placed near the text matter it illustrates. This is not always possible in typewritten or mimeographed papers and reports where several visual aids may be part of a discussion within a paragraph. If a visual aid is relatively small, it can be integrated into the appropriate paragraph of text. A full-page visual aid might be placed either facing or immediately after the page where it is discussed. Graphic material should not precede its discussion in the text because it will then only confuse the reader. In typed and mimeographed reports, illustrations which cannot be integrated within the text page are placed for convenience in a separate section, usually in the appendix. All visual aids, whether integrated within the text page or collected at the end of the article or report in an appendix, must be referred to in the text itself. The reader should be directed by explanatory sentences to the specific illustration or table so that he may better correlate, interpret, and understand the data being communicated. If an illustration or table supplements, clarifies, confirms, analyzes, or reveals conclusions, such explanations must be clearly stated in the text. The more complex the illustration, the more explanatory material is necessary in the text. Directions for interpretation of illustrations may be included in the text or in captions below the illustrations. Illustrations need to be as simple as possible. Text matter and lengthy explanations are kept out of the illustration proper.

Lists

A list is a series of names, words, numbers, sentences, or similar data arranged, enumerated, or catalogued in a sequence demanded by a logic encompassing the grouping. Each column of a table could be considered a list. Lists identify data in some sort of sequence. While one item of a list might be compared with another, the purpose for the listing is identification rather than immediate comparison with equivalent data. Occasions for using lists are numerous; among the more frequent are catalogs, directories, tables of contents, indexes, parts lists, specifications, directions, and summaries. Often, within a paper or report, it is convenient to structure a series of parallel statements in a numbered, indented, or tabulated form. It is a good display device for securing emphasis.

I have found it convenient to use lists throughout this text. The process of listing was an efficient device for summarizing major points of a subject. In chapter 2, for example, after discussing the "problem concept" and its significance to scientific investigation, I listed four guides in defining and formulating a problem. In chapter 6, before explaining the elements of a formal report, I listed them.

Tables

Consider a paragraph of text which contains the following data:

As of December 31, the Consolidated Electric Company had a total of 7,863 employees. Of these, 1,989 (or 25 percent) had less than five years' service; 1,590 (or 20 percent) had five to ten years' service; 1,275 (or 16 percent) had ten to fifteen years' service; 784 (or 10 percent)had fifteen to twenty years' service; 931 (or 12 percent) had ten to thirty years' service; 1,294 (or 17 percent) of the employees had thirty or more years' service with the Consolidated Electric Company.

The data, while fairly brief, became rather formidable for a reader to digest. Not that the information is complex or difficult, but it is full of numbers arranged in horizontal lines of text, a factor which makes the information difficult for the reader to assimilate and interpret. Let us rearrange this information by tabulating it. All of the data in the paragraph concern the Consolidated Electric Company's employees' length of service. That subject, then, could become the title of the table. The rest of the information could fit neatly, accurately, and in easily readable form into three columns. The first column might have the heading, "Years of Service." The second column would have the heading, "Number of Employees"; the third column, "Percentage of Total." Under "Years of Service" would be listed: under five, five to ten, ten to fifteen, fifteen to twenty, twenty to thirty, thirty or more. Under "Number of Employees" would be listed: 1,989, 1,590, 1,275, 784, 931, and 1,294. Under "Percentage of Total" would be listed: 25, 20, 16, 10, 12, 17. The final item in this table would be "Total as of December 31"; under "Number of Employees" would be listed "7,863"; and under "Percentage of Total," 100 percent.

CONSOLIDATED ELECTRIC COMPANY EMPLOYEES' LENGTH OF SERVICE

YEARS OF SERVICE	NUMBER OF EMPLOYEES	PERCENTAGE OF TOTAL
Under 5	1,989	25
5 to 10	1,590	20
10 to 15	1,275	16
15 to 20	784	10
20 to 30	931	12
30 or more	1,294	17
Total as of December 31	7,863	100 percent

The table permits the reader to see clearly at a glance all of the numerical information and permits him to interpret and generalize. He can see that, while numerically a fourth of the employees have less than five years of service, more than half have been with the company for more than ten to as many as thirty years. The table lends itself very easily to comprehension and permits the reader to compare the various data groupings.

Tables offer a convenient means for presenting characteristics of things, processes, and concepts. Tables offer the most precise way to present experimental data in a compact arrangement of related facts, figures, and values in orderly sequence, usually in lines and columns for convenient reference.

Tables are boxed or framed within a page when the data are self-sufficient and self-explanatory. Where significance and meaning of a table are dependent upon explanatory material preceding and succeeding the information in the columns, the table is usually not boxed, as in the following example:

The average surface wind velocities and direction at the Seattle Weather Station during the swimming season (for 20-year and 38-year averages) are as follows:

	June	July	August	September
Average wind velocity m.p.h.	4.2	3.9	3.7	3.9
Predominant direction	S	N	N	S

Thus, in summer, with light northerly or southerly winds, we could expect that the water in the swimming areas would be exchanged by surrounding lake water in from two to four hours.

Tables, such as the one immediately preceding do not have numbers and titles. The amount of data they present is limited. They are called *dependent* tables.

Independent tables are those which are self-sufficient and have numbers and titles which often answer the questions, Who, What, Where, When, and How. For example:

TABLE XXI
U. S. POPULATION CHANGE BY SIZE OF COUNTY, 1965 TO 1975

TABLE 102
MORTALITY FROM LEADING TYPES OF ACCIDENTS
IN CANADA BY SEX AND AGE
1965–1975

TABLE V
LABORATORY Z ANALYTICAL DATA OF THREE PREPARATIONS OF
HEPARIN ISOLATED FROM RAT SKIN

Tables within a paper or report are numbered consecutively. The title states where the material was obtained and what it means. The number and title of the table are placed above the table. Subtitles and/or headnotes are frequently used. Headnotes give additional details such as condition under which data were recorded, limitations of the data, accuracy of source, and other explanatory matters. For example, the title of a table might be as follows: Absences during 1974 of Consolidated Electric Company Employees. The headnote under the title in parentheses, might be "(Figures represent man-days lost)." The first column at the left of a table is known as the stub. It serves to identify the horizontal line of data. The column headings to the right sometimes have a second tier of heads for subclassification. The table itself should be arranged as simply as possible. Footnotes should be used wherever necessary for explanation to avoid complicating the title or mixing the words and figures in the table itself. It is usually desirable to use symbols (asterisks and daggers, for example) for footnote references rather than numerals, so that the reader will not confuse the references with the tabular data. Double lines, or heavier lines, are used in the table to indicate division. Totals are placed at the bottom or at the right or at the top and left. Blank spaces within the table should be avoided. Either writing "0" in the space or indicating by a dash that no data were taken for that item is preferable. Tables, like any other visual aid, should be integrated with the text matter. The text discussion should consider the table, even though the table's information may be independent of the text.

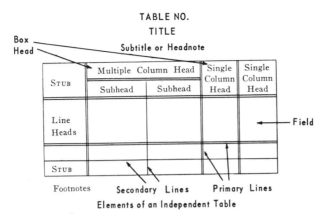

Elements of an Independent Table

Checklist of Fundamentals and Considerations for Constructing Tables

1. When four or more items of statistical information or data are to be presented, the material will be clearer in tabular form.

2. Qualitative, descriptive, and comparative data are more readily comprehensible in table form.

3. The data of a table should be crystallized into a logical unit. Extraneous data should be excluded; the table should be self-explanatory. Though self-contained, the table should be integrated with text matter for fuller explanation and interpretation.

4. The table should have both a number and title. Tables may be numbered in Roman numerals or in Arabic. The title should be concise, yet clearly identify the contents. A subtitle may be used for providing precise details.

5. Each vertical column and, as necessary, each horizontal line should have an identifying head.

6. Standard terms, symbols, and abbreviations should be used for all unit descriptions. The same unit system of measurement should be used for comparable properties or dimensions; for example, in linear measurements feet and meters should not be intermingled.

7. If all the numbers in the table are measurements in the same units, then the unit is stated in the title.

8. Data to be compared should be placed in a horizontal plane.

9. If headings are not self-explanatory, footnotes should be used. If an item is repeated several times in a table, it should be removed from the data in the table and placed in the title, in a footnote, or in a column or line head.

10. Footnotes should be numbered or identified in sequence, line by line, from left to right, across the table.

11. Figures in columns are aligned to similar digits — ordinarily the right digit. However, when the data set up in a column are composed of different units, they should be centered in the column or aligned on the

left. For example, in a table providing data of performance characteristics of an instrument, succeeding line heads would call for different units of data:

Frequency Range	0 to 10.0 kmcs
Power Range	1 mw to 1 watt
Input VSWR	1.15 maximum
Accuracy	± 2%
Reading time	1 minute manually

12. Fractions should be expressed in decimals. Decimal points are aligned in a column. When the first number of a column is wholly a decimal, a cipher is added to the left of the decimal point, e.g., 0.192.

13. Column and line headings should be used to group related data.

14. Tables containing similar data should be set up in the same manner within a given report.

15. Whenever possible, a table should be designed and structured so that it can be typed on one page. If the data cannot be made to fit one page, a continuation page should be used. The word "continued" should be placed at the bottom of the first page to indicate that the table has not been completed, as well as at the top of the second page. Column heads must also be shown on the second page. Subtotals, when appropriate, should be shown at the bottom of the first page and at the top of the second page. Subtotals should always be clearly identified as such.

16. Only significant or summary tables should be placed in the main body of a paper or report. Supporting tables or those of record interest are placed in the appendix.

Example of Experimental Data in Table Form

TABLE II (4:46)

RESULTS OF COLORADO STATE UNIVERSITY LABORATORY TESTING
OF POTENTIAL SEALANTS FOR THE COACHELLA CANAL

Sample No.	Material	Grit Content %	Colloidal Yield %	Wall Building Filter Loss (cc)	Cake (in)	Viscosity (centipoises)
Sl-1	Coyote Well	1.3	53.5*	40	3/32	3
Sl-2	Ackins Claim	12.1	42.9*	189	8/32	2
Sl-3	Thermo Claim	2.5	48.9*	88.5	1/8	2
Sl-4	Burslem Claim	20.7	28.9	69	3/16	2
Sl-5	Armaseal	4.2	65.2	41	1/16	<4
Sl-6	Maas Clay	5.7	60.1	38	1/16	1
Sl-7	Western Clay (Utah)	17.5	55.5	28.5	1/16	6
Sl-8	Western Clay (Utah) reserves	5.0	41.3	33.3	1/8	3
Sl-9	Bent. Corp. (Utah)	4.1	84.6	14.5	5/64	8
Sl-10	Baroid (Wyo) crushed	4.8	89.4	16.5	1/8	22
Sl-11	Baroid (Wyo) 200 mesh	2.9	88.2*	16	3/32	23

*Dispersant (sodium tripolyphosphate—0.75 gms.) added where tendency for flocculation noted.

Example of Statistical Data in Table Form

TABLE II

BATHING BEACH BACTERIOLOGICAL DATA*
KING COUNTY, JUNE 25-SEPT. 2, 1974

	Green Lake		Lake Washington		Puget Sound Golden Gardens	Lake Sammamish State Park
	West Beach	East Beach	Seward Park	Juanita		
No. of values	113	105	112	113	112	113
Median value†	23	230	230	60	230	60
Range of values	0 to 2400	0 to 24,000	0 to 13,000	0 to 7000	0 to 7000	0 to 2400
Avg. No. bathers	130	104	110	110	43	114
Avg. water temp.	70.3	71.1	69.0	70.5	58.2	71.9
Coliform per bather‡	0.18	2.2	2.1	0.54	5.4	0.53

*Most probable number of coliform bacteria per 100-ml sample, computed from Seattle-King Co. Health Dept. data.

†Less than 45 proportioned equally among 0, 15, 30, and 45.

‡Coliform per bathers = $\dfrac{\text{Median}}{\text{Avg. No. Bathers}}$

Example of Comparison of Characteristics in Table Form

COMPARISON OF COMMONLY USED FREQUENCY MEASUREMENT DEVICES

TYPE OF INSTRUMENT	FREQUENCY RANGE (mc/sec)	APPROXIMATE LOADED Q	FREQUENCY MEASUREMENT ACCURACY	RELATIVE CONVENIENCE OF USE	TYPICAL APPLICATIONS
Tuned Lumped Constant Circuits	300 to 1200	200	2%	1	Quick checking of frequency in unshielded systems.
Cavity Meters Sealed	300 to 40,000	10^2-5×10^4	0.01 to .05%	1	General laboratory and field work in conjunction with microwave measurements. Setting of systems frequency.
Unsealed	300 to 40,000	10^3-5×10^4	.05 to .1%	1	General laboratory work in conjunction with microwave measurements.
Slotted Sections	300 to above 40,000	—	0.1 to 5%	2	Checking of frequency while making VSWR measurements.
Frequency Standards Crystal Reference Standards	300 to 40,000	10^4-10^6	.02 to .0001%*	3	Setting of systems frequency. Extremely precise scientific investigations. Calibrating less accurate instruments.
Spectral Lines	Fixed Points	10^5-10^6	.0001%*	4	Stable reference frequencies. Spectroscopic measurements.

*Limited by associated measurement apparatus.

Example of Process in Table Form

TROUBLES AND REMEDIES BRAKE SYSTEM (3:27)

Symptom and Probable Cause	Probable Remedy
Pedal Spongy a. Air in brake lines.	a. Bleed brakes.
All Brakes Drag a. Mineral oil in system. b. Improper pedal to push rod clearance. c. Compensating port in main cylinder restricted.	a. Flush entire brake system and replace all rubber parts. b. Adjust clearance. c. Overhaul main cylinder.
One Brake Drags a. Loose or damaged wheel bearings. b. Weak, broken or unhooked brake retractor spring. c. Brake shoes adjusted too close to brake drum.	a. Adjust or replace wheel bearings. b. Replace retractor spring. c. Correctly adjust brakes.
Excessive Pedal Travel a. Normal lining wear or improper shoe adjustment. b. Fluid low in main cylinder.	a. Adjust brakes. b. Fill main cylinder and bleed brakes.
Brake Pedal Applies Brakes but Pedal Gradually Goes to Floor Board a. External leaks. b. Main cylinder leaks past primary cup.	a. Check main cylinder, lines and wheel cylinder for leaks and make necessary repairs. b. Overhaul main cylinder.
Brakes Uneven a. Grease on linings. b. Tires improperly inflated. c. Spring center bolt sheared and spring shifted on axle.	a. Clean brake mechanism; replace lining and correct cause of grease getting on lining. b. Inflate tires to correct pressure. c. Replace center bolt and tighten "U" bolts securely.
Excessive Pedal Pressure Required, Poor Brakes a. Grease, mud or water on linings. b. Full area of linings not contacting drums. c. Scored brake drums.	a. Remove drums — clean and dry linings or replace. b. Free up shoe linkage, sand linings or replace shoes. c. Turn drums and install new linings.

Photographs

Photographs offer not only realistic and accurate representations but also dramatic and artistic effects. Effective photographic illustrations require thoughtful planning, so that all desired details are shown at the most favorable angle. Poor photographs have such faults as cluttered backgrounds, inadequate lighting, poor camera angle, and improper focus. Good photographs are sharply detailed and do not contain distortions of any essential elements. Proper lighting eliminates or reduces strong shadows which destroy fine details.

To give the reader an idea of the equipment size or to lend human interest to the picture, include in the photograph, whenever appropriate, human beings who might be logically associated with the equipment; or include portions of the body, such as the hand or head, in proper relation to the equipment in use. This technique is illustrated by the photographs on pages 158 and 159.

In addition to the techniques of the ordinary still camera, those of the color photograph, the aerial photograph, the X-ray photograph, the photograph taken through an electron microscope (see photograph on page 161)

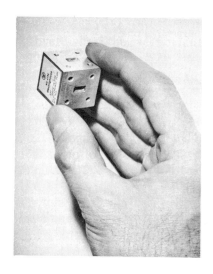

This tiny discriminator cavity (WX-4334) is a simple, rugged frequency stabilizing device for use with parametric amplifier pumps, navigation systems, beacon radars, or any system that requires a stable high-frequency source. It is a dual mode, transmission type cavity designed for operation at a fixed frequency of 35 kilomegacycles. The cavity is vacuum sealed and incorporates invar construction to minimize frequency shift with ambient temperature change (2:25).

or through a telescope, and a photo micrograph can lend impressive visual aid. *Callouts* are frequently added to the photograph to help provide distinctive details and identification. Callouts are labels with leaders that point to individual items in an illustration; the labels identify or give information about the item. When there are many elements to be identified, numbers or letters replace the callouts. Each letter, then, is referenced to the identifying caption. The photo cutaway view on p. 160 has callouts for identifying parts.

Drawings

The drawing is one of the oldest forms of symbolic representation. Writing and drawing were identical in prehistoric Egypt and in early Greece. The Greek word *graphein* means both writing and drawing. Drawings are made with pencil, pen, crayon, or brush. There are many kinds, from a

Wide-angle view from above of the new National Bureau of Standards 12-million pound testing machine. The machine is believed to be the largest testing device in the world and capable of applying forces of 12 million pounds in compression and 6 million pounds in tension. Designed primarily for testing full-scale structural components, the machine will also be used to apply forces for calibrating large capacity force-measuring devices such as those used to measure rocket thrust (10:261).

simple freehand sketch to detailed engineering and architectural drawings with elaborate minutae of detail. Drawings offer more flexibility than photographs for showing the inner movements of equipment, cross sections, and relationships. The more complex the subject matter of a drawing is, the greater the necessity for identifying callouts. Letter symbols with keys to identify the symbols are required.

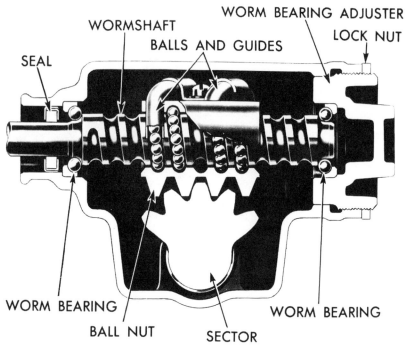

WORM BEARING ADJUSTER

WORMSHAFT   LOCK NUT

BALLS AND GUIDES

SEAL

WORM BEARING   WORM BEARING

BALL NUT   SECTOR

The Saginaw recirculating ball nut gear is used on a large number of American vehicles today. It is basically a screw-nut type with recirculating balls between the thread grooves and the nut. This nut has rack teeth engaging with sector teeth on the pitman shaft. This gear provides increased efficiency to accommodate the increased steering demands caused by heavier vehicles and lower tire pressure (7:5).

The three examples on pp. 163-65 illustrate how drawings may be integrated with text to help explain a technical device. In the first example, drawings illustrate the design of a device, its components, and how they work. These drawings permit cutaway and cross-sectional views of schematics which clarify the explanation in the text. In the second example, the explanation of movement and action offered by the text is made clearer and more vivid by drawings. The third example of how text and drawings may be effectively integrated involves a freehand crayon drawing that is on the decorative side but still appropriate to the exposition of the text (13:79; 26;215).

Diagrams

Diagrams are a subclassification of drawings. Diagrams are representations of abstractions. They are symbolic configurations. Some times they are charts or graphs, explaining or illustrating ideas or statistics. Some texts

This is rust, as photographed through an electron microscope. Part of a research project on corrosion of iron, this photo was taken to demonstrate the effect of water vapor. In an atmosphere of pure water vapor, the surface erupts into thin blade-shaped crystals, which reach a density of nearly one billion per square inch (1:26).

use the words "drawing" and "diagram" interchangeably. Drawings attempt to represent the likeness of their subject; diagrams attempt to show the operation of the subject. Diagrams present an analysis of a symbolic or conventional representation of an actuality. The verbal equivalent of a diagram is an outline. Therefore, the diagram outlines or sketches, rather than represents an actuality. Among the more frequently used diagrams are schematic diagrams of which there are two major types.

To those schooled in the art of lightning measurement, this interesting figure is a film record of the electric field intensity developed by a thundercloud. The measurement was made with a specially developed Klydonograph capable of recording voltages about three times those obtainable with conventional Klydonographs. In essence, the record is produced on a sheet of photographic film that is placed between a probe electrode and a grounded plate — the greater the field intensity, the larger the figure. In this example, the record shows that an electric field intensity of approximately 2500 volts per inch existed just prior to the lightning stroke (5:33).

1. *Schematic diagrams:*

   a. One type is a cutaway (cross-sectional view) drawing of the interior of a mechanical device or equipment. (See page 163).

   b. The second type is a graphic representation in which symbols and lines are used to depict various components and their connections. This type of schematic is the equivalent of the blueprint. Of this second type of schematic diagram there are two subtypes:

      (1) *Complete:* this type shows the arrangements of the components of an entire equipment or a major unit.

      (2) *Stage:* this type is sometimes called a partial or functional schematic. It is a functional segment of usually one stage of a complete schematic diagram.

2. *Wiring diagrams:* These are graphic representations of the relative physical location of all component parts of an electronic equipment. This type of diagram identifies all wires or interconnections between components. Leads are identified by a colored code; cable numbers and type of wire used are charted. Graphic symbols are not used in wiring diagrams. Parts are shown either pictorially or by simple rectangular boxes. The layout of the components follows the actual chassis arrangement.

ADJUSTMENT DEVICES

FRICTION RELIEF HANDCRANK

The handcrank can be provided with an adjustable friction drive, by using a cup spring (or a helical spring) to apply pressure upon a gear bearing against a wood or composition disk. Adjustment of a clamp nut provides the means of varying the pressure to obtain the degree of friction required. If pressure becomes greater than the friction imposed upon it, the gear will slip and protect associated gearing from strain or damage.

ADJUSTABLE HOLDING FRICTION

The same handcrank may be provided with cork disks, a collar, and a bushing. This assembly puts a drag on the handcrank, keeps it positioned, and prevents motion from backing out through the handcrank.

POSITIONING PLUNGER

We can carry the design of the handcrank still further and add a plunger for the purpose of holding the shaft in either of two positions: an IN position and OUT position. In changing position, the shaft and the drive gear move in relation to the adapter housing. The plunger is pulled out and the handcrank pushed or pulled to its new position. When released, the plunger is returned by a spring and enters a hole in the bushing, locking the assembly in a particular position.

Moving the handcrank to the in or out position will cause it to engage or become disengaged, or this arrangement can be used to drive one or the other mechanism. By using a wide gear, this drive can be kept in engagement all the time, the in and out position being to control the drive of another gear. Thus, it is possible to drive one gear at all times, alone, or in conjunction with another. Further, we

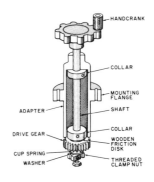

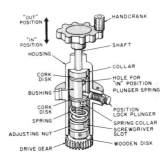

would include a switch actuated by the in or out position of the shaft.

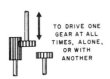

TO DRIVE ONE GEAR AT ALL TIMES, ALONE, OR WITH ANOTHER

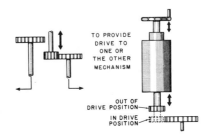

3. *Pictorial diagrams:* These visuals are graphic representations which show components of equipment as they appear normally to the eye. They may be drawn orthographically, isometrically, angularly, through projection or perspective. Schematic drawings often are used to depict the design of an apparatus or process.

4. *Block diagrams:* These diagrams show the function of an entire unit. All stages or functions are represented by rectangular boxes or similar symbols and are arranged in the order of signal flow. Each block is labeled clearly. Notes are frequently used to aid the explanation of the diagram.

SECTOR

MULTIPLIERS

Some multipliers operate on a geometric principle slightly different from that previously described. The computing portion of one such multiplier is shown to the right. It consists of a pivoted sector arm along which a multiplier pin can be moved by a lead screw. The A input is applied as the angle between the sector arm and the reference line, and the B input as the distance between the pivot and the multiplier pin carried on lead screw. The output is taken as the vertical height of the multiplier pin above the reference line. The input and output quantities form the right triangle shown next to the sector arm. Solving the right triangle gives: Output = B sin A. However, for small angles up to about 1/3 radian (20 degrees), the sine is nearly equal to the angle in radians. Hence, the equation for the output can be rewritten as Output = AB. This means that within its range the mechanism acts as a multiplier (to a close approximation).

The additional mechanism shown to the right in heavy line is provided merely to pick off the linear output of the A sector arm and convert it to an angular quantity. The mechanism is set up so that the radius arm, parallel arm, and C sector arm form a parallelogram with the result that the slot in the parallel arm always remains parallel to the reference line. The multiplier pin of the A sector arm projects up through the slot and moves the parallel arm an amount equal to output AB. This linear output is applied to the C sector arm causing the arm to rotate through output angle C. As shown in the diagram, linear motion AB, angle C, and the constant K(distance between C sector pivot and hinge pin) form a right triangle. Solving the triangle gives: K sin C = AB. Again, for small angles sin C is almost equal to C.

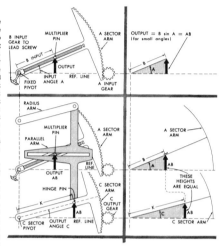

Hence, the equation can be rewritten as KC = AB. The constant K is accounted for in the external gearing. The complete sector multiplier in the form shown here is actually used in computers.

MULTIPLYING

BY

A CONSTANT

Any of the complete multipliers that have been described can be used to multiply a variable by a constant as well as to multiply two variables. However, for multiplying by a constant, the use of such complex devices is uneconomical in both cost and space utilization. There are many simpler and more compact devices available for the handling of constant factors. All of these devices are based on a fixed ratio between the variable input and the output. A number of different types of simple constant multipliers are shown at the right. These and many other similar devices are employed extensively throughout Naval computers.

scale factor The simplest way to multiply by a constant is to introduce the ratio into the scale or dial from which the value of the product is read.

gears A pair of gears is a device that can be used as a constant multiplier. The constant is the ratio of the number of teeth on the driving gear to the number of teeth on the driven gear.

$RATIO = \frac{30}{10} = 3$

levers When pivoted levers are used, the constant factor is established by the ratio of the arm lengths as measured from the fixed pivot. Two examples of levers are shown.

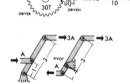

bell cranks These cranks are essentially the same as levers. They allow the input and output to move in different directions. The bell crank can have any desired angle and shape.

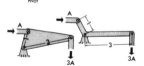

164

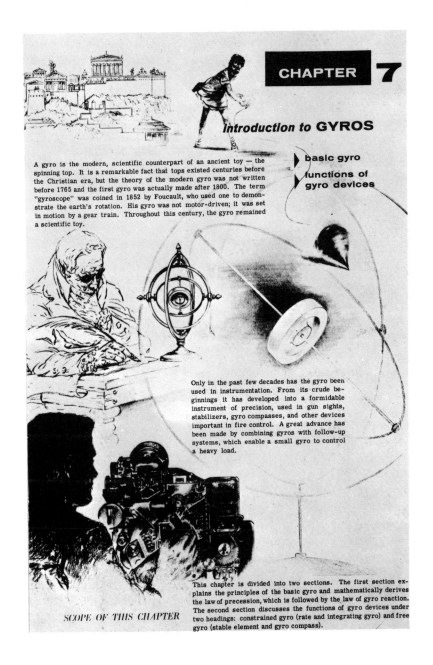

CHAPTER 7

Introduction to GYROS

▶ basic gyro
▶ functions of
gyro devices

A gyro is the modern, scientific counterpart of an ancient toy — the spinning top. It is a remarkable fact that tops existed centuries before the Christian era, but the theory of the modern gyro was not written before 1765 and the first gyro was actually made after 1800. The term "gyroscope" was coined in 1852 by Foucault, who used one to demonstrate the earth's rotation. His gyro was not motor-driven; it was set in motion by a gear train. Throughout this century, the gyro remained a scientific toy.

Only in the past few decades has the gyro been used in instrumentation. From its crude beginnings it has developed into a formidable instrument of precision, used in gun sights, stabilizers, gyro compasses, and other devices important in fire control. A great advance has been made by combining gyros with follow-up systems, which enable a small gyro to control a heavy load.

SCOPE OF THIS CHAPTER

This chapter is divided into two sections. The first section explains the principles of the basic gyro and mathematically derives the law of precession, which is followed by the law of gyro reaction. The second section discusses the functions of gyro devices under two headings: constrained gyro (rate and integrating gyro) and free gyro (stable element and gyro compass).

165

A Schematic Drawing

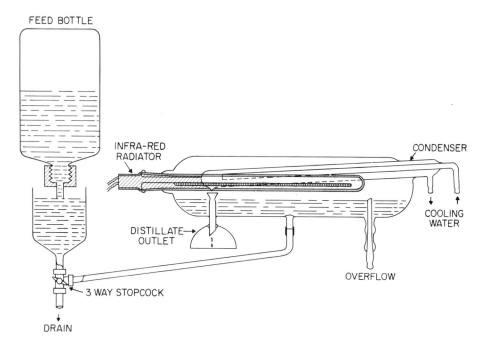

Schematic of a sub-boiling still used at the National Bureau of Standards in the preparation of ultrapure mineral acids and water (11:104).

A Schematic Diagram

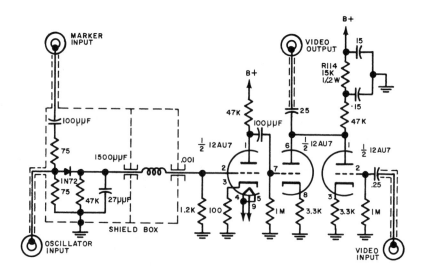

Figure 1.
Schematic diagram of the marker injection circuit.

Stage or Partial Schematic Diagram

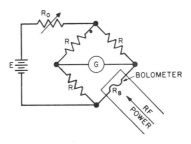

Figure 2.
The basic dc Wheatstone bridge circuit which is commonly used to measure the microwave power in a bolometer.

A Pictorial Diagram for an Electronic Instrument

Excellent integration of text matter and pictorial perspective diagrams to explain the operation of a system are shown below (13:130-131).

BASIC MANUAL SERVO

A SERVO PROBLEM

One of the many types of problems which a servo may be required to solve is the movement of a heavy gun mount in response to an order generated by a computer, and indicated by the reading on the dial.

A SIMPLE SOLUTION

The simplest type of servo which will solve a problem like this is the manual servo. This type consists of a man reading the computer dial and turning the gun mount by hand until the mount position is the same as the ordered position.

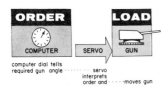

computer dial tells
required gun angle ········· is servo
interprets
order and ······ moves gun

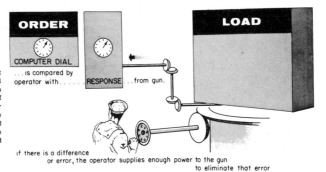

To position the gun mount accurately, the operator will need an indicator to show him the exact present position of the mount. Such an indicator may be a simple dial. The reading on the dial is called the response of the gun. The difference between order and response is called error.

...is compared by operator with...... RESPONSE ...from gun.

if there is a difference
or error, the operator supplies enough power to the gun
to eliminate that error

Although the Basic Manual Servo is simple compared to the complicated servos found on a modern ship, nevertheless, it has all the essential properties of a servo.

BASIC AUTOMATIC SERVO
need for automatic servo · · · · · · · ·

making servo automatic

The necessary speed and accuracy for modern servo use can be attained by mechanization. In order to mechanize a manual servo, we must mechanically duplicate each manual function.

These functions, whether performed manually, or automatically, are essential in any servo, and may be considered as the basic principles of all servos.

1. COMPARISON of order and response to determine error.

2. CONTROL of power by the error.

In a manual servo, these functions are performed by the operator. In an automatic servo, they are performed by mechanisms.

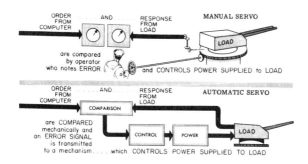

The specific devices used as components to fill these "black boxes" can take many forms. The function of the comparison device is to subtract the response from the order. A differential can accomplish the required subtraction. The power supply may be a motor, controlled by the error, through a switch.

The manual servo does not have the necessary speed and accuracy for many servo applications. A heavy object such as a gun mount can be best controlled by an automatic servo with a source of high power.

The devices contained in the "black boxes" need not take a mechanical form. It is very common to have servos which operate electrically, hydraulically or pneumatically.

Signals may be transferred from one mechanism to another by any of these means.

TRANSFER OF SIGNALS

The order and response signals of a servo need not be mechanical. For example, servos using the same principles as discussed here may control such diverse quantities as missile direction, height of water in a tank, machine shop operations, pressure and temperature. In fact, the household thermostat is a common form of automatic servo.

SUMMARY

Both the Basic Manual and Basic Automatic Servos function according to the same principles. These principles apply to all servos discussed in this chapter.

They are: comparison of order and response to determine error; and control of power by error.

Block Diagram of an Electronic Instrumentation

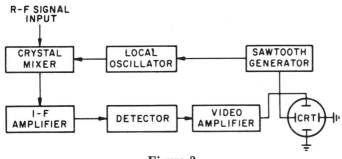

Figure 3.
Block Diagram of the Spectrum Analyzer.

Exploded Views

Exploded views (see p. 170) can be either photographs or drawings of a device or equipment in which the parts are shown in a disassembled state but arranged to simulate a perspective view. The parts are arranged in sequence in their respective axis of assembly. Each part is called out by name or by a reference designation. (See also page 000, chapter 10, for exploded view of Oldham Coupling.)

Maps

The map (see p. 171 for an example) is a symbolic, conventionalized representation of reality. Maps are most frequently used to show geographic or spatial distributions or relationships, representing areas of land, sea, or sky. Among the types used in technical writing are: contour, profile, historical, linguistic, political, or demographic.

Maps make use of color and shading, as well as conventionalized symbols, to depict details of information and to indicate relationships. The maps should include a key to help explain the symbols and to enable the reader to read or interpret the maps satisfactorily. Scales are included where geographical distances are important.

Graphs, Charts, and Curves

Graphs, charts, and curves are terms which are used interchangeably for the diagraming, mapping, and presentation of statistical information. These devices compare pictorially changes in value or interrelationships of variable quantities. Charts, curves, and graphs simplify statistical aspects of information and aid in their interpretation. Some types of visual aids within these groupings are: the curve or line chart, the pie chart, the sur-

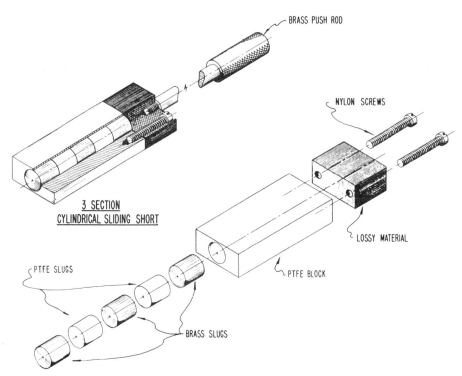

BRASS PUSH ROD

NYLON SCREWS

3 SECTION
CYLINDRICAL SLIDING SHORT

LOSSY MATERIAL

PTFE SLUGS

PTFE BLOCK

BRASS SLUGS

Exploded and assembled views of a 3-section cylindrical sliding-short, a new design which permits greater ease in making precision phase-shift measurements at microwave frequencies. PTFE block prevents brass slugs from contacting inside wall of waveguide and provides a stable mechanism to slide within the waveguide. Brass slugs maintain a light press fit within PTFE block; PTFE slugs maintain optimum spacing between brass slugs (8:17).

face chart, the flow chart, the pictorial chart, the organization chart, the bar graph, and the line graph.

The line chart or graph is the simplest and most commonly used visual aid. Generally, the independent variable is plotted on the abscissa—the horizontal axis or scale—and the dependent variable is plotted on the ordinate—the vertical axis or scale. The most common independent variables include time, distance, voltage, stress, load; they are plotted on the horizontal or abscissa scale. Temperature, money, current, and strain—common dependent variables—are plotted on the vertical or ordinate scale.

Theoretically, each scale begins at "0." The zero is placed where the two scales intersect. However, this is not always possible. Occasionally, the abscissa or horizontal line needs to represent a variable where zero is meaningless. Examples of such a variable would be hours, months, or years during which an event occurred. The ordinate or vertical axis may also need to represent values wherein the zero is not appropriate—for example, where costs may begin at millions or more dollars. This problem is met by putting some value higher than zero where the scales intersect or by putting a break in the scale. This is frequently done by starting the scale at zero

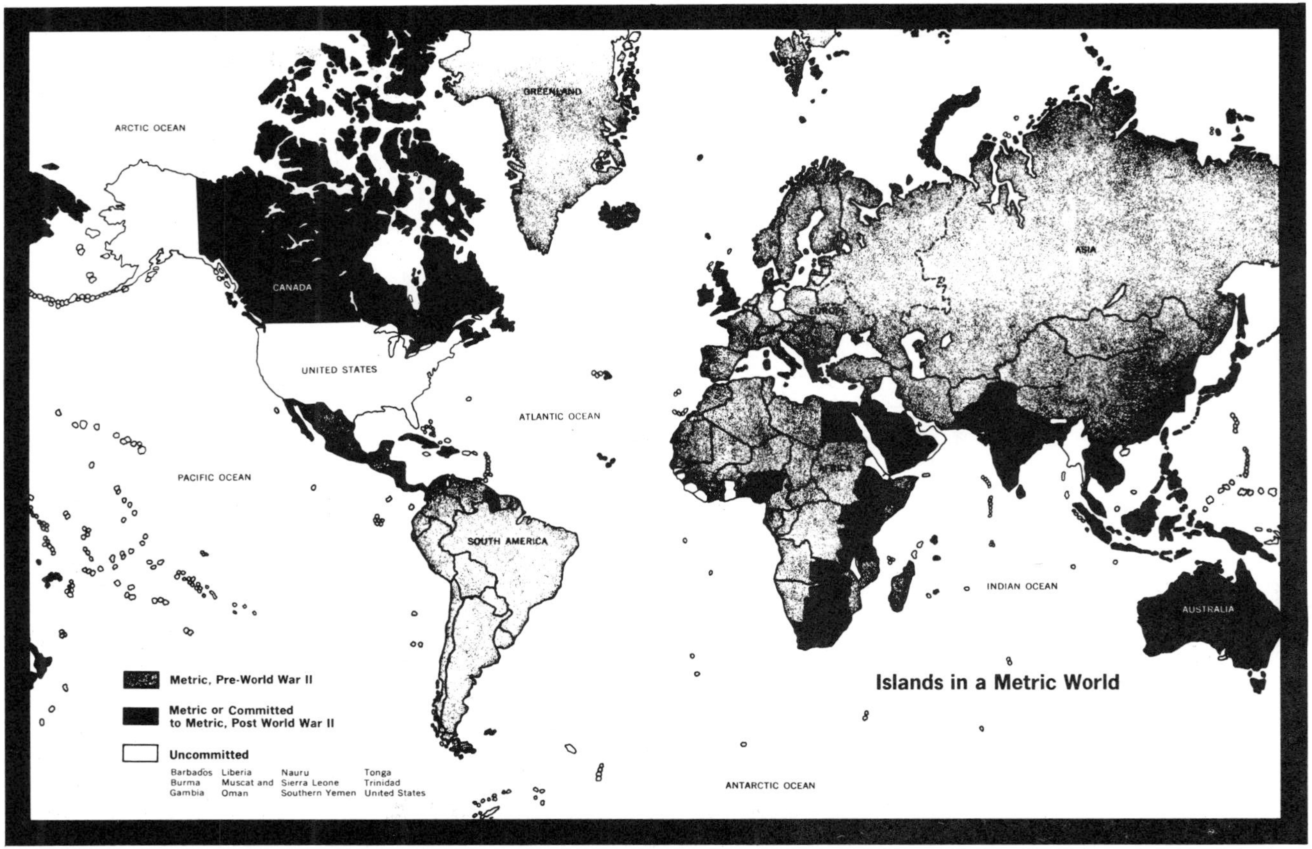
Islands in a Metric World
ARCTIC OCEAN
CANADA
GREENLAND
UNITED STATES
ATLANTIC OCEAN
PACIFIC OCEAN
SOUTH AMERICA
ASIA
EUROPE
AFRICA
INDIAN OCEAN
AUSTRALIA
ANTARCTIC OCEAN
Metric, Pre-World War II
Metric or Committed to Metric, Post World War II
Uncommitted
Barbados
Burma
Gambia
Liberia
Muscat and Oman
Nauru
Sierra Leone
Southern Yemen
Tonga
Trinidad
United States

Islands in a Metric World (9:223).

and then depicting the fracture that represents values between zero and the first value drawn to follow it on the scale.

Choosing the proper scale values are important. Identical data presented on two graphs each of which has a different scale may give the viewer two entirely different impressions.

Various types of charts are used to present information to the reader for convenient and easy comprehension:

1. A *line chart* or *line graph* is used to show comparative trends or values over a long period of time. Line graphs may be used in combination with bar charts or may be used to represent the profile curve of continuous or changing data. See the combination line chart and line graph on page 173.

2. The *bar chart* is used to show relative quantities by vertical or horizontal bars of varying lengths. Causal relationships cannot be shown by a bar chart. Multiple factors can be shown by bar charts by changing the appearance of the bars through crosshatching, utilizing different colors and filling in the bars in different ways. A logarithmic horizontal bar graph is illustrated on page 174.

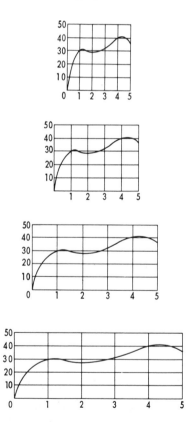

Effect of Choice of Scale.

3. The *pie chart* or circle graph is an excellent device to partition or classify a whole into various parts or elements. See the typical pie chart on page 176.
4. The *pictorial chart* is a visual aid in which units are represented by a picture symbol. Each symbol may stand for a stated number of units. Pictorial charts are illustrated on page 175.
5. A *surface chart* is a variation of the line chart. It can be used to show very clearly cumulative totals of two or more components; relative sizes of various components are dramatically, although not always accurately, shown through this charting device. The surface chart is made by shading or crosshatching the areas between index lines, so that the areas are differentiated. The areas at the bottom

UPPER ARKANSAS

RED RIVERS REGION (12:104)

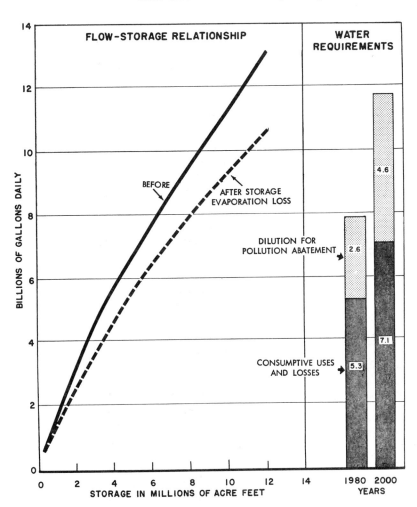

are shaded more darkly; the upper areas become progressively lighter. Surface charts are not practical when index lines cross each other. See the surface chart on page 176.

6. The *organizational chart,* like the block diagram, is a directional chart. Blocks are arranged to simulate the sequence of the line of authority in an organization. Instead of functions of components, the titles of persons or the names of departments are printed in the boxes. Reading is from top to bottom. Equal functions are drawn

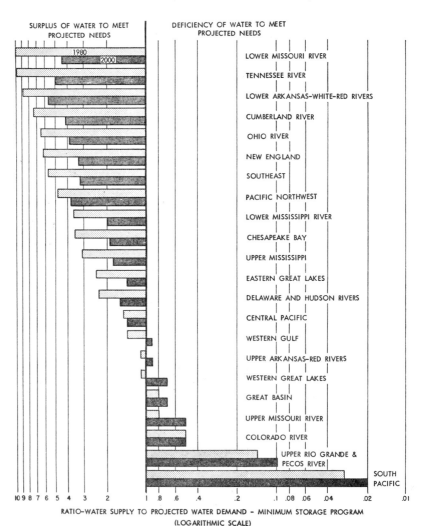

RATIO OF MAXIMUM OBTAINABLE SUPPLY TO

MINIMUM FLOW REQUIREMENT

1980 and 2000 (12:127)

A Bar Graph

from left to right. Direct relationships are indicated by solid lines connecting the boxes, indirect relationships by broken lines. A typical organizational chart is illustrated on page 177.

7. The *flow chart* traces pictorially the movement of a process from beginning to completion of an action, or from the raw material to a finished product. Processes may be pictured by simplified drawings or conventional symbols to represent the operations. Arrows are used to indicate flow of movement. See the flow chart on page 178.

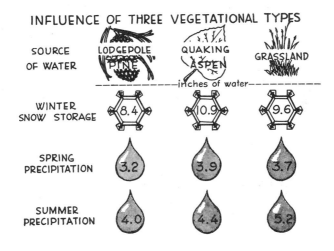

A Pictorial Table Graph

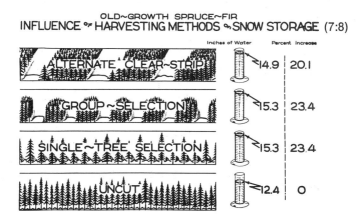

A Pictorial Chart

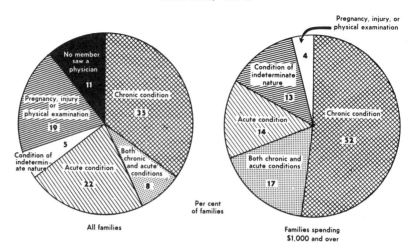

Reasons for Seeing a Physician by Expenditures for Health*
United States, 1957-58

Pregnancy, injury, or physical examination

No member saw a physician — 11
Chronic condition — 35
Pregnancy, injury, or physical examination — 19
Condition of indeterminate nature — 5
Acute condition — 22
Both chronic and acute conditions — 8

Condition of indeterminate nature — 13
4
Chronic condition — 52
Acute condition — 14
Both chronic and acute conditions — 17

Per cent of families

All families

Families spending $1,000 and over

A Circle Graph or Pie Chart

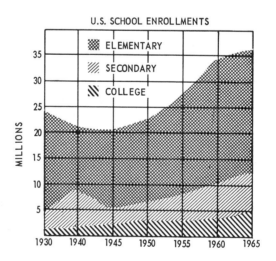

U.S. SCHOOL ENROLLMENTS

▒ ELEMENTARY
▨ SECONDARY
▧ COLLEGE

MILLIONS

35
30
25
20
15
10
5

1930 1940 1945 1950 1955 1960 1965

Source: Industrial College of the Armed Forces

A Surface Chart

176

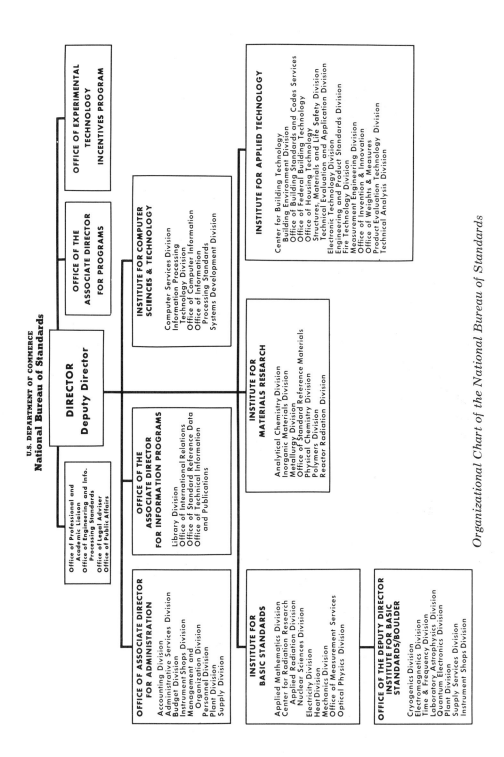

DIRECTOR
Deputy Director

Office of Professional and Academic Liaison
Office of Engineering and Info. Processing Standards
Office of Legal Adviser
Office of Public Affairs

OFFICE OF EXPERIMENTAL TECHNOLOGY INCENTIVES PROGRAM

OFFICE OF THE ASSOCIATE DIRECTOR FOR PROGRAMS

OFFICE OF THE ASSOCIATE DIRECTOR FOR INFORMATION PROGRAMS

Library Division
Office of International Relations
Office of Standard Reference Data
Office of Technical Information and Publications

INSTITUTE FOR COMPUTER SCIENCES & TECHNOLOGY

Computer Services Division
Information Processing Technology Division
Office of Computer Information
Office of Information Processing Standards
Systems Development Division

INSTITUTE FOR APPLIED TECHNOLOGY

Center for Building Technology
Building Environment Division
Office of Building Standards and Codes Services
Office of Federal Building Technology
Office of Housing Technology
Structures, Materials and Life Safety Division
Technical Evaluation and Application Division
Electronic Technology Division
Engineering and Product Standards Division
Fire Technology Division
Measurement Engineering Division
Office of Invention & Innovation
Office of Weights & Measures
Product Evaluation Technology Division
Technical Analysis Division

OFFICE OF ASSOCIATE DIRECTOR FOR ADMINISTRATION

Accounting Division
Administrative Services Division
Budget Division
Instrument Shops Division
Management and Organization Division
Personnel Division
Plant Division
Supply Division

INSTITUTE FOR MATERIALS RESEARCH

Analytical Chemistry Division
Inorganic Materials Division
Metallurgy Division
Office of Standard Reference Materials
Physical Chemistry Division
Polymers Division
Reactor Radiation Division

INSTITUTE FOR BASIC STANDARDS

Applied Mathematics Division
Center for Radiation Research
Applied Radiation Division
Nuclear Sciences Division
Electricity Division
Heat Division
Mechanics Division
Office of Measurement Services
Optical Physics Division

OFFICE OF THE DEPUTY DIRECTOR INSTITUTE FOR BASIC STANDARDS/BOULDER

Cryogenics Division
Electromagnetics Division
Time & Frequency Division
Laboratory Astrophysics Division
Quantum Electronics Division
Plant Division
Supply Services Division
Instrument Shops Division

Organizational Chart of the National Bureau of Standards

177

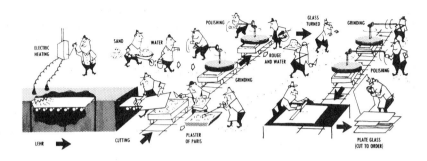

Simplified "Flow Diagram" of Flat Glass Manufacture (14:38-39).

Some Final Words on Graphic Presentation

In summary, the most effective graphic aids are those which serve the reader's needs, supplementing and clarifying the sense of the text. Illustrations for embellishment alone are a distraction. Poorly executed drawings or those illogically planned or placed confuse and frustrate the reader. Graphic aids should be functionally integrated into the discussion. Illustrations and tables should not be cluttered with excessive details or notations. They should bear proper identification by means of a table or figure number, title, and caption. Figures are numbered independently of tables and in sequence. Titles are identification labels; captions are brief but complete explanatory annotations.

Illustrations should be expertly rendered. Poor execution will distort the facts they are required to communicate. If you, the writer, are inexpert, make freehand sketches with ample explanatory notes and instructions for submission to an experienced technical illustrator. In your discussions with him, keep in mind reproduction requirements. Illustrations should be planned to have a uniform reduction ratio for reproduction. Recommended reductions are 1.5 to 1 for photographs and 2 to 1 for line drawings. Uniformity and consistency in such details as direction of arrowheads in callouts, in nomenclature, symbols, weight of lines, lettering, and similar matters give a professional quality to the printed paper or report.

As a closing word, you will find that an open mind and a willingness to accept advice and suggestions from the professional illustrator will frequently enhance the clarity and effectiveness of your ultimate product — the reproduced or published report.

Discussion Problems and Exercises

1. Refer to the four different reports you located under problem 5 of chapter 4. Examine also a representative number of technical periodicals. Select effective and ineffective examples of graphic presentations. What relationships do you

find between subject-matter and the types of graphic aids? Are there any new or unusual combinations or uses of graphic aids? Write an analysis discussing the effectiveness or ineffectiveness of the visual aids used in four representative items of technical writing.

2. Check the report you are writing. Determine where graphic presentation can effectively promote the communication of your data. On the basis of the considerations discussed in this chapter, determine which types of graphic aids you will need. First make rough sketches. As you deem necessary, obtain advice and help from more expert friends or teachers in the execution of your drawings and photographs. Construct the tables required. Lastly, rewrite your text to integrate the graphic presentations of your report.

References

1. "A Basic Theory about Rust," *Westinghouse Engineer,* January 1961.

2. "Checking Out Electrical Systems," *Westinghouse Engineer,* January 1961.

3. *Chevrolet Passenger Car Shop Manual,* Detroit, Michigan: General Motors Chevrolet Division, 1954.

4. Dirmeyer, R. D., Jr., and R. T. Shen, *Sediment Sealing of Irrigation Canals,* Fort Collins, Colorado: Colorado State University, July 1960.

5. "Estimating Lightning Performance," *Westinghouse Engineer,* January 1961.

6. *The Fraser Experimental Forest ... Its Work and Aims,* Forestry Station Paper No. 8, Fort Collins, Colorado: U.S. Department of Agriculture, Revised June 1960.

7. Marquis, Donald P., "How Fundamentals are Applied to the Design of Safe, Efficient Automotive Steering Systems," *General Motors Engineering Journal,* October-November-December 1957.

8. *National Bureau of Standards Technical News Bulletin,* January 1971.

9. *National Bureau of Standards Technical News Bulletin,* September 1971.

10. *National Bureau of Standards Technical News Bulletin,* November 1971.

11. *National Bureau of Standards Technical News Bulletin,* May 1972.

12. *Dater Resources Activities in the United States,* Washington, D.C.; U.S. Government Printing Office.

13. *Weapons Systems Fundamentals, Basic Weapons Systems Components,* NAVWEPS OP3000, Vol. 1, Washington, D.C.: U.S. Government Printing Office, July 15, 1960.

14. *Westinghouse Engineer,* March 1960.

PART TWO

8

Semantics and the Process of Communication

Efficient communication of scientific thought or technical information—whether conveyed by a technical writer, a scientist, or an engineer—does not just happen. Effective technical writing is a craft, never an accident. The poet, the playwright, the novelist must master his art before he can give form and meaning to what he has to say. So must the conveyor of technical information. If he is to reach the minds of other men, he must use language with precision, clarity, and grace. Technical writing is a form of exposition in which one explains facts, expounds theory and principles, and analyzes concepts, objects, processes, events, and data. The chief end of technical writing is to inform the reader and to evaluate the information.

The purpose of this text is to teach you how to communicate technical information effectively. To begin with, you need to know how language and semantics affect the communication process and how their various elements operate.

Origin and Nature of Language

Linguistic scholars and anthropologists are not in agreement as to how language originated. However, there must have been a time when language started, just as there was a time when man began to use fire. In *An Introduction to Philosophy*, W. A. Sinclair claims that language originated and remains as a "symbolic pointing." He says:

> Speech is simply gesture made audible. . . . Speech is an extremely complex system of more or less standardized and conventionalized noises, and writing is an even more highly standardized and conventionalized system of visible marks upon a surface, but in principle speech and writing are as much gesture as is pointing with the finger.
>
> The origin of language appears to have been roughly as follows: Our remotest human ancestors when they attempted to draw the attention of others of their kind to anything in particular (or when they behaved in a way which did in fact draw attention to it, intention or no intention) pointed and gesticulated in its direction. These movements were accompanied by various movements in the flexible tissues and other parts of the body, especially when any high degree of vigor was put into action. Among these were movements in the tongue, lips, windpipe, and associated parts. This was further accompanied in certain cases by contraction of the walls of the chest and movements of the diaphragm which led to the expulsion of air from the lungs, through the windpipe, over the tongue, and between the teeth and lips, thereby creating noises. The gestures and the noises together resulted in drawing attention to the object or situation in question, and then in course of time the noises alone served to do so. Later, after thousands upon thousands of years, conventionalized marks upon surfaces came to be employed to represent the conventionalized noises.
>
> Consider an example. When a biologist writes and you read, that the hind leg of a horse is homologous with your own leg . . . what is happening? What he does is to arrange the printer's black marks on paper so cunningly that you, sitting in your chair by the fire, have your attention drawn to certain matters which he could otherwise have drawn your attention to only by leading you up to a horse and pointing with his hands. . . .
>
> If the biologist in our example were unskilled in pointing, you might have difficulty in understanding what he was driving at when he pointed. To be effective, he needs some natural aptitude and a good deal of acquired skill gained by long practice. If he were unskilled in handling the English language, you might similarly have difficulty in understanding what he was driving at when he wrote. Here also to be effective, he needs both some cultural aptitude and a good deal of acquired skill gained by long practice. . . . For some purposes, there are suitable words available in our vocabularies. For others there are not. Since man through countless generations has conventionalized and developed the system of symbolic pointing which we call language, we can all generally find suitable words to direct other people's attention to what we wish, provided that it is the commonplace things of life that we wish to direct their attention to. On the other hand when we wish to direct their attention and our own to less plain and obvious matters, we fall into difficulty, because we have to use for that

purpose words which were originally developed and used for other and more matter-of-fact purposes, i.e., we have to use words as metaphors. The parallelism between words to write with and fingers to point with is very close (9:113-119).

Anthropologists tell us that language brought the birth of civilization. Whether the Neanderthal man was anthropoid or human depends less on his cranial capacity, his upright posture, or even his use of tools and fire than on whether or not he spoke or communicated with his fellows. Aldous Huxley (3:13) observed that human behavior as we know it became possible only with the establishment of relatively stable systems of relationships between things and events on the one hand and words on the other. Behavior is not human in societies where no such relationship has been established, that is to say, where there is no language. Only language makes it possible for man to build up the social heritage of accumulated skill, knowledge, and wisdom and enables man to profit by the experience of past generations as though it were his own. Man's mastery over reality, both technical and social, depends on his knowledge of how to use words. This is true of craftsmanship within a primitive community or in our own highly technological society. Knowledge of the name of a "gizmo" is frequently the direct result of knowledge of how to use a "gizmo." The right words to describe the activity of a craft or trade or the right words to describe techniques and abilities assume meaning to the extent that the person becomes the master of them or acquires the ability to carry out the proper action. Such a mastery of putting an ability or craft or technique into communicable words serves as the basis for cooperative action. Civilization depends on cooperation made possible by communication.

Communication

The word *communication* is derived from the Latin *communis*, meaning common. When we communicate, we are trying to establish a *commonness* with someone. Dictionaries define the process as "the giving and receiving of information, signals, or messages by talk, gestures, writing, signals." Communication involves the sharing of experience with others. While I want to stress the *human symbolizing* activity of language (which is the basis of human thought), I want you to be aware also:

1. That within human experience there is the capability of communication without recourse to formal language through signs, signals, symbols, etc.
2. That other living beings — animals and insects, for example — also communicate.

Nonverbal Communication

Two lanterns hung from old North Church were a sign to Paul Revere to get on his horse. A runny nose, watery eyes, a hoarseness, and a cough communicate to a mother, despite the protestations of the eight-year-old, that the boy is better off in bed. The impatient horn of the automobile behind you as the traffic light turns green tells you to get your car moving. Blasts on a ship's whistle from the lead ship tell others in a convoy in which direction to maneuver. A school bell tells the teacher and the students that classes are ready to begin, classes are over, recess has begun, recess is over, or that there is a fire in the building or a fire drill is under way. The pattern on the screen of an oscilloscope communicates to the technician or engineer the performance characteristics of an electronic instrument being tested. The rate of pulse beat or heart beat gives a physician informative data. So man communicates or receives communication from natural phenomena in his experience and from devised conventional signs like gestures, grimaces, pictures, codes, flags, lights, bells, signs, punched paper tape, and other signals of all sorts.

Insects, animals, and birds communicate. A bee returning from flowers with a load of nectar will perform a dance which instructs other bees where to obtain the nectar. The length of time taken by the dance indicates the number of flowers there are and the number of workers needed to transport the nectar. If the bee dances for a few seconds only, a few bees will go out. If for two minutes, twenty or more may go. Long dances indicate a very rich food supply. The direction for the bees to go is indicated by the direction of the dance (2:36-40).

Animals know and read signs. Overcast skies inform birds that it will rain. Thirsty cattle know that low, thick foliage means water. Dogs communicate their wants by barking, whining, scratching, and running. Wild geese in a V-flight-formation coordinate and communicate their next movement to each other so that their V-formation persists with precision throughout the flight.

Biologists tell us that by means of chemical, optical, auditory, tactile, electrical, and other signals there are communications not only between organisms of the same species and between organisms of different species, but also between cells of the same organism and between parts of the same cell. Without inter- and intra-cellular communications, life would not be possible.

However, language is the unique characteristic of man which has enabled him to rise above the insects, birds, and animals. As Irving Lee has observed:

> A bird builds a nest. A man builds an engine. Each uses time, effort, and materials, but soon their purposes have been served. A new generation comes. The new bird builds again and the next nest is not unlike it. With the new man, however, come new possibilities, new searchings, new ways of looking, new experimenting, so that when the engine is built, it is built anew. To the work of the past something is added from the present. For with the symbol-using class of life, the power to achieve is reinforced by past achievement. . . (5:15).

Human Communication

Now let us examine this great human achievement which has made civilization and progress possible. There are three basic elements in communication: the communicator, the thought or message communicated, and the person receiving the message. A diagram, Figure 1,[1] will help us understand the process. Though we will borrow terminology from a radio or telephone circuit, the theoretical concepts of a radio or telephone communication system are based upon the human communication system.

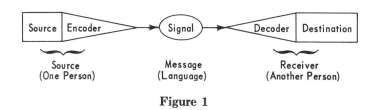

Figure 1

There are three basic elements in human communications—the source, or sender; the message, symbols, or signal; and the destination or receiver. The source may be a person speaking, writing, typing, drawing, gesturing, or it may be a communication organization such as a magazine, book publisher, radio or television station, or motion picture studio. The message may be in the form of writing, Braille, printed characters, diagrams or pictures on paper, sound waves in the air, impulses in an electrical current, a gesture or a grimace, lights, bells, flags, or smoke signals in the air, or any form of signal capable of being interpreted meaningfully. The destination may be a person listening, watching, reading, or perceiving by any of the senses.

What happens when the source tries to establish commonness of experience with his intended receiver? The answer is diagramed in Figure 2, (8). First, the source (the person) encodes his message; that is, he takes the information, feeling, thought, or idea he wants to share and encodes or expresses it in a form that can be transmitted. The coding process is the thought or formulation process which takes place in the mind of the source for translation into a signal message or symbolic means to enable sending. The message or signal is sent in the form of gestures, grimaces, speech, written words, drawn diagrams, pictures, other graphic illustrations,

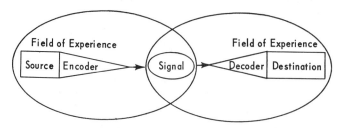

Figure 2
Commonness of experience is requisite for communication.

---

[1]Figures 1, 2, and 4 are based on diagrams in Wilbur Schramm's excellent article in reference 8.

printed means, or radio pulses. The message is received by the decoder in the human communication system, which is, of course, the perception system of the individual receiving the message. The decoder can translate the message only within the framework of his experience and knowledge.

The source can encode and the destination can decode only in terms of the experience each has had. If the circles have a large area in common, communication is easy. If the circles do not meet—if there is no common experience—communication is impossible.

Plainly, the communication system can be no better than its poorest link. To use engineering terms, there may be noise (a disturbance which does not represent any part of the message), filtering (abstraction or abridgment), or distortion (perversion of meaning) at any stage in the communication process. If the person sending the message does not have adequate or clear information, the message is not formulated or encoded adequately, accurately, or effectively in transmittable signs. If these are not transmitted accurately enough despite interference and competition to the desired receiver, if the message is not decoded in a manner that corresponds faithfully to the encoding, and if the person receiving the message is unable to handle the decoded message so as to produce the desired response, then there is a breakdown in the communication system.

Look at Figures 1 and 2 again. Let us say an electronics engineer is working in a laboratory designing a new instrument. A technician is assisting him. The engineer is testing a circuit. If it works, it will solve an important problem in the design of his instrument. At a critical moment he tells his assistant to connect a lead to a scope. His assistant asks, "How?" "Use an alligator clip," the engineer answers. "Quick!" The technician connects the alligator clip to the lead of the instrument and then to the terminal of the scope. "Good boy!" says the engineer and watches the trace of the circuit on the scope.

Now, while it is difficult to imagine an electronics engineer not being able to communicate to his technician-helper what an alligator clip is, let us suppose, for the sake of illustration, that an engineer, through incompetence, emotional pressure, or difficulty of speech, was unable to formulate the proper message about the alligator clip to his technician; or that the engineer's helper was inexperienced or had auditory or other perception difficulties or did not speak or understand the language that the engineer spoke. Then noise, filtering, and distortion would occur at the signal encoding or decoding elements in the communication process. The source can encode and the destination can decode only in terms of the experience each has had. If the circles in Figure 2 have a large area in common, communication is easy. If the circles do not meet, there is no common experience, and communication is impossible. The most important factor about a system such as that pictured in our diagram is that the receiver and sender must be in tune. This, of course, is clear in the case of a radio transmitter and receiver, but a little more difficult to understand when it means that the human receiver must be able to understand the human sender.

If we have never learned any Russian, we can neither encode nor decode in that language. If we have had no experience with a "gizmo," like an alligator clip, unless the message includes a proper description, explanation, and background, we might then think that the alligator clip is a device which clips the claws of an alligator. *The source must always be aware of the experiential background of the receiver* so that he may encode his message in such a way that the destination will be in tune with the message and be able to relate it to the parts of his experience which are most like those of the source.

In actuality, the communication system can get much more complicated than the preceding two simple diagrams have indicated.

Figure 3 represents a situation that frequently occurs before the intended report reader (destination) receives a report about a problem in which he has been interested. The source, let us say, is the engineer working on a problem. He is also the transmitter. The transmitting agent may be in the form of words used to formulate the message about the problem. The first receiver in this case might be a dictabelt on a dictaphone into which the engineer dictates his message. The dictabelt becomes the first relay. A stenographer listening to the belt transcribes the message into rough draft. The transmitting agent is the rough draft. A technical writer may be the receiver. Reading the rough draft, he is also the destination. After he has polished up the message or report, he becomes relay No. 2.

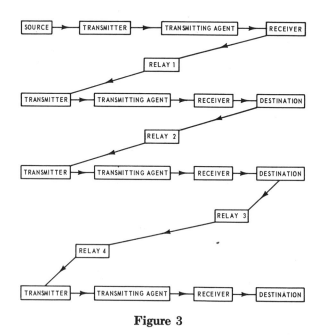

Figure 3

He (the transmitter), in turn, dictates his polished version (transmitting agent) to a stenographer (receiver or destination). The new typed copy is relayed by the stenographer to a supervisor. The supervisor becomes relay No. 4. He goes over the polished version and, in turn, acts upon it.

It may then receive another typing and a new chain is started over again. The final typing may be transmitted directly to the intended destination, or it may be turned into reproduction copy or printed copy, where again another chain is started. Within any of these subcommunication systems, noise, filtering, or distortion may take place so that the final message or report may have elements of incomprehensibility, inaccuracy, or distortion in some or all of its elements. Breakdowns in the message may occur at any of the diagramed links within the communication chain. Conversely, the report may be better for all the collaboration it received en route.

Figure 3 shows the links in the communication chain and points out the importance of watching for any weaknesses in the communication chain.

Each person in the communication process is both an encoder and decoder. He receives and transmits. The communicator gets feedback of his own message by listening to his own voice as he talks or by reading his own message as he writes (see Figures 4a and 4b). Thus, we are able to correct mispronunciations as we listen to ourselves talk or catch mistakes in our writing as we read what we have written or add information as we recognize omissions. The process of feedback enables us to catch weaknesses in the communication link when they occur.

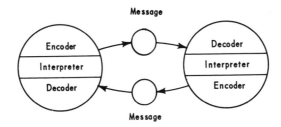

Figure 4a (8)

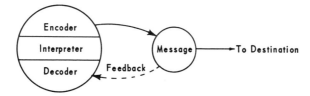

Figure 4b (8)

Semantics

Words as Symbols

Semanticists caution us that words and language are not the nonverbal phenomena they represent. Semanticists say we live in two worlds which should not be confused. Irving Lee called them "a world of words" and "a world of not-words." "If a word is not what it represents, then whatever you might say about anything will not be it. If in doubt, you might try eating the word *steak* when hungry, or wearing the word *coat* when cold. In short, the universe of discourse is not the universe of our direct experience" (5:16). Semanticists also caution us about symbols. Symbols are not the things they are supposed to represent. The picture of Uncle Sam, although representing the idea of the United States, is not the United States. *Au* is the chemical symbol for gold. It may do a proper job in a lab notebook, but it will not fill a tooth or replace currency.

Symbols in human experience have a physical aspect. They are able to cause sensations within us. They permit us to deal with things that are absent as though they are present. Our minds, through the use of symbols, can handle dollars, pounds of apples, oceans of water, and billions of stars without having these things physically present. We are able to do this because, as humans, we have developed an elaborate substitute system for human experience and thought in signs, symbols, signals, words, and language. Unfortunately, reactions or sensations to a particular symbol or word are not the same for one man as they are for another. For example, the symbol of Uncle Sam represents one thing to an American, another thing to a Russian, and a third thing to a Britisher. For good or bad, words and symbols have acquired emotional associations. (See Figure 8.)

Words and symbols can be manipulated independently of what they represent. Semanticists point out that the structure of the language we use does not always correspond to the structure of the world. To make their point, semanticists frequently use the simile of the map. A road map which puts St. Louis north and west of Denver would not be much use *if* you were driving from New York to St. Louis. Language is a map — unless it corresponds to the territory, the driver is in trouble.

Process of Abstraction

All words and symbols are an abstraction of reality; that is, they are removed from life, as the definition of the word signifies. When people — you and I — use words, we use them as arbitrary symbols agreed upon by word users to represent reality. The word *chair*, for example, is an arbitrary symbol representing many different types of a piece of furniture on which people sit. If I wanted or needed to be more specific in a situation, I might say "easy chair." To be more specific, I might say "the brown, overstuffed easy chair my wife, Margaret, purchased at Macy's Department Store in New York City, which I use every night to sit on and relax in after I get home from work." The word *chair* symbolizes or is an abstraction of the particular chair to which I refer. The word *chair* can also refer to any number of chairs existing in reality or capable of existing in the minds of any number of people. The word *chair* is more abstract than my particular chair. While the word *chair* is more specific than the word *furniture*, it still might be considered a term of higher order abstraction. Semanticists have constructed what they call "abstraction ladders." Abstraction ladders are graphic devices which help us understand how to make words more accurate maps of the fact territories they represent.

Here is a typical abstraction ladder which I have borrowed from Stuart Chase. To get at the most accurate map in the abstraction ladder, we begin at the top and work downward:

Mountains. What can be said about mountains which applies in all cases? Almost nothing. They are areas raised above other areas on land, under the sea, on the moon. The term is purely relative at this stage: something higher than something.

Snow-capped Mountains. Here on a lower rung, we can say a little more. The elevations must be considerable, except in polar regions — at least 15,000 feet in the tropics. The snow forms glaciers which wind down their sides. They are cloud factories, producing severe storms, and they require special techniques for climbing.

The Swiss Alps. These are snow-capped mountains about which one can say a good deal. The location can be described, also geology, glacier systems, average elevation, climatic conditions, first ascents, and so on.

The Matterhorn. Here we can be even more specific. It is a snow-capped mountain 14,780 feet above the sea, shaped like a sharp wedge, constantly subject to avalanches of rock and ice. It has four faces, four ridges, three glaciers. It was first climbed by the Whymper party in 1865, when four out of seven were killed — and so on. We have dropped down the ladder to a specific space-time event (1:143-144).

I have illustrated the process of abstraction as it applies to the concrete term *mountain*. A concrete term or word designates something that exists in the physical world. A concrete term may be specific; that is, it designates only one specific object, such as the Matterhorn, Pike's Peak, Cache de la Poudre River, the General Motors Corporation, the Charles E. Merrill Publishing Company, Albert Einstein, my daughter, Abbi, my son, Harlan; or, it may be general or designate a class of objects, such as mountains, rivers, corporations, scientists, children, baseball players. An abstract term usually designates something which does not exist in the physical world. It applies either to classes of things or to attributes or relationships. It does not have a specific reference against which its meaning can be checked. Examples of such terms are: culture, Americanism, beauty, art, philosophy, evil, honesty. There are abstract words which are more definite. These are based on concepts which are derived from relationships which our minds associate with elements operating in our experience. Among such words are: mass, force, magnetism, dimensions, absence, profit, rigid, cost, ripe, red, confused, equal.

These words are related to another set of words which grammarians call "relative" words. These words name qualities. Their meaning is dependent upon their relation or reference to the experience and/or intention of the user of the word. The word *dark*, for example, implies total or partial absence of light. We can say "brown is a dark color" or "a dark shade of red at sunset" or "it is getting dark." We can say that someone's mood was "dark," meaning gloomy. We can say that "the room is dark." We can say a person is "in the dark," if he is ignorant about something. We can say that someone's "intentions were dark," which can mean either insidious or obscure or secret. The word "dark" used in any of these examples would offer the reader or auditor some difficulty as to the specific meaning the writer or speaker intends. Meaning depends on a frame of reference or the context in which words are used, as we shall soon see.

Words representing colors, degrees of warmth, weight, and pressures are also relative terms. The technical writer is obligated to be precise in his communication and must therefore exercise much care in his choice of terms. In those instances when he needs to refer to a color, to a weight, or to a pressure, the technical writer will describe colors by their chromatic or achromatic scales; will indicate that an object is not just heavy, but that it weighs exactly 163 pounds, or 1 milligram as the case might be; will specify that a force exerts a thousand pounds of pressure per square inch; or that a pulse has a frequency beat of 100 kilocycles.

Problems in Achieving Meaning

Since man invented words and language to represent reality and to help human existence, endeavor, and progress, an easy conclusion might be: "Why can't we devise our language so that we stabilize the meanings of words in the way a mathematician has stabilized the quantities of his science and in the way the chemist has stabilized the symbols of his elements? We could avoid so much confusion if we established or found words to mean *one* thing and one thing *only*; we should not allow one word to mean one thing in one context and another thing in a different context. Language should be perfected so that each thing would have its own meaning."

This seems ideal. However, there is an insurmountable problem because there are an infinite number of things and experiences in human reality. While each person may have universal experiences, each experience is unique for that person. To structure a language with an infinite number of words for the infinitely complex variety of life facts is to build impossibility of communication into language. People would be inventing words for each spontaneous situation, as F. C. S. Schiller has pointed out:

> For if a word has a *perfectly* fixed meaning, it could be used only once, and never again; it could be applied only to the situation which originally called for it, and which it uniquely fitted. If, the next time it was used, it retained its original meaning, it could not designate the actual situation but would still hark back to its past use, and this would disqualify it for all future use.... It is evident that the intellectual strain of continuously inventing new words would be intolerable, and that the chances of being understood would be very small.... Thus the price of fixity would be unintelligibility. The impracticability of such a language might not be regarded as its "theoretical" refutation; but after all languages are meant to be used, and human intelligence, at any rate, could make no use of it (7:56-57).

Wilbur M. Urban, in his book *Language and Reality*, describes effectively the character and obligation of language:

> In the animal world the meanings of things are largely instinctive and the signs of these meanings are adherent to the thing signified. In human society, on the contrary, all is mobile, so a language is required which makes it possible to be always passing from what is known to what is yet to be known. There must be a language whose signs — which cannot be infinite in number — are extensible to an infinity of things.
>
> The mobility of signs, the tendency of a sign to transfer itself from one object to another is characteristic of language. This phenomenon of transference forms the central fact of linguistic meaning.
>
> Every meaning-situation . . . has two . . . components, the sign and the thing signified — that which means and that which is meant. Meaning in its most elementary form is, then, this relation. . . (10:107-115).

Urban further points out that there is no semantic meaning situation which does not involve both speaker (source) and hearer (destination) as necessary components. Words are signs (signals) but they are expressive signs. As such, they imply communication, either overt or latent (10:115-116).

Some diagrams would be helpful at this point in understanding what occurs when a source successfully or unsuccessfully communicates a message to a receiver destination. Figure 5a represents the communication of a simple idea. By substituting the alligator clip (Fig. 5a) for the square (Fig. 5a), you can clearly see the extent of loss that can occur in a communication situation. Figure 5b represents the extent of loss in the communication of a complex idea.

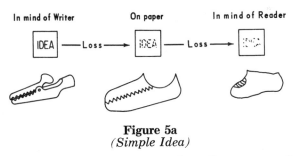

Figure 5a
(Simple Idea)

If we reverse the diagrams, we can see how the process of encoding and sending the signal to the receiver — the communication process — can aid

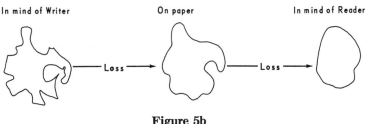

Figure 5b
(Complex Idea)

in formulating an effective message which the receiver receives and in his own decoding process helps clarify. In Figures 6a and 6b, we have a situation in which the source may begin with only a vague idea. In the process of encoding, that is, formulating it, the idea begins to take shape. The sender conveys the message in a formulated concept or in words and sentences which the receiver decodes in accordance with his experience, or the receiver acts upon the message in such a way that when he receives it, it is decoded in a clear, definite statement. The idea or concept, as the receiver has decoded it by his own added thought process, becomes complete, clear, concise, and meaningful. The receiver, using his experience and associations which the signals he has received stimulated, organizes through his cerebral processes the signals received and makes the proper sense dictated by the requirements of the message situation. This has been illustrated operation-

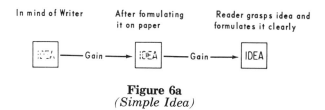

Figure 6a
(Simple Idea)

ally by Figure 6b. *This gain process is,* of course, *highly unusual.* The diagrams help us visualize what frequently occurs in the revision stages of writing. Initially, the writer might begin with a blurred idea in a first or second draft. As he writes out his idea, it becomes clarified (see also Fig.

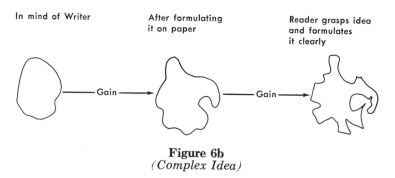

Figure 6b
(Complex Idea)

4b), so that by the time the writer's third or fourth draft reaches the reader, the idea has been totally clarified in the writer's mind.

Operation of Meaning

When we speak or write to one another, we transmit our thoughts through physical signals. Properly structured and organized, these signals become messages of orderly selected signs and symbols. The signs, then, are the physical embodiments of our messages or thoughts.

There are three elements involved in the conveyance of meaning:

1. A person having thoughts
2. A symbol
3. A referent

These three elements are frequently represented by three corners of a triangle as in Figure 7.

The symbols or words have no direct relationship to their referents and cannot be identified with the physical fact they symbolize except indirectly through the reader's mental processes. Line AC of the triangle in Figure 7 becomes a direct relationship only after the reader pronounces the word symbols of the fact; the symbolization (the sentence "Acid turns blue litmus paper red") becomes identical (means the same thing in the mind of the reader) with the fact it represents.

Words represent facts. A person perceives the fact that acid turns blue litmus paper red. The person selects the proper words to form a message which will designate (mean) the fact or thing perceived (the referent).

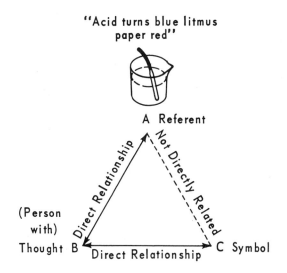

Figure 7
*Triangle of Meaning Operating in the Mind of the Encoder
of a Communication Message*

Words, then, refer through the mind to facts. Past experience through memories and external environment influences the proper selection of the symbolic response to the referent by the person encoding and sending the message. Meaning to the receiver emerges from the operation indicated in Figure 2.

We need to recognize that there are three types of meaning:

1. what the communicator or source intends to indicate,
2. what is suggested to a particular receiver (destination) by the message,

3. and the more or less general habit of using a given symbol to indicate a given thing.

What a communicator intends and what a receiver understands both
depend heavily on meaning 3, which frequently is influenced by the frame
of reference or context in which the words are used.

Bridging the Gap Between Words and Thought

Language-use is the occupational hazard of technical writers who must
communicate knowledge clearly and precisely; language is the link between
the reality of the outside world and the concepts or thoughts in people's
minds. Because each human nervous system is different, each person sees
reality differently. This can cause problems, as illustrated in Figure 8.

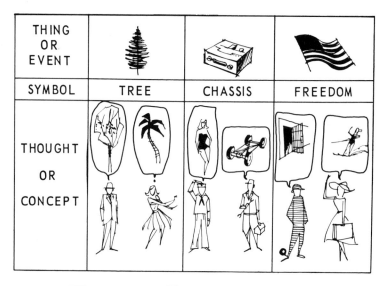

Figure 8

Effective technical writing is clear, precise, and accurate writing. A
knowledge of semantics can help the technical writer and can help his
reader.[2] It guides the writer in the correct choice and use of both technical
and nontechnical terms, and it guides readers in the correct interpretation
of the symbols and terms used. A knowledge of semantics also makes the
technical writer realize that words and terms are subject to emotional

---

[2]We have had time for only an introduction to semantics and communication theory
in this chapter. A deeper and more comprehensive study would be rewarding for you.
The bibliography at the end of this chapter lists some excellent references. In the area
of semantics, I suggest your reading Hayakawa's *Language in Thought and Action* and
Irving Lee's *Language Habits in Human Affairs*. Urban's *Language and Reality* is difficult but comprehensive. Ogden and Richards' *The Meaning of Meaning* is outstanding
for an examination of meaning. Colin Cherry's *On Human Communication* provides
both a mathematical and nonmathematical introduction to communication theory.

associations. Therefore, care should be taken to avoid words and terms placed in a context which will make such terms or words emotionally "loaded." Finally, technical writers must keep their readers' background always in mind so that terms, words, and other communication symbols used will be common or familiar to the reader.

Role of Writing in the Scientific Method

Because technical writing is as much a way of thought as a way of express-ing thought, the following observations by the linguist Benjamin Lee Whorf form an appropriate close to this chapter on semantics and communication:

> The revolutionary changes that have occurred since 1890 in the world of science — especially in physics but also in chemistry, biology, and the sci-ences of man — have been due not so much to new facts as to new ways of thinking about facts. The new facts themselves, of course, have been many and weighty; but, more important still, the realms of research where they appear — relativity, quantum theory, electronics, catalysis, colloid chem-istry, theory of the gene, Gestalt psychology, psychoanalysis, unbiased cul-tural anthropology, and so on — have been marked to an unprecedented degree by radically new concepts, by a failure to fit the world view that passed unchallenged in the great classical period of science, and by a groping for explanations, reconciliations, and restatements.
>
> I say new ways of thinking about facts, but a more nearly accurate state-ment would say new ways of *talking* about facts. It is this *use of language upon data* that is central to scientific progress. Of course, we have to free ourselves from that vague innuendo of inferiority which clings about the word "talk," as in the phrase "just talk": that false opposition which the English-speaking world likes to fancy between talk and action. There is no need to apologize for speech, the most human of actions. The beasts may think, but they do not talk. "Talk" *ought to be* a more noble and dignified word than "think." Also, we must face the fact that science begins and ends in talk; this is the reverse of anything ignoble. Such words as "analyze, compare, deduce, reason, infer, postulate, theorize, test, demonstrate" mean that, whenever a scientist does something, he talks about this thing that he does. As Leonard Bloomfield has shown, *scientific research begins with a set of sentences which point the way to certain observations and experiments, the results of which do not become fully scientific until they have been turned back into language, yielding again a set of sentences which then become the basis of further exploration into the unknown*... (11:220).

Discussion Problems and Exercises

1. The philosopher George Santayana once wrote: "Looking at the moon one may call it simply a light in the sky; another prone to dreaming while awake, may call it a virgin goddess; a more observant person, remembering that this luminary is given to waxing and waning, may call it the crescent; and a fourth, a full-fledged astronomer, may say (taking the aesthetic essence before him

merely for a sign) that it is an extinct and opaque spheroidal satellite of the earth, reflecting the light of the sun from a part of the surface. All of these descriptions envisage the same object..." (6:176-177).

Which of these descriptions would you consider the most appropriate? A technical writer always must consider the audience he is writing for. Using each of the descriptions as a starting point, write separate but appropriate descriptions for a ten-year-old boy lying on his back looking at the night sky; a college girl studying the poetry of the English romantic period; a layman who is interested in knowing about man-made satellites; a college student in a beginning course in astronomy.

2. The Arabic language is said to have almost six thousand terms to describe camels. In other languages of the world, there are no terms for camels at all. What generalization about languages can be made from this? Would you consider the automobile the English vocabulary equivalent to the camel in that respect? Check your answer in an unabridged dictionary. What other terms in the English language might approach the camel in that respect? If you know a foreign language, can you think of a word, idiom, or expression which cannot be adequately translated into English and vice versa? Why should there be such a difficulty? Does this mean that there are things, processes, and concepts that a technical writer can never adequately describe? Justify your answer.

3. George Bernard Shaw once said, "England and America are two countries separated by the same language." Give examples of his observation which apply to the technical field. How would you overcome the situation you call attention to in your examples?

4. When technical writers forget or ignore the fundamentals of the communications process, "noise" or "interference" occurs. Indicate the type of "noise" that may occur in writing for various levels of readership. One of the best ways for overcoming interference is to use "the principle of feedback." How does the principle of feedback operate for the technical writer? List several types.

5. Harold Laswell defined the act of communication operationally as "to answer the following questions:
   Who
   Says what
   In which channel
   To whom
   With what effect?" (4:37)
   Explain this definition's operation and applicability within the technical writing situation.

6. Library research will be necessary for the following exercises:
   (a) Distinguish between *semantics* and *general semantics*.
   (b) Find definitions within the communication context of
      (1) *Message*
      (2) *Signal*
      (3) *Sign*
      (4) *Symbol*
   (c) Distinguish between *language* and *code*. How are codes used in human and machine communication processes?
   (d) Distinguish between *phonetics, phonemics,* and *graphemics*.

(e) Find at least three theories of the origin of language. Of what value is the knowledge of language origin to the technical writer? Explain.

(f) Distinguish and relate
   (1) *Rhetoric*
   (2) *Linguistics* and *psycholinguistics*
   (3) *Logic*

(g) Look up (not in a dictionary but in an encyclopedia or reference source):
   Semantics
   Semiosis
   Significs
   Pragmatics
   Syntactics
   and differentiate their meanings.

7. Do we know the meaning of a word, a term, or a concept when we know to what the word, term, or concept refers? Explain.

8. In poetry and in puns, ambiguity — the intentional use of multiple meanings of a word or phrase — is employed by the writer because the effectiveness in what he writes is served by the several meanings offered the reader. In technical writing, however, uncertainty as to which of two or more possible meanings to assign to a word, term, or phrase impedes or destroys the effectiveness of the message. Other failures of meaning might be due to words or phrases being incorrect, vague, or meaningless. Carelessly placed modifiers cause ambiguity through unclear reference:

When properly pickled, no customer can resist a green tomato.
Sizzling slightly, the technician changed the damaged resistor.
Throw the horse over the fence some hay.

How would you correct the preceding sentences?

9. Read one of the following books and write a review. Your review should be directed to the technical writer. Include a brief summary of the contents, an indication of the purpose and scope, and an evaluation of the contents:

Ogden, C. K. and I. A. Richards, *The Meaning of Meaning*, New York: Harcourt, Brace & Co., 1936.
Cherry, Colin, *On Human Communication*, New York: The Technology Press of Massachusetts Institute of Technology and John Wiley & Sons, 1957.
Sapir, E., *Language*, New York: Harcourt Brace & Co., Inc., 1939.
Morris, Charles, *Signs, Language and Behavior*, New York: Prentice-Hall, Inc., 1946.
Urban, Wilbur M., *Language and Reality*, New York: The Macmillan Company, 1939.
Wiener, Norbert, *Cybernetics*, New York: The Technology Press of Massachusetts Institute of Technology and John Wiley & Sons, 1948.

References

1. Chase, Stuart, *Power of Words*, New York: Harcourt, Brace and Company, Inc., 1954.

2. Haldane, J. B. S., "Communication in Biology," in *Studies in Communication*, Communication Research Centre, University College, London, London: Martin Secker and Warburg, 1955.

3. Huxley, Aldous, *Words and Their Meaning*, Los Angeles: The Ward Ritchie Press, 1940.

4. Laswell, Harold, "The Structure and Function of Communications in Society," in *The Communication of Ideas*, Lyman Bryson, Editor, New York: Institute for Religious and Social Studies, 1948.

5. Lee, Irving J., *Language Habits in Human Affairs*, New York: Harper and Brothers Publishers, 1941.

6. Santayana, George, *Scepticism and Animal Faith*, London: Constable and Company, Ltd., 1923.

7. Schiller, F. C. S., *Logic for Use*, New York: Harcourt, Brace and Company, 1930.

8. Schramm, Wilbur, "The Nature of Communication Between Humans," in *The Process and Effects of Mass Communication, Revised Edition*. Wilbur Schramm and Donald F. Roberts, Editors, Urbana, Illinois: University of Illinois Press, 1971.

9. Sinclair, W. A., *An Introduction to Philosophy*, London: Oxford University Press, 1944.

10. Urban, Wilbur M., *Language and Reality*, New York: The Macmillan Company, 1951.

11. Whorf, Benjamin Lee, "Linguistics as an Exact Science," in *Language, Thought and Reality, Selected Writings of Benjamin Whorf*, John B. Carroll, Editor, New York: John Wiley and Sons and Massachusetts Institute of Technology Press, 1956.

9

Special Expository Techniques in Technical Writing—Definition

Because the first aim of technical writing is communication — an activity requiring at least two participants, a writer and a reader — effective writers are always aware of this dualism and direct their writing to a particular reader-audience. Ways of bringing understanding to the reader vary with the particular situation and reader background and interest. The fundamentals and elements of effective communication are the same in the technical field as in other fields. However, the special and basic techniques which characterize technical writing are expository. The present and following chapter are concerned with exposition. The specialized style characteristics of technical writing which aid and promote effective communication will be dealt with in chapter 12.

Exposition is explanation and/or instruction. It is the term given to the kind of writing we use when we need to explain facts or ideas. The aim of exposition is to bring about an understanding of something. The expository techniques of technical writing may be grouped into three major categories or methods: definition, description, and analysis. In explaining facts and ideas of a specific or technical nature, the technical writer will use these types more often in combination than separately. However, for purposes of illustration and understanding, we will examine these methods as individual types, but you will probably notice that among the illustrations there is an overlapping of method.

Nature of Definitions

A definition is an explanation of an object or idea that distinguishes it from all other objects or ideas. Definition is basic to knowledge. The infant, for example, learns about his environment by gradually defining the objects in it. A parent points to himself and says, "Daddy." After a time, the infant will begin, in imitation and understanding, to point to Daddy. Using this process, the parent will teach by definition through pointing to the infant's food, to common articles of usage, and to common experiences. Certain dangerous or delicate articles become "no-no's"; others are "nice-nice." Through this rudimentary instructional process begins the child's perception and definition of his environment and the world.

Similarly, adults attach importance to the names of things and ideas in their environment. What we are familiar with and what we can name, we seem to understand. Anthropologists tell us that man's belief in the potency of names was one of his earliest human traditions. As a new activity achieves maturity, it also attains a nomenclature. Systematically, every component is named. In order for a technician, an engineer, or a scientist to understand a mechanism, he usually must know the nomenclature of its various parts.

However, let us not oversimplify. Involved in the knowing or understanding of an instrument, species, or concept is more than just the ability to give its name or to list its parts. We must be able to relate it to essentially similar classes of devices, species, or concepts and to be able to distinguish it by its significant, identifying characteristics which differentiate it from other members of its group. Definition is the process whereby we become better aware of the exact nature of things which exist, or are capable of existing in our world.

The derivation of the word *definition* will help us understand its essential meaning. The Latin word *definire* is its origin; *de-* means from, and *finire* means to set a limit to or to set a boundary about. To define, then, is to delimit the area of meaning of a word, term, or concept. Definitions are verbal maps which indicate or explain what is included within a term and what is excluded. The latter is often communicated by implication.

Methods of Developing Definitions

The Formal Definition

A definition is said to be formal when it has a prescribed form consisting of three parts. The first part is the *term* — the word or object, idea or concept, to be defined. The second is the *genus* — the class, group, or category in which the term belongs. The third part is the *differentia* — the distinctive characteristics which distinguish the term from other members of the group (genus). The differentia excludes all other members of the class except the term being defined.

1. "A *rectangle* is a four-sided figure having all its angles right angles, and thus its opposite sides equal and parallel."

    "Rectangle" is the *term*; "four-sided figure" is the *class*; "having all its angles right angles and thus its opposite sides equal and parallel" is the *differentia*.

2. "An *emetic* is a substance or drug that induces vomiting, either by direct action on the stomach or indirectly by action on the vomiting center in the brain."

    "Emetic" is the *term*; "substance" is the *class*; "that induces vomiting either by direct action on the stomach or indirectly by action on the vomiting center in the brain" is the *differentia*.

3. "An *electric cell* is a receptacle containing electrodes and an electrolyte, used either for generating electricity by chemical reactions or for decomposing compounds by electrolysis."

    "Electric cell" is the *term*; "receptacle" is the *genus* or *class*; and "containing electrodes and an electrolyte used for generating electricity by chemical reactions or for decomposing compounds by electrolysis" is the *differentia*.

4. "*Hydrology* is the science that deals with the processes governing the depletion and replenishment of the water resources of the land areas of the earth" (15:1).

    "Hydrology" is the *term*; "science" is the *class*; "that deals with the processes governing the depletion and replenishing of water resources of the earth" is the *differentia*.

The formal definition is not an academic exercise. It is closely related to the scientific process of classification; it utilizes a logical method of analysis to place the subject to be defined into a general class (genus) and then to differentiate it from all members of its class. The term to be defined should not be repeated in the genus or differentia. For example, to define chess as a game played on a chessboard tells the reader very little about the game. To be completely satisfactory, the genus must identify the term precisely and completely. Too general a genus complicates the identification by necessitating the expansion of the differentia. The differentia must be broad enough to include everything the term covers and specific enough to exclude everything that the term does not cover. The differentia fences in the meaning of the term defined by listing qualities, giving quantities, making comparison, and/or itemizing elements or components.

Definitions should be stated in simpler or more familiar language than the term being defined; otherwise the purpose of the definition is defeated. A classic example of this fault is Samuel Johnson's definition of a cough: "A convulsion of the lungs, vellicated by some sharp serosity." Too, definitions should not be phrased in obscure or ambiguous language. Definitions should be impersonal, objective, and should *describe*, not praise or condemn, the matter being defined. Samuel Johnson's definition of oats as "a grain eaten by horses and in Scotland by the inhabitants" reflects a personal bias. These stated principles apply, of course, not only to the formal but also to all methods of definition that follow.

The Informal Definition

There are situations in which the strict format of the formal definition is not appropriate or efficient. The *informal definition* uses the shortest and simplest means for identifying or explaining the matter to be defined. The method may be the substituting of a short, more familiar word or phrase for the unfamiliar term or using special expository devices such as antonyms or illustrations that give the reader quick recognition of the term. Informal definitions are used in less formal writing, in writing for the general public, not for the technical specialist.

Definition by Synonym and Antonym

If you were asked, "What is a microbe?" you probably would answer, "a germ." *Germ* means the same as *microbe*. The two words can be and are used interchangeably. They are synonyms. *Double* means *twice*; paleography is the technical term for ancient writing; *helix* means *spiral*. A known word is substituted for an unknown word in definition by synonym.

Closely associated with this technique is the use of an antonym for aiding definition. *Down* is the opposite of *up*. *Direct* means not *deviating* or not *roundabout*; *foreign* means the opposite of *indigenous*; *heterodox* means not *orthodox*.

Both synonyms and antonyms are simplified forms of definitions which frequently set rough, approximate boundaries of meaning rather than complete and exact limits. They are techniques which are more appropriate for impromptu and informal situations than for more formal requirements. A child or lay person does not usually require the complete and specialized details for his normal requirements and understanding. For example, if a child or an uninformed layman asks, "What is penicillin?" his informational needs would be satisfied with "a substance which kills germs." He might be actually confused by fuller and technically more accurate information such as:

> Penicillin is an antibacterial substance produced by microorganisms of the *Penicillium chrysogenum* group, principally penicillium notatum *NRRL 832* for deep or submerged fermentation and *NRRL 1249, B21* for surface culture. Penicillin is antibacterial toward a large number of gram positive and some gram negative bacteria and is used in the treatment of a variety of infections (13:1206).

Definition by synonyms and antonyms finds use in the technical article aimed at the general reader.

Definition by Illustration

Offering an illustration is a primary means for aiding definition of an unfamiliar thing or concept. We have seen at the beginning of this chapter that adults use the system of pointing as an effective way of explaining things to children: "*This* is an oak leaf." "*This* is a Phillips screwdriver." "*This* is a ten-penny nail." Definition by "pointing" or illustrating makes the task easier for both adult and child.

Dictionaries, too, resort to definition by illustration. For instance, in defining *emu*, the dictionary says it is "a large, nonflying Australian bird, like the ostrich, but smaller." The dictionary finds it necessary to include a drawing of the bird which gives the reader details of the bird's appearance. In a like manner, the dictionary defines *trapezium* as a "plain figure with four sides, no two of which are parallel." The dictionary finds it necessary to include an illustration of this geometric configuration.

Definition by Stipulation

In chapter 6 of Lewis Carroll's *Through the Looking Glass*, Humpty Dumpty announces:

> "There's glory for you!"
>
> "I don't know what you mean by 'glory,'" Alice said.
>
> Humpty Dumpty smiled contemptuously. "Of course you don't — till I tell you. I mean 'there's a nice knockdown argument for you.'"
>
> "But 'glory' doesn't mean 'a nice knockdown argument,'" Alice objected.
>
> "When I use a word," Humpty said in rather a scornful tone, "it means just what I choose it to mean — neither more nor less."
>
> "The question is," said Alice, "whether you *can* make words mean so many different things."
>
> "The question is," said Humpty Dumpty, "which is to be master — that's all." (3:214)

Humpty Dumpty is using definition by stipulation — that is, he attributes to a word or term a specific meaning he wants it to have. Notice how the writer of the National Science Foundation Report, *Science, Technology, and Innovation* attributes special meaning to the term *event* and stipulates particular meanings to types of *events*:

> Throughout this report the term "event" is used in a special and technical sense. The innovative process comprises myriad occurrences, some of which happen sequentially, and some concurrently at different places. From these occurrences, one can identify some that appear to encapsulate the progress of the innovation. These special occurrences are the "events" in the technical sense just referred to. Their selection reflects the best judgment of the investigators, and is necessarily somewhat arbitrary.
>
> To clarify further how the study proceeded, other terms associated with the "events" are defined below.
>
> A significant event is an occurrence judged to encapsulate an important activity in the history of an innovation or its further improvement, as re-

ported in publications, presentations, or references to research. Generally these events follow one another in historical sequence, along channels of developing knowledge. Significant events include other classes of events.

A decisive event is an especially important significant event that provides a major and essential impetus to the innovation. It often occurs at the convergence of several streams of activity. In judging an event to be decisive, one should be convinced that, without it, the innovation would not have occurred or would have been seriously delayed.

Since science and technology lie at the focus of the investigation, the great majority of significant events are technical in nature. However, a few events that did not involve science and technology were important enough to be included among the significant events. These events are termed nontechnical.

A nontechnical event is a social or political occurrence outside the fields of science and technology. For example, a war or natural disaster would be a nontechnical event, in contrast with a management venture decision within a technical organization, which would be classed as a technical event (11:2-3).

The Operational Definition

In the scientific and technical fields, the operational definition has become a very useful technique. It offers meaning of a term, not by classifying it into genus and then isolating its distinguishing characteristics from other elements of its class, but by describing the activities, procedures, or operations within which the term operates.

Anatol Rapoport, in an article called "What Is Semantics?" appearing in the *American Scientist* of January 1952, explains this approach:

> An operational definition *tells what to do* to experience the thing defined. Asked to define the co-efficient of friction, the physicist says something like this: "If a block of some material is dragged horizontally over a surface, the necessary force to drag it will, within limits, be proportional to the weight of the block. Thus the ratio of the dragging force to the weight is a constant quantity. This quantity is the co-efficient of friction between the two surfaces." The physicist defines the term by telling how to proceed and what to observe. The operational definition of a particular dish, for example, is a recipe (9:128-29).

Here are some further examples of operational definitions:

> The sound [f] is a voiceless, labiodental fricative continuant, formed by placing the lower lip lightly against the upper teeth, closing the velum, and forcing breath out through the spaces between the teeth, or between the upper teeth and the lower lip. We call [f] labiodental because of the contact between the lower lip and the teeth; a fricative, because a characteristic of the sound is the audible friction of the breath being forced past the teeth.
>
> [f] occurs in the beginning, middle, and end of words, and is spelled *f, ff, ph,* or *gh,* as in *fife, offer, photograph* and *cough.* Occasionally in sub-

standard speech it may be replaced by the first consonant of *thin*, most commonly *trough*, but sometimes in other words as well. Ordinarily [f] causes little difficulty to the native speaker of English. In some foreign languages, notably Japanese, [f] is formed by the lips alone without the aid of the upper teeth. This weaker [f] may sound slightly queer in English, or the listener may fail entirely to hear it (12:38).

Torque, in a motor, is simply a measure of how much load it can turn or "lift." On small motors and synchro receivers, torque is measured in inch-ounces. A simple way to determine torque is to wrap a cord around a pulley secured to the shaft, and then add washers or small weights until the total weight is such that the motor is not capable of lifting the load... (4:106).

The Expanded Definition — Amplification

Sometimes a synonym, antonym, or formal definition is adequate to explain the meaning of a thing or idea. At other times, especially if an idea or a complex object is being explained, it is necessary to develop the definition further by making use of details, examples, comparisons, and other explanatory devices. The expanded definition is frequently an amplification of the formal definition. How it is developed depends on the nature of the concept or term and the writer's own approach to it. Most expanded definitions follow the structure of the paragraph. They will begin with a topic sentence which may be structured as a formal definition or as a statement of the topic for discussion. Many expanded definitions are not longer than a paragraph, but some may require several hundred or several thousand words. In the example below, a formal definition begins the explanation of *innovation*. Not a simple process, the term's formal definition requires amplification through the use of several expository devices:

> Innovation is a term that describes certain activities by which our society improves its productivity, standard of living, and economic status. Basic to the progress of innovation are the tools, discoveries, and techniques of science and technology ...
>
> When inventions or other new scientific or technological ideas are conceived, they do not immediately enter the stream of commercial or industrial application. In fact, many never get beyond the stage of conception, while others are abandoned during the period of development. But some go through a full course of gestation, and finally emerge as new and useful commercial products, processes, or techniques. Such advances are called innovations.
>
> Innovation should be distinguished from scientific discovery, although relevant discoveries may be incorporated into the innovation. Innovation should also be differentiated from invention, although an invention frequently provides the initial concept leading to the innovation. Nor is innovation merely a marginal improvement to an existing product or process. Rather it is a complex series of activities, beginning at "first conception," when the original idea is conceived; proceeding through a succession of

interwoven steps of research, development, engineering, design, market analysis, management decision making, etc.; and ending at "first realization"*, when an industrially successful "product", which may actually be a thing, a technique, or a process, is accepted in the marketplace. The term "innovation" also describes the process itself, and, when so used, it is synonymous with the phrase "innovative process."

As so defined, innovation extends over a bounded interval of time (the innovative period) from first conception to first realization. Implicitly, therefore, we have defined two other periods of time — the "preconception" period, which precedes the time of first conception, and the "post-innovative" period, which follows the time of first realization. During the preconception period science and technology develop the foundation for the innovation. In the post-innovative period improvements of the innovation are made and marketed, and the technology diffuses into other applications. The post-innovative period is sometimes called the period of "technological diffusion", although the two terms are not synonymous, because diffusion is only one of the activities of this period.

Let's look at an example — one of the innovations studied — the Heart Pacemaker. For this innovation, the preconception period saw advances in electricity, especially electrochemistry, in cardiac physiology, and in surgery and intracardiactherapy techniques. But first conception did not occur until 1928 when Dr. Albert S. Hyman conceived the idea of periodic electrical stimulation of the heart by means of an artificial device, an idea for which he filed a patent application in 1930. The innovative process for the pacemaker proceeded between 1928 and 1960; during this period, batteries were upgraded, the transistor was invented, materials technology enjoyed rapid development, and surgical techniques were advanced. The year 1960 marked the first implantation of a pacemaker in a human patient, and marketing of the device began soon thereafter; 1960 is therefore the date of first realization for this innovation. Since that date — in the post-innovative period — heart pacemakers have become more sophisticated, and work on further improvements continues.

Our study concentrated on the innovative period. But to understand the background from which innovation evolves, and to appreciate the impact of innovation on society, we considered also the preconception and post-innovative periods.

The preconception period presents problems, because it is historically open ended. At what point in history should one start? Electrical stimulation of muscular activity is important to the development of the heart pacemaker. Should one then go back to Galvani's experiments with frogs' legs? Since some time horizon had to be chosen, we selected 1900. The continuity of history makes it difficult to close one's eyes completely to pre-1900 events, so some especially significant scientific and technical events that occurred before 1900 are included in the historical record (11:1-2).

The Expanded Definition — Example

Closely related to definition by illustration is definition by example. In the scientific and technical fields, examples are necessary to bring under-

---

*Also termed "culmination" in the historical accounts.

standing to the uninformed. They are especially useful in explaining abstract and conceptual matters. In his book *Electrons, Waves, and Messages*, John R. Pierce used some very common, everyday experiences to exemplify and define quickly some very abstract processes:

> If you touch a steam radiator, your hand is heated by *conduction*; if you hold your hand over the hot air arising from the hot air register, your hand is heated by *convection*; if you hold your hand in the beam of heat from a reflector-type electric heater, your hand is heated by *irradiation*; it is warmed despite the fact that the air surrounding it is cool (8:181-182).

Even though conduction, convection, and irradiation are complicated processes, they enter into our everyday experience and can be identified for us by instances within our experience. However, to define another type of abstract concept, in the same book, Pierce turns to another exemplifying illustration to help define communication theory, a very difficult concept:

> In communication theory, information can best be explained as choice or uncertainty. To understand this, let us consider a very simple case of communication. For instance, if you want to send a birthday greeting by telegraph, you may be offered a choice of sending one of a number of rather flowery messages, perhaps one from a list of 16. Thus, in this form of communication, the sender has a certain definite limited choice as to what message he will send. If you receive such a birthday telegram, there is some uncertainty as to what it will say, but not much. If it was chosen from a list of 16 standard messages, it must be one among the 16. The received message must enable the recipient to decide which among the 16 was chosen by the sender (8:245).

This, of course, is not the whole story of communication theory, but the example gives the reader a generalized view of the essential characteristics.

The Expanded Definition — Explication

Explication is explanation or interpretation of the terms used in a definition. This is illustrated in the following definition:

> The rather odd term, optical pumping, means just what it says. In general, "pumping" is a process of raising matter from lower to higher energy. For example, raising the potential energy of water by moving it from an underground well to an elevated tank. In this article, we shall be concerned with the pumping of individual atoms from lower to higher states of internal energy. The word "optical" refers to the light energy that is the source of power for the pump (1:72).

The Expanded Definition — Derivation

Some terms may be explained effectively by discussing the origin of the term. I have used this approach in this text at several points. My discussion of "definition" as an expository technique began with the etymological origin of the word. Here is another example:

> The word *biology* is derived from two Greek words, *bios,* "life" and *logos,* "word" or "discourse," and so "science." In its narrower sense, biology may be defined as the science of life, that is, the science which treats of the theories concerning the nature and origin of life; in a broader sense, it may be defined as the science of living things. In this broader sense, biology is the sum of zoology (Greek *zōon,* "animal," + *logos,* "science"), or the science of animals, and botany (Greek *botonē,* "plant"), or the science of plants. It is in this broad sense that the word is used in this volume (10:1).

The Expanded Definition — History

Frequently, a knowledge of the origin or the history of an object or concept is helpful to the understanding of it. Consider the following definition of *eutrophication:*

> During the past few years the word *eutrophication* has continually cropped up in the daily press and in popular articles on the environment in magazines. It has usually been supposed to describe the process by which a beautiful lake or river became converted into a body of water covered with decomposing blue-green algae. That the word should mean the process of becoming well fed merely illustrates once again the horrid dichotomy between starvation and excess that is apt to characterize much of contemporary society. It has therefore seemed timely to examine what the word may mean and how the ideas that it is used to express have developed during the sixty-six years since the German adjectival form *eutrophe* was first introduced into scientific discourse.
>
> What do we mean by eutrophication? The word is due to C. A. Weber, who, in 1907, described the nutrient conditions determining the flora of German peat bogs as "nährstoffreichere (eutrophe) dan mittelreiche (mesotrophe) und zuletzt nährstoffarme (oligotrophe)," a sequence that expressed the changes as a bog built up and was raised above the surrounding terrain, so being more and more easily leached of its nutrients.
>
> At first the bog vegetation was what Weber called *eutraphent,* requiring high concentrations of essential elements in the soil solution; at the end of the process an *oligotraphent* flora covered the bog, composed of species tolerating very low nutrient concentrations. In Weber's case the process that took place in going from a less elevated, less leached to a more elevated, more leached bog would now be called oligotrophication, if anyone had occasion to use the word.
>
> In 1919 Einar Naumann, who knew of Weber's work, employed the terms in a discussion of the phytoplankton of Swedish lakes. He originally used them to describe water types, so that springs, streams, lakes, or bogs could contain oligotrophic, mesotrophic, or eutrophic water, according to the concentration of phosphorus, combined nitrogen, and calcium present. Originally no estimates, however, could be given as to what these concentrations were. Naumann throughout his works gives the impression that he

liked to draw limnological conclusions, expressible in schematic terms, merely from looking at lakes. Weber's original words *oligotrophic* and *eutrophic* were, in fact, now redefined for limnologists in terms of the appearance of a lake in summer. A lake containing eutrophic water is *"sehr stark getrübt oder sogar vollständig verfärbt"* as the result of a very dense population of algae. The unproductive oligotrophic lake did not support such a population and in consequence remained much less turbid and either blue, if it were unstained by peat, or brown in peaty montane areas. Even today, transparency and color are the simplest indicators of the nutrient condition of a lake, though of course they must be used with great discretion if no other information is available.

Much of the work that was being done by other limnologists in the first third of the twentieth century contributed to the development and ultimate confusion of Naumann's typological scheme.

The great difference between the lakes of the lands bordering the Baltic and those of more mountainous districts, notably the Swiss Alps, had been known for some time. Wesenberg-Lund, perhaps the greatest of limnological naturalists, in the general part of his work on the biology of the fresh-water plankton of Denmark, had discussed the biological characteristics of the Baltic lakes, of which the phytoplankton consisted largely of diatoms and blue-green algae.

At the same time the English algologists W. and G. S. West were studying the lakes of northern and western Britain, finding a diversified desmid flora in the phytoplankton. Teiling observed comparable assemblages in the mountains of Scandinavia and spoke of a Caledonian type of phytoplankton in contrast to the Baltic type. It was reasonable to regard these two types of plankton as regional expressions of the eutrophy of the waters of the Baltic Lakes and the oligotrophy of those of the mountainous parts of Europe (5:269).

The Expanded Definition — Analysis

A reader can grasp a complex subject better if it is broken up for him into its component parts. He can then digest one portion before proceeding to the next. This approach is effective in defining various steps in a process, especially if succeeding steps increase in complexity. This is seen in the following analytical explanation of intermittent drives:

> An intermittent drive is a mechanical arrangement which will automatically deliver power at definite intervals, or automatically and selectively engage or disengage mechanisms having different limits of operation

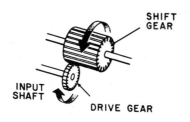

and being driven from the same line of gearing. A device enables the output shaft to be locked upon disengagement from the drive, and held immobile until picked up later at the very same position at which it was cut out.

The simplest form of intermittent drive would be a hand-operated assembly. For example, to disengage input or re-engage at a desired point, we could utilize a slidably mounted broad-faced gear between the input and output gears. With gears in mesh, power would be delivered to the output shaft.

Pushing on the shaft of this broad-faced gear would serve to disengage the drive.

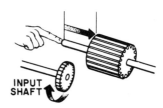

INPUT SHAFT

To lock the output in the particular position at disengagement, the shaft can be grasped with the hand, or locked with a clamp or wedge. In place of these crude methods, however, we can use a lever and cam assembly. Thus, power would be delivered to an output gear for a definite period, then stopped, and kept inactive for another definite period. The input would operate continuously. In certain mechanical equipment, such an assembly would be used to provide a regular rhythm of power application and idle time.

If we put a spring on the shaft, removal of the pressure would allow the gear to slide back into driving position (4:177).

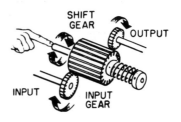

SHIFT GEAR   OUTPUT

INPUT   INPUT GEAR

The Expanded Definition — Comparison and Contrast

What is unknown and unfamiliar can often be very effectively defined by comparing or contrasting it with something known and familiar. Warren Weaver uses this method of comparison to define the difficult concept of statistical mechanics or "disorganized complexity" as he phrases it:

Subsequent to 1900, and actually earlier, if one includes heroic pioneers, such as Josiah Willard Gibbs, the physical sciences developed an attack on nature of an essentially and dramatically new kind. Rather than study problems which involve two variables, or at most three or four, some imaginative minds went to the other extreme and said: "Let us develop analytical methods which can deal with two billion variables." That is to say, the physical scientist, with the mathematicians often in the vanguard, developed powerful techniques of probability theory and of statistical mechanics to deal with what may be called problems of *disorganized complexity*.

The last phrase calls for an explanation. Consider first a simple illustration in order to get the flavor of the idea. The classical dynamics of the nineteenth century was well suited for analyzing and predicting the motion of a single ivory ball as it moves about on a billiard table. In fact, the relationship between positions of the ball and the times at which it reaches these positions forms a typical nineteenth-century problem of simplicity. One can, but with a surprising increase in difficulty, analyze the motion of two or even three balls on a billiard table. There has been, in fact, considerable study of the mechanics of the standard game of billiards. But, as soon as one tries to analyze the motion of ten or fifteen balls on the table at once, as in pool, the problem becomes unmanageable, not because there is any theoretical difficulty, but just because the actual labor of dealing in special detail with so many variables turns out to be impractical.

Imagine, however, a large billiard table with millions of balls rolling over its surface, colliding with one another, and with the side rails. The great surprise is that the problem now becomes easier, for the methods of statistical mechanics are applicable. To be sure the detailed history of one special ball cannot be traced, but certain important questions can be answered with useful precision, such as: On the average how many balls per second hit a given stretch of rail? On the average how far does a ball move before it is hit by some other ball? On the average how many impacts per second does the ball experience?

Earlier it was stated that new statistical methods were applicable to problems of disorganized complexity. How does the word "disorganized" apply to the large billiard table with a million balls? It applies because the methods of statistical mechanics are valid only when the balls are distributed, in their positions and motions, in a helter-skelter, that is to say, a disorganized way. For example, the statistical methods would not apply if someone were to arrange the balls in a row parallel to one side rail of the table, and then start them all moving in precisely parallel paths perpendicular to the two in which they stand. Then the balls would never collide with each other nor with two of the rails, and one would not have a situation of disorganized complexity.

From this illustration, it is clear what is meant by a problem of disorganized complexity. It is a problem in which a number of variables is very large, and one in which each of the many variables has a behavior which is individually erratic, or perhaps totally unknown. However, in spite of this helter-skelter, or unknown, behavior of all the individual variables, the system as a whole possesses certain orderly and analyzable average properties (14:537).

Dr. J. Bronowski develops an expanded but clearly luminous definition of creativity by the methods of comparison and contrast:

> The most remarkable discovery made by scientists is science itself. The discovery must be compared in importance with the invention of cave-painting and of writing. Like these earlier human creations, science is an attempt to control our surroundings by entering them and understanding them from inside. And like them, science has surely made a critical step in human development which cannot be reversed. We cannot conceive a future society without science.
>
> I have used three words to describe these far-reaching changes: discovery, invention and creation. There are contexts in which one of these words is more appropriate than the others. Christopher Columbus discovered the West Indies, and Alexander Graham Bell invented the telephone. We do not call their achievements creations because they are not personal enough. The West Indies were there all the time; as for the telephone, we feel that Bell's ingenious thought was somehow not fundamental. The groundwork was there, and if not Bell then someone else would have stumbled on the telephone as casually as on the West Indies.
>
> By contrast, we feel that *Othello* is genuinely a creation. This is not because *Othello* came out of a clear sky; it did not. There were Elizabethan dramatists before Shakespeare, and without them he could not have written as he did. Yet within their tradition, *Othello* remains profoundly personal; and though every element in the play has been a theme of other poets, we know that the amalgam of these elements is Shakespeare's; we feel the presence of his single mind. The Elizabethan drama would have gone on without Shakespeare, but no one else would have written *Othello*.
>
> There are discoveries in science like Columbus's, of something which was always there: the discovery of sex in plants, for example. There are tidy inventions like Bell's, which combine a set of known principles: the use of a beam of electrons as a microscope, for example. In this article I ask the question: Is there anything more? Does a scientific theory, however deep, ever reach the roundness, the expression of a whole personality that we get from *Othello*?
>
> A fact is discovered, a theory is invented; is any theory ever deep enough for it to be truly called a creation? Most nonscientists would answer: No! Science, they would say, engages only part of the mind — the rational intellect — but creation must engage the whole mind. Science demands none of that ground swell of emotion, none of that rich bottom of personality, which fills out the work of art.
>
> This picture by the nonscientist of how a scientist works is of course mistaken. A gifted man cannot handle bacteria or equations without taking fire

from what he does and having his emotions engaged. It may happen that his emotions are immature, but then so are the intellects of many poets. When Ella Wheeler Wilcox died, having published poems from the age of seven, *The Times* of London wrote that she was "the most popular poet of either sex and of any age, read by thousands who never open Shakespeare." A scientist who is emotionally immature is like a poet who is intellectually backward: both produce work which appeals to others like them, but which is second-rate.

I am not discussing the second-rate, and neither am I discussing all that useful but commonplace work which fills most of our lives, whether we are chemists or architects. There are in my laboratory of the British National Coal Board about 200 industrial scientists — pleasant, intelligent, sprightly people who thoroughly earn their pay. It is ridiculous to ask whether they are creators who produce works that could be compared with *Othello*. They are men with the same ambitions as other university graduates, and their work is most like the work of a college department of Greek or of English. When the Greek departments produce a Sophocles, or the English departments produce a Shakespeare, then I shall begin to look in my laboratory for a Newton.

Literature ranges from Shakespeare to Ella Wheeler Wilcox, and science ranges from relativity to market research. A comparison must be of the best with the best. We must look for what is created in the deep scientific theories: in Copernicus and Darwin, in Thomas Young's theory of light and in William Rowan Hamilton's equations, in the pioneering concepts of Freud, of Bohr, and of Pavlov (2:59-60).

The Expanded Definition — Distinction

The essence of definition is differentiation. An extended definition may be composed entirely of the distinguishing characteristics between concepts or things, as is the case with the following consideration of what constitutes published information and what constitutes unpublished information.

> The traditional system of formal scientific publication constitutes the most highly-organized channel of scientific communication, having evolved over the years into a complex integrated series of operations intended to insure that a given piece of information, once deposited in the system, is permanently available to all who need and look for it. The sequence for new information begins with *primary* publication of the original paper; *secondary* publication as an abstract follows; and appropriate reference to the paper made in standard indexes to the literature, represent what may be called *tertiary* publication.
>
> This orderly sequence is the formal channel for disseminating new knowledge, and is the traditional mainstay for all scientific communication. However, there have always existed other less formal channels through which information flows and by which information is disseminated and exchanged

among scientists, practitioners, teachers and students in the biomedical sciences. In the past few decades, these alternative and supplementary channels have multiplied in numbers and are handling increasing volumes of information.

Biomedical information transmitted by any method other than the formal system of scientific publication has been termed "unpublished information." This terminology assumes a restricted and special definition of publication that differs from common usage. Most scientists regard any information recorded in the form of the written word and disseminated more or less widely as having been published.

By the lawyer, publication is construed more broadly, and does not necessarily require that information be printed. Librarians and documentalists, in contrast, use the term in a much more restricted sense. To them, information is "published" only when it has appeared in those scientific periodicals covered by the standard lists of scientific periodicals and indexed and abstracted by the standard bibliographic services, or in books obtainable by the usual library purchasing or borrowing practices. Several of the more conservative scientific societies and journals also honor the strict definition and allow authors to cite as references only information to which these criteria apply.

Printed information not falling in this category is often referred to as "ephemeral" or "exotic" and is considered "unpublished." Since this distinction is in essence, an operational definition of publication, what is included in the category of unpublished information changes as practices of libraries and bibliographic services change. Defining "published" information in this way, i.e., in terms of ready availability and ease of finding emphasizes an important point — if information is to be useful for more than a very short time, it must be incorporated into some system capable of finding and supplying it to a user when it is needed. Using this restrictive definition, however, causes confusion in many minds when information is classified as "unpublished," although it has been printed and distributed widely. For want of better terminology, the adjective "informal" will therefore be used here to distinguish all those media methods for transmitting biomedical information other than the conventional scientific periodicals and monographs, of other bound volumes, issued by scientific societies and commercial publishers. These "informal" channels will be contrasted with the "formal" channels — the traditional system of scientific publication described above. This distinction roughly parallels the documentalists delineation of "published" from "unpublished" information (7:7-8).

The Expanded Definition — Elimination

In some instances, definitions may be strikingly developed by the process of elimination — demonstrating what something is by the enumeration of what it is not. Sir Julian Huxley does just this in his essay "War as a Biological Phenomenon":

> War is not a general law of life, but an exceedingly rare biological phenomenon. War is not the same thing as conflict or bloodshed. It means

something quite definite: an organized physical conflict between groups of one and the same species. Individual disputes between members of the same species are not war, even if they involve bloodshed and death. Two stags fighting for a harem of hinds, or a man murdering another man, or a dozen dogs fighting over a bone, are not engaged in war. Competition between two different species, even if it involves physical conflict, is not war. When the brown rat was accidentally brought to Europe and proceeded to oust the black rat from most of its haunts, that was not war between the two species of rat; nor is it war in any but a purely metaphorical sense when we speak of making war on the malaria mosquito or the boll weevil. Still less is it war when one species preys upon another, even when the preying is done by an organized group. A pack of wolves attacking a flock of sheep or deer, or a peregrine killing a duck, is not war. Much of nature, as Tennyson correctly said, is "red in tooth and claw"; but this only means what it says, that there is a great deal of killing in the animal world, not that war is the rule of life.

. . . So much for war as a biological phenomenon. The facts speak for themselves. War, far from being a universal law of nature, or even a common occurrence, is a very rare exception among living creatures; and where it occurs, it is either associated with another phenomenon, almost equally rare, the amassing of property, or with territorial rights (6:26-27).

Definition by Combination of Methods

The more complicated a term or concept is, the more involved may be the explanation necessary. No one system of definition may be adequate, and a combination of several techniques may be called for. We have already seen this in some of the previous examples. In the following definition of *waves*, negative details, history, example, listing of characteristics, explication, illustration, reiteration, operational analysis, comparison, as well as description, are interwoven to give a very graphic definition:

What are waves? They are not earth, or water, or air; steel, or catgut, or quartz; yet they travel in these substances. Nineteenth-century physicists felt constrained to fill the vacuum of space with an ether to transmit electromagnetic waves, yet so arbitrary a substance seems more a placebo to quiet the disturbed mind than a valid explanation of a physical phenomenon. When we come to the waves of quantum mechanics, the physicists do not even offer us a single agreed-upon physical interpretation of the waves with which they deal, although they all agree in the way they use them to predict correctly the outcome of experiments.

Rather than asking what waves are, we should perhaps ask, what can one say about waves? Here there is no confusion. One recognizes in waves a certain sort of behavior which can be described mathematically in common terms, however various may be the physical symptoms to which the terms are applied. Once we recognize that in a certain phenomenon we are dealing with waves, we can assert and predict a great deal about the phenomenon even though we do not clearly understand the mechanism by which the waves are generated and transmitted. The wave nature of light was understood, and many of its important consequences were worked out long before the idea of an electromagnetic wave through space was dreamed of. Indeed, when the true explanation of the physical nature of light was proposed, many physicists who recognized clearly that light was some sort of wave, refused to accept it.

We can study the important principles of waves in simple and familiar examples. As we come to understand the behavior of these waves, we can abstract certain ideas which are valid in connection with all waves, wherever we may find them. Such a study is the purpose of this chapter.

Suppose we watch the waves of the sea from a pier. Let us imagine that today the waves are particularly smooth and are very regular in height. We see a certain number of crests pass us each second — let us say a number f. This number f is the *frequency* of the waves. Frequency is reckoned in *cycles per second*, or *cycles* for short. The cycle referred to is simply a complete cycle of change; the departing of a wave crest, the passing of the trough, and finally, the arrival of the next crest. As a complete wave, from trough to crest again, passes us, the height of the water goes through a complete change, from high to low to high again.

A cycle is a complete cycle of change, at the end of which we are back to the original state. It is the same in the case of 60-cycle electric power. The 60-cycle electric current alternates in the direction of flow and goes through a complete cycle of change 60 times a second. Broadcast waves reach a receiver about a million crests per second; some television waves a hundred million crests a second, and radar waves leave the radar antenna and are reflected back again at the rate of billions of waves or cycles per second.

The waves in the ocean each take several seconds to pass us, so the frequency of the ocean waves is a fraction of a cycle per second. We can if we wish measure, instead of the frequency, the time between the passing of the two crests; this is the *period* of the wave, which we will call T. We see that

T is the *reciprocal* of f, that is $T = \dfrac{1}{f}$.

Looking out at the waves, we may estimate or measure the distance between the crests of the waves; this is the wave length, which is always denoted by the Greek letter λ (lambda). Among radio waves from radar to broadcast, the wave length ranges from a little over an inch to around 1,000 feet.

The time between the passage of wave crests is T. In this time the next crest must travel just one wave length λ, to reach the position of the preceding crest. Thus, the wave travels with a velocity, v, which is the distance of travel λ, divided by the lapsed time, T, so that

$$v = \frac{\lambda}{T} = \lambda f$$

Thus, we can express λ in terms of f and f in terms of λ by using the velocity, v:

$$\lambda = \frac{v}{f}$$

$$f = \frac{v}{\lambda}$$

Light waves and radio waves are both electromagnetic waves, and for such electromagnetic waves traveling through space, the velocity, v, is the velocity of light:

$$v = 186,000 \text{ miles per second}$$
$$v = 3 \times 10^8 \text{ power meters per second}$$

Waves may have various shapes. We have been considering smooth rollers which come in one after another, regularly spaced. We can also have a single wave or a short *train*, or waves such as those caused by throwing a single stone into a pond. There is a reason, however, for considering a particular regular, smooth persistent kind of wave called a *sinusoidal* wave.

The waves we consider are waves of what we call *linear* systems. We will see what this means later on. Now, we will merely say that, while for some linear systems a wave of any form travels along preserving that same form, in many other linear systems a wave of arbitrary form will change form as it travels. Consider, however, a wave such that as it passes a given point, the height of the water or the magnitude or *amplitude* of some other significant quantity, varies *sinusoidally* with time with some frequency f. If this is so at any point in any linear system, the wave will also vary sinusoidally with time, with the same frequency f, at any other point. Strictly, term *frequency* should be applied to sine waves only.

A sinusoidal variation can be understood in terms of a crank on a shaft which rotates at a constant rate, f turns per second, as shown in Figure 1. The height, h, of the end of the crank above the level of the center shaft varies sinusoidally with time. If we plot height or amplitude vs. time in seconds, as in Figure 2, we get *sine curve* or *sine wave*.

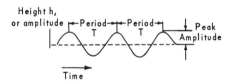

Figure 1

This is the way we will talk about the way waves vary with time. As long as the wave travels with constant velocity, this is also the way the height above the mean or zero level, which is called the *amplitude*, varies with distance.

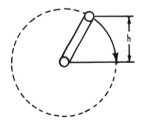

Figure 2

In instances in which a wave other than the sine wave will change in form as it travels, sine waves of different frequencies (and hence of different wave lengths) travel with different velocities... (8:82-85).

From the many various examples quoted, it can be seen that there are many methods of achieving definition. The effective writer does not choose his method haphazardly, but for appropriateness in bringing understanding to his particular reader. In report writing, as in other types of writing, definitions are given in the introductory section in order to give the reader the understanding necessary to follow the main body of the writing.

Summary of General Principles of Definition

1. Definitions should include everything that the term means and exclude everything the term does not mean.
2. Definitions should not include the term to be defined or any variant form of it in the genus or differentia.
3. Definitions should include the essential qualities of the term defined.
4. Definitions of terms in which magnitudes are the essential differentia should give the essential quantities involved.
5. Definitions not measurable by quantities but which are nevertheless limited or bounded by other terms that are closely related or similar should include essential similarities and differences of the term defined and the terms bordering it.
6. Definitions should be stated in simpler or more familiar language than the term defined.
7. An expanded definition should employ expository devices, such as examples, contrast, comparison, cause and effect, details, analysis, analogy, history, or other such devices as will promote clarity, meaning, and interest. Each expanded definition should contain a logical sentence definition of the term being amplified.
8. Definitions should not be phrased in obscure or ambiguous language.
9. Definitions should describe, not praise or condemn, the matter being defined.

Discussion Problems and Exercises

1. Find five formal definitions in any of your textbooks. Chart the definitions into their three parts.
2. Write logical definitions for the following terms:

   a. gimlet (the hand tool)   k. IQ
   b. gimlet (the cocktail)   l. work
   c. beer cheese   m. host
   d. syllogism   n. symbology
   e. entropy   o. *in vivo*
   f. Oedipus complex   p. mesosphere
   g. Cubic equation   q. fault
   h. Calibration curve   r. glacier
   i. Isocline   s. dielectric
   j. Thermistor   t. hypothesis

3. Write three expanded definitions of about 200-300 words each of any of the terms in exercise 2, using the expository devices of explication, example, analogy, comparison and contrast, analysis, history, operational, descriptive details, or cause and effect.

4. a) Find two examples of stipulative definition and cite authors, titles, and pages. Are these definitions arbitrary or do they have logic behind their intended usage?
   b) Write a stipulative definition for a definite purpose, communicating a meaning that can be defended by logic and usage.

5. Man has been defined as "metazoan, triploblastic, chordate, vertebrate, pentadactyl, mammalian, eutherian, primate." Write an expanded definition explicating these terms.

6. An old maxim attributed to the Chinese says, "A picture is worth a thousand words." What qualifications, if any, would you put on this assertion?

7. Check two consecutive pages of the dictionary adopted for use by your college English department. How many rules listed in the summary at the end of this chapter are violated? What generalizations should you infer?

8. Definitions range in length from a single explanatory word to entire monographs of many hundred pages. Definitions in reports, papers, and books may be located in the text itself, in footnotes, in a special section of the introduction, or in a glossary at the end. Check the placement of definitions in this text. Why were some incorporated into the lines of the page and others placed in footnotes? If you were to compose a glossary for this text, which terms would you include in it? What method of definition would you use in the glossary? Why?

9. Write an expanded definition of any of the following terms for a class of entering high school students who have not been exposed to an exact science:
   a. erosion
   b. polymerization
   c. feedback
   d. gyro
   e. galvanometer
   f. pressure
   g. resonance
   h. capacitance

10. Write an expanded definition for a college graduate who has had a year's work in one of the sciences, for one of the terms in problem 9.

References

1. Bloom, Arnold L., "Optical Pumping," *Scientific American*, October 1960.

2. Bronowski, J., "The Creative Process," *Scientific American*, September 1958.

3. Carroll, Lewis, *The Complete Works of Lewis Carroll*, New York: Random House, Inc., [n.d.]

4. Chief of the Bureau of Weapons, *Weapons Systems Fundamentals, Basic Weapons Systems Components,* Washington, D.C.: U.S. Government Printing Office, 1960.

5. Hutchinson, G. Ewelyn, "Eutrophication," *American Scientist*, May-June 1973.

6. Huxley, Julian, "War as a Biological Phenomenon," in *On Living in a Revolution*, New York: Harper and Brothers, 1942.

7. Orr, Richard H., "Communication — 'Unpublished Information'," *Industrial Science and Engineering*, October 1960.

8. Pierce, John R., *Electrons, Waves and Messages*, Garden City: Hanover House, 1956.

9. Rapoport, Anatol, "What is Semantics," *American Scientist*, January 1952.

10. Rice, Edward Loranus, *An Introduction to Biology*, Boston: Ginn and Company, 1935.

11. *Science, Technology, and Innovation*, Columbus, Ohio: Battelle Columbus Laboratories, February 1973.

12. Thomas, Charles Kenneth, *An Introduction to the Phonetics of American English*, New York: Ronald Press, 1947.

13. *Van Nostrand's Scientific Encyclopedia*, New York: D. Van Nostrand Company, Inc., 1958.

14. Weaver, Warren, "Science and Complexity," *American Scientist*, October 1948.

15. Wesler, Chester O. and Ernest F. Brater, *Hydrology, 2nd Edition*, New York: John Wiley and Sons, Inc., 1959.

10

Special Expository Techniques in Technical Writing—Description, Explanation, and Analysis

Technical Descriptions

The primary aim of science, it has been said, is the concise description of the knowable universe. The technical writer will therefore be concerned with exact representations of phenomena. To achieve exactness, the technical writer must often use drawings that enable the reader to visualize what is being described, but drawings and photographs cannot give complete representation. They help the reader see what the thing looks like, but he has additional questions: What is it? What is it for? What does it do? How does it do it? What happens after it does it? What is it made of? What are its basic parts? How are they related to make it do what it does?

To answer these questions fully for the reader, the technical writer must be able to visualize clearly the thing or process being described. He must have a thorough command of the appropriate nomenclature. He must know the components and their interrelationships and proportions. Finally, he must know the "why," the purpose of the thing or organism in question, the end which it serves.

Complete knowledge of the thing described is fundamental to writing a technical description. Being able to draw the matter to be described indicates ability to visualize it completely and comes only with intimate familiarity and understanding. Nomenclature is important and inherent in technical descriptions. Every part has a name and should be correctly labeled. Part identification is best accomplished through labeling on the drawing. However, mere drawing is not full description because the sketch or diagram cannot indicate many essentials and attributes, nor can it always show relationships or structure.

It should be apparent that the matter to be described must be at hand during the writing. You, the writer, need to have the subject before you in order to be able to give complete details. Drawings are two-dimensional; the writer's description needs to be three-dimensional. You must be concerned with not only proportions but also shape, materials, essences, finishes, connections, relationships, purposes, and actions of the various components. Size, proportions, materials, finishes, and compositions of essences are fairly straightforward. Connections and relationships, however, need precise definition. Here the writer needs to examine very carefully how an element is fastened or connected. Is one component riveted, bolted, screwed, coupled, molded, wired, etc., to another?

The precise location of components in relation to each other is important. At what angle is one element joined to another? Is it under or above? Inside or outside? Beside, parallel to, diametric to, etc.? These details are significant to the *what* and the *how* of the matter described.

In such analysis, one other element remains: the *why*, or the purpose. What does the thing or organism do? How and why does it do it? What is the significance of its action? The combination of the *what*, *how*, and *why* becomes the complete technical description.

The Written Description

After you have become familiar with the mechanism or organism, your task is to organize the data into a written description. Written description usually falls into an arrangement of major sections. Section I is the introduction; it consists of a general description of the mechanism, phenomenon, or organism. Section II contains the main functional divisions of the subject, its main parts, and the principle under which they operate. Section III, the concluding section, shows how the subject operates by taking it through a cycle of operation. The three sections with their components might be considered as an outline for a technical description.

I. Introduction
   A. Definition of the device
   B. Purpose
   C. Generalized description, including perhaps an analogy or comparison with a familiar matter
   D. Division of the device into its principal functional portions
II. Principle or theory of operation, if the device is a complicated or complex one. Detailed description of the portions, major assemblies, sections, etc.

    A. Division One — what the part is, its purpose and appearance, including an analogy or comparison or divisions into subparts

        1. Subpart One

           a. Purpose

           b. Appearance

           c. Detailed description

               (1) Shape, size, relationship to other parts, method of connection, material, and finish

               (2) Function

        2. Subparts Two, Three, etc.

    B. Division Two, etc.

III. Brief description of the device in a cycle of operation

    A. How each division achieves its purpose

    B. Causes and effects of the device or organism in operation

    C. Summary or other generalized conclusions

This outline considers the essentials which are included in a technical description. It is not prescriptive. Some elements of this outline might require enlarged treatment. Major headings and subheadings are helpful road signs to the reader and serve as transition devices.

The Introduction

Technical descriptions usually begin with a formal definition of the matter to be described. The definition is followed with a statement of the purpose or use of the matter and then with a very general description of its appearance. This initial description is general; its purpose is to give the reader a visual image of the subject by describing its size, shape, and appearance. The use of an analogy can help the reader to become familiar with something that might otherwise seem quite alien to his experience. The major components or assemblies of the subject are then listed. The listing serves to lead the reader into the middle portion.

Major Divisions of the Object and its Principle of Operation

A technical description of more complex devices often includes the principle or theory of operation. The description of the major components can be arranged in three ways. One arrangement may follow the order in which a viewer's eye might see the matter described. For example, in the case of a claw hammer, the viewer might first see the head, then the peen, the face, the neck, and finally the handle. Each part, then, would be described in sequence from the viewpoint of how the eye sees the hammer.

    Another way of arranging the description of the parts may be according to the sequence in which the parts are assembled. For example, a car-door lock cylinder might be described in accordance with the way its various parts are arranged in the car door; the retaining clip, the pawl, the cylinder housing, the lock cylinder, the cylinder cap springs, the cylinder cap, and the cylinder housing scalp.

A third method of arrangement that might be appropriate is the order in which the parts act in operation, as in the example of a duo-brake on pages 229-231.

Which order of parts listing to use depends on your reader's requirements. If the reader is interested only in a general knowledge and understanding, the first order—from the viewpoint of how the reader's eye sees the thing—would be appropriate. If the reader needs to assemble the object, the second order should be used. If the reader needs to know how to operate the mechanism, obviously the third order is called for.

When all the parts have been fully described, your next step is to show how the complete unit functions by taking it through a cycle of operation. The concluding section is in effect a description of a process in condensed form (see pp. 231-240, Explanation of a Process). This section should show how each division achieves its purpose. Emphasis is placed on the action of the parts in relation to one another. Special applications, variations, cost, and other pertinent general details are appropriate for the terminal section.

Illustrations

Drawings and photographs can indicate precisely the size and shape of things. They need to be integrated with the text. The following suggestions are offered for the use of illustrations in technical descriptions:

1. All drawings should be made to scale and fully dimensioned.
2. All illustrations should be properly numbered and each should have its own captioned title.
3. Even where there is only one illustration, it should have a title.
4. Text matter in the illustration should be kept to a minimum.
5. Descriptive matter needed to explain an illustration should be incorporated.
6. Standard abbreviations should be used.
7. Where space permits, words should be spelled out completely.
8. Reference letters or numbers in illustrations make it easier for the reader to correlate the drawing with the text.
9. Where the purpose of a technical description is to enable the reader to reproduce or fabricate a device, complete dimensions should be indicated in drawings and critical dimensions given in the text description.

Style in Technical Descriptions

The two major requirements in a technical description are completeness and clarity. The requisites of technical style discussed in chapter 12 are certainly necessary. Conciseness, precision, and objectivity are the keystones. Description of an existing mechanism or organism is usually written in the present tense. On occasion, a more appropriate tense is the future or the past, but whatever tense is used, it should be consistent and logical throughout the description. A writing pitfall in taking a mechanism or organism through a cycle of operation is lack of parallel grammatical structure (see pp. 340-41 in chapter 13). Finally, always keep in mind the reading audience. Are you writing for experts? For well-informed individuals? Or for the general public? Does the reader need to know only in a general way what the matter described looks like and how it functions, or does he need to be taught how to manufacture or to operate the device? Your purpose is going to determine the principle or organization and the amount of detail necessary. In a generalized approach, for example, the organizational divisions of the text are going to be the same, but the details are going to be fewer and not as exacting. Drawings will be generalized and functional rather than specific and detailed.

Examples of Technical Description

The following description of the venturi carburetor by a student in a technical writing class is an example of an effective technical description, meeting, on the whole, the requirements set forth in this chapter. It would be enlightening to check this description against the outline of a detailed, typical technical description.

THE SIMPLE VENTURI CARBURETOR[1]

The simple venturi carburetor is used primarily as a fuel metering device for small, internal combustion engines such as are found on modern, self-powered lawn mowers. This type of carburetor, which is only remotely similar to a complicated automotive carburetor, consists basically of a venturi, a fuel supply line and needle jet, a butterfly valve, and sufficient mounting lugs so that the unit may be adequately secured to the engine on which it is to be used.

DESCRIPTION OF PARTS

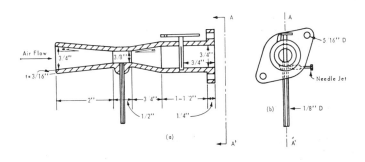

Figure 1

[1] A student paper by Joe Marcus written to meet an assignment for writing a technical description.

The Main Casting

The main casting, the cross-hatched area, which is formed of 195-TS-62 aluminum, is 5 inches in length, and has machined surfaces on all of the inner areas. The venturi entrance angle 0 is 5.38 degrees and the throat exit angle α is 7.12 degrees. The mounting flange is ¼-inch thick, and two 5/16-inch diameter holes are drilled in this area for mounting bolts, as shown in Figure 1-b. Butterfly shaft holes are drilled to a size of ⅛-inch in diameter, and are located in the unit's central, vertical plane, ¾ of an inch inward from the flange mounting surface. The fuel pick-up tube mounting hole is ⅛-inch in diameter and is also located in the central, vertical plane of the carburetor. This hole is drilled through the bottom of the casting at the mid-point of the throat section. The needle jet hole is radially drilled into the pick-up tube mounting lug, as shown in Fig. 1-b. This hole, which is 7/32-inch in diameter, is then threaded with a ¼-inch National fine tap which corresponds to the needle jet's adjustment threads.

The Pick-up Tube

This unit is a ⅛ + 0.001 inch diameter (outside) brass tube which is two inches long, and is press fit into the casting due to the 0.001 inch interference. The inside diameter is 1/16-inch, and a 1/32-inch diameter hole is drilled radially through one wall of the tube 7/16 of an inch down from the top surface which eventually is flush with the inside surface of the throat. This hole is the one through which the needle jet point protrudes so that the air-fuel mixture may be altered.

The Needle Jet

Figure 2

The needle jet, as shown in Fig. 2, is formed of brass material employing the dimensions shown. The threads indicated are ¼-inch fine to correspond with the threads cut in the main casting needle jet hole.

The Butterfly Valve

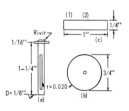

Figure 3

The butterfly valve consists of a ⅛-inch diameter by 1¼-inch long brass shaft with a diametral slot cut in the location shown. This slot and the circular butterfly plate are 0.020 inches thick and ¾ of an inch in height and diameter respectively. The hole "A" in Fig. 3 is drilled after the shaft is installed in the main casting, and it corresponds with the hole in the butterfly plate. A rivet then joins the plate and shaft, employing hole "A" which is 1/16-inch in diameter. A small shank is machined in the shaft which provides a retaining shoulder for the throttle linkage, Fig. 3-c. The stub extending from this shank is 1/16 of an inch in diameter and ⅛ of an inch tall; it is used as a rivet to secure the throttle linkage to the shaft. This throttle linkage plate is 1/16-inch thick steel plate 1-inch long and ¼-inch wide. Holes (1) and (2) are 1/16-inch in diameter, and are used for throttle rod attachment and mounting to the shaft respectively.

The Mechanism in Use

The nature of the engine with which this fuel metering device is to be used is such that a flow of air will almost continuously be drawn through the carburetor's throat. The amount of flow entering the carburetor is also a function of how much the butterfly valve has been opened. From fluid mechanics it can be shown that as a fluid (gas or liquid) passes through a venturi, its velocity increases: but its pressure decreases to some value below atmospheric. This negative pressure is greatest at the point in the throat where the fuel pick-up tube is located. This differential pressure, P (atmospheric) minus P (throat), will cause the gasoline, into which the tube has been submerged, to be forced up the tube and into the air stream. The needle jet, previously described, is used to accurately control the amount of gas flow through the pick-up tube. As the gasoline enters the air stream, it is vigorously mixed with the air flow and is therefore made compatible with combustion requirements.

The following is a more generalized description, that of a duo-servo brake taken from an automobile service manual. It would be interesting for you to check it carefully with the specifications for a technical description in this chapter. Note that the description is more generalized. Dimensions are offered only where the knowledge is useful to the reader—the serviceman. Note, too, that the descriptions of the brake and its components otherwise are quite detailed. The writing is directed to aid the garage mechanic to service and repair the brake. You may note a grammatical slip in the writing — a dangling participle (first sentence last paragraph).

THE DUO-SERVO BRAKE

General Description

The brakes used on both front and the rear of all 1955 Chevrolet models are the Duo-Servo, single anchor type, which utilize the momentum of the vehicle to assist in the brake application. The self-energizing or self-actuating force is applied to both brake and shoes in each wheel in both forward and reverse motion.

Each brake (Fig. 4) has one wheel cylinder located near the top of the brake flange plate, just below the anchor pin. Each wheel has two shoes with a pull-back spring installed between each shoe and the anchor pin, to hold the upper ends of the shoes against the anchor pin when the brakes are released. The lower ends of the shoes are connected by a link and helical spring. The link is made up of an adjusting screw, riding a socket at one end, and threaded into a pivot nut at the other. The outer ends of

1. *Backing Plate*
2. *Anchor Pin*
3. *Guide Plate*
4. *Secondary Shoe*
5. *Pull Back Spring*
6. *Primary Shoe*
7. *Pull Back Spring*
8. *Hold Down Spring*
9. *Hold Down Pin*
10. *Adjusting Screw*
11. *Adjusting Screw Spring*

Figure 4
Duo-Servo Brake

the socket and the pivot nut are notched to fit the webs of the brake shoes, providing freedom of motion between the link and the shoes. The spring is stretched from one shoe web to the other, crossing over the notched head of the adjusting screw. It bears against one of the notches in the head, and thus acts as a lock for the adjusting screw. Bonded brake linings are used, and brake drums are 11″ in diameter. The front brakes are 2″ wide, while the rear are 1¾″.

In each brake assembly, the linings for the front and rear shoes differ in length, the secondary facing being 2½″ longer than the primary, because in operation, a greater force is applied to the secondary facing than to the primary.

The brake flange plate has six bearing surfaces, three for each shoe, against which the inner surfaces of the shoes bear to maintain alignment. Slightly below the center of each shoe web is a hole through which a hold-down pin is inserted. A spring, fitted over the outer end of the pin holds the shoe against the braking surfaces. At the top of the brake, where the shoes butt against the anchor pin, a guide plate separates the pullback springs from the shoe webs, and assists in keeping the shoes properly aligned. The brake mechanism is effectively sealed against the entrance of dirt or mud by the joint between the brake flange plate and the drum. The outside edge of the flange plate fits over the edge of the drum which has annular grooves located between two flanges.

The outer flange is of a larger diameter than the inner one, so that dirt and moisture which collect in the groove are thrown off the larger flange by the centrifugal force of the rotating drum, thus keeping foreign matter away from the drum-to-flange plate joint.

Operation

When the brakes are applied, the pistons in the wheel cylinder, acting on the brake shoes, through the connecting links, force the shoes against the drum. Since the shoes float free in the brake, the force of friction between the shoes and the rotating drum, turns the entire assembly in the direction of the wheel rotation. The front or primary shoe moves downward, and the back or secondary shoe is carried upward until its upper end butts against the anchor pin. The friction between the moving drum and the stationary shoes now tends to roll both shoes toward the drum with increased pressure. The secondary shoe pivots on the anchor pin at the top, and the primary shoe tends to turn out the adjusting link at the bottom which is held stationary by the secondary shoe. This self-energizing effect greatly increases the pressure of the shoes against the drum and reduces the physical force required on the brake pedal.

Inasmuch as the brake shoes are freely connected at the bottom by the adjusting link, the self-energizing or friction force which is applied to the primary shoe by the brake drum is transmitted to the secondary shoe through the link. The effectiveness of the secondary shoe is nearly doubled, because the total force applying this shoe becomes the sum of the force which is received from the primary shoe, and the self-energizing effect that is derived from the rotating drum.

When backing the car, the brake action is reversed. The rear shoe becomes the primary shoe, and the front shoe becomes the secondary, butting against the anchor pin during braking and being forced against the drum with great pressure (3:5-12).

Explanation of a Process

Explaining a process is describing, narrating, or instructing how something is done, how an occurrence has happened, or how an effect has taken place. Certain events or activities are connected in a significant sequence and create an observable or measurable effect which, if repeated with the same quantitative and qualitative ingredients and in the same sequence, should produce the same or similar results. There are actions and activities in which no human operator is involved; the process may be due to interaction of natural, psychological, or sociological forces. Processes may be industrial, psychological, physiological, economic, mathematical, or sociological. They are the steps or sequence of events by which occurrences take place or by which products are made. Technicians, engineers, biologists, physicians, geologists, social scientists, and home economists are concerned daily with processes. Frequently they have to explain them so that others may understand the same or a similar result.

One of the more effective explanatory devices is the analogy. The writer makes use of something familiar or known to the reader to help explain the workings of something little known or complex. The process of how the debilitating disease of multiple sclerosis afflicts a patient is graphically and clearly portrayed by the analogical explanation below:

Close-up of a Neural Network

Imagine for a moment a complicated telephone switchboard in a large organization. Hundreds of wires run from the central console to all floors. These wires carry messages from the outside world and from one office to another. Data, reports, instructions — converted into electrical impulses —

race along the wires all day, tying the many units of this complicated enterprise together in a communications network. Most of the wires are protected by a sheath of insulating material that keeps the electrical impulses from leaking away or being short-circuited.

Let us suppose that, for some unknown reason, patches of this insulation begin to disintegrate and sections of bare wire are exposed. Very slowly, as the disintegration progresses, strange things start happening in the network. Some offices get a scratchy buzz when the receivers are raised. There are complaints of garbled messages. Messages transmitted from the switchboard wind up at the wrong offices. In some sections of the building, the phones go dead. The functioning of the entire system becomes erratic, and the baffled repairmen are unable to explain what has got into the insulation.

This is a highly simplified mechanical analogy for an advanced case of multiple sclerosis — an analogy that conveys none of the human anguish caused by each attack of the disease.

Multiple sclerosis attacks the insulation around the message-bearing nerve fibers of the most complicated communication network known, the central nervous system of man. Almost all the information he receives from the outside world and from inside his own body, as well as all the conscious and unconscious control he has over his own movements and behavior, man owes to this system that consists of the brain, the spinal cord, and the various nerves that connect the brain and the spinal cord with the sense organs, glands, and muscles of his body. Although such chemicals as hormones and such purely physical factors as temperature help to integrate the various tissues and systems of the body into a smoothly operating unit, the central nervous system is the most important means by which the body coordinates its activities and responds to environmental changes.

The easiest way to understand how multiple sclerosis affects this communication system is to take a microscopic look at its enormously complex tissues. Such a close-up reveals that the central nervous system is built of cells, as are other parts of the body. The basic units are the nerve cells, or neurons. In addition, there are many neuroglial cells, or glia, which correspond to the connective tissue cells found in other organs. Together with the blood vessels, the glia pervade the substance of the brain and spinal cord, and serve as the "glue" and structural supports that hold the delicate nerve network together. There are many more glia than neurons, but it is the neurons that conduct messages to and from the brain and from one part of the brain to another.

A neuron is a highly specialized unit with a central body containing a nucleus and cytoplasm. What distinguishes the neuron from other cell types are the various threadlike appendages or processes extending from it. Each neuron has relatively short processes called dendrites, which have many fine extensions like the branches of a tree, and a single fiber called the axon, which usually does not branch out except at the very end, although it may be two or three feet long. Both the dendrites and the axon have a property not shared by most other cells in the human body: the capacity to transmit electrical impulses or messages. The dendrites, transmitting nerve impulses toward the cell body, and the axon, conducting the impulses to the dendrites of other nerve cells or to muscles and glands, are in effect the wiring of the body's communication system.

The Myelin Sheath

If a section of an axon is properly stained and placed under a microscope, two parts become apparent — a central core, or axis cylinder, surrounded by a sheath of fatty tissue called *myelin*, which is formed by processes of neuroglial cells that wrap around the nerve's fibers. Unstained, myelin is whitish in appearance. Even to the naked eye, certain areas of the brain and spinal cord appear to be white; this *white matter* contains the nerve fibers with their myelin sheaths. What is popularly known as a *nerve* is really a bundle of axons with their myelin sheaths, all bound together by a packing of connective tissue much as a bunch of wires is wrapped together to form an electric cable.

Because the fatty material is indeed a good insulator, scientists presume that the sheath restricts a nerve impulse to a distinct pathway and prevents the leakage of electric potential to surrounding tissue. However, since the chemical composition of the myelin is much more complex than an insulator needs to be, some researchers suggest that the sheath also helps to nourish or maintain the integrity of the axon, which stretches far from the nucleus so vital to the life of the entire cell.

There is also evidence that the sheath speeds up the transmission of the nerve impulse. Myelinated nerve fibers transmit at the rate of 100 to 120 meters per second, or 225 miles an hour — fast enough for a message to travel from head to toe of a six-foot man in 1/50 of a second. Along a nonmyelinated nerve, the impulse travels at a speed of only 10 to 20 meters per second.

The living web of the central nervous system is in ceaseless activity, with passages of sensory information from the outside and from all parts of the body along sensory nerves to the spinal cord and brain, and an unending stream of return messages speeding along motor nerves to the muscles and glands. The brain is the central switchboard. The spinal cord, a somewhat flattened cylindrical bundle of fibers running from the base of the skull to the lower back through a bony canal formed by the vertebrae, is the main trunkline for sensory and motor impulses.

If the central nervous system of a person suffering from multiple sclerosis could be spread out for inspection and the impulses made visible as light traveling along the nerves, it would be easy to spot discolored patches where the myelin sheath has disappeared or been destroyed, exposing the nerve fibers. The disease is termed *multiple* because the patches, or *plaques* as they are called, often are widely scattered and affect many areas of the central nervous system — the spinal cord, the cerebellum, the brain stem, the cerebrum.

The plaques, which vary greatly in size and shape as well as location, appear as clearly defined islands where the soft myelin has degenerated and later will be replaced by hard, semitransparent scar tissue. These are islands of *sclerosis*, from a Greek work meaning "hardening."

At first only the sheath is affected. The axons still transmit messages, although there may be malfunctions indicating that a weakening or partial blockage of the nerve impulse has occurred. As the plaques develop and scar tissue forms, the impulses find it increasingly difficult to travel past these hard spots; ultimately, nonconducting scar tissue may completely block the impulses (2:2-5).

Effect of Demyelination

Normal   Myelin   Myelin Sheath
Myelinated   Breakdown   Totally
Nerve Fiber   Begins   Disrupted

The explanation of a process in which the reader must take part requires more specific details than the explanation of an activity or occurrence in which the reader does not take part but needs to understand the occurrence. In other words, the written description of the process depends, as does every type of writing, on the intended reader's requirements. If the reader needs to perform the process, it will be necessary for the technical writer to consider every detail. If the reader needs only a general knowledge of the principles involved and will not need to perform the process or to supervise its performance, the writer need not go into specific details which may only confuse the reader, but will instead emphasize the broad principles and give a generalized account of the steps and sequences.

Intimate knowledge of the process to be described is fundamental. This involves not only complete understanding of how the process works or an ability to perform it but also a knowledge of the appropriate terminology and a logical organizational plan for the explanation. The organizational pattern for explaining an industrial process might be as follows:

I. Introduction
  A. Definition of the process — general information as to why, where, when, by whom, and in what way the process is performed, carried out, or occurs
  B. Fundamental principle of operation, unless explained above
  C. List of the main steps, events, or sequences in the action
  D. Requirements for the performance of the process or the occurrence of the event
    1. Principal tools, apparatus, and supplies
    2. Special conditions required
II. Description of the steps or analysis of the action
  A. First main step (or sequence of events)
    1. Definition
    2. Special materials, apparatus, etc.
    3. Division into substeps
      a. Details of the first substep — action, quality of action, reasons for results of the action
      b, c, etc.,—Description of the other substeps, as needed

B, C, D, etc.,—Descriptions of other main steps or sequences, as required

III. Conclusion (a summary statement about the purpose, operation, evaluation of the whole process)—the process is taken through a complete cycle of operation or through to its final stage

Introduction

The Introduction serves to define what is being done or what happens. Inherent within it are other questions: What exactly is the process? Who performed it or how does it happen? When and why is it being performed or when and why does it happen? What are the chief elements or steps? Not all of the questions need necessarily be answered, nor is the order of the listing significant. Reader level and reader requirements determine the depth of detail used or omitted.

As in the case of the technical description, the explanation of the process may begin with a definition; then it is followed by a generalized description to give the reader an immediate visual image of the character of the process.

The Introduction should include any special circumstances, conditions, and personnel requirements. Frequently, it is necessary to explain why the process is being described. Sometimes the reason is contained in the definition. Sometimes the purpose is so obvious that it is unnecessary to explain, as, for example, the opening of a jar of pickles or the stapling of a sheaf of papers.

The Introduction might also include an indication of necessary materials, tools, and apparatus required for the process. The tabular form lends convenience and clarity. It is helpful to the reader to get a bird's-eye view of the general procedure before he gets into the details; the Introduction, accordingly, will end with a listing of the chief steps in the process.

In an explanation of a simple process, the Introduction may consist of no more than a paragraph or two. In a complex process, the principle of operation or theory may, of itself, occupy several hundred words. Drawings are frequently used to help clarify tools, instruments, or material used.

Main Steps

The main body of the explanation may well have a paragraph of details for each major step. Each step is taken in chronological order, defined, described, and explained as a whole. Materials, conditions, tools, and apparatus necessary in the performance of the steps are included in the explanatory unit of each step. Each descriptive unit has a topic sentence indicating the subject of the step. The details clarify the process and action in their relation to the process as a whole. All essential details of action should be included in the explanation. Whenever a nut and bolt are removed, the explanation must be complete enough to show their replacement. Each essential action should be specified in the terms of who, what, why, and how. Each substep within a major step constitutes a process within itself. Substeps should be properly introduced and, if necessary, subdivided so that the reader will understand and will be able to perform the entire action. In the description of any action, the reader needs to understand and to visualize the activity. Qualitative conditions are important in the performance of an activity. Precise communication of the qualitative factors is essential to obtain the required result.

Concluding Section

After the step-by-step explanation of the process is completed, it is desirable to summarize the whole process so that the reader can see the activity as a whole and not as a series of separate steps. Requirements of the reader may necessitate additional information, such as an evaluation of the process, discussion of its importance, a repetition of important points, cautions to consider, or perhaps, alternative steps which might be taken if certain difficulties are encountered.

Drawings

The reader will frequently be helped by drawings which show different stages of the process. Drawings can verify relationships and show shapes which are difficult to describe precisely in language. Dimensions are most conveniently indicated in a drawing. In complex processes or those in which several steps or actions are going on simultaneously, a flow diagram will be necessary to show the relationship of the several phases. Drawings must be integrated with the text and referred to specifically by figure number to help the reader follow the complete process in text and illustration.

Style

The style considerations discussed in chapter 12 are applicable and appropriate to the explanation of a process. Depending upon the point of view and the formality of the writing situation, processes may be explained in the first, second, or third person, in the active or passive voice. Many processes are described for the express purpose of directing others to duplicate the particular activity. In those instances, the active voice and imperative mood are required. However, the style must be consistent. If the explanation of the process is part of a larger report or a handbook of instructions, it must be integrated stylistically with the larger work.

Summary of Principles in Explaining the Process

The explanation of a process, especially one that involves the fabrication of a product or the conducting of an experiment or activity which leads to a specific result, should include the following information:

1. The definition of the process
2. A description of the time, setting, performers, equipment, and preparations
3. An indication of the principles behind the operation
4. The listing of the major steps in chronological order
5. A step-by-step account of every action and, if appropriate, inclusion of description of apparatus, materials, and special conditions
6. Details under major steps, arranged in chronological order
7. A concluding section which may be a summary or an evaluation of the process
8. Drawings to aid in the explanation of crucial steps and in the vividness of the description of an action

Examples of Explanations of Process

The following explanation of a technical process written by a student exemplifies rather well the principles discussed in this chapter. It would be a valuable learning exercise for you to check this student paper against the outline and text material to see which guidelines were followed and which were not.

How to Obtain a Velocity Profile of Water Flowing in a Pipe

The process of obtaining a velocity profile of water flowing in a pipe consists of taking velocity measurements by means of a U-tube manometer at specified distances across the pipe. The velocity is measured by a pitot tube. This is a tube which is situated in the pipe so that it is pointing in the direction of flow. The flowing water is brought to a stop in front of the tube. This results in an increase in pressure of the fluid in the tube. This increased pressure is transmitted inside a rubber tube to a U-tube manometer where the pressure differential pushes the water level down one leg and up in the other. The difference in the height of these columns of water in the U-tube manometer is an indication of the velocity of the water in the pipe.

The process of measuring the velocity of water flowing in a pipe requires the use of the following pieces of equipment:
U-tube manometer
Rubber tubing
Pitot tube assembly with adjustment screw, pointer and scale
Pipe flanges
Rubber gaskets
Wrenches
Data paper and pencil
French curve
Scale

The steps to be followed are: (1) placing the pitot tube assembly in the pipe, (2) assembling and checking the equipment, (3) taking the velocity data, (4) plotting the results.

Placing the Pitot Tube Assembly in the Pipe

1. Choose a joint in the pipe and screw the pipe flanges on the pipe at this point. Bolts through the flanges will hold the pipe together with the pitot tube assembly between them.

2. Insert the pitot tube assembly in the pipe making sure that the pitot tube is pointed upstream. See Figure 1.

3. Slip the bolts through the holes in the flanges and pitot tube assembly. See Figure 5.

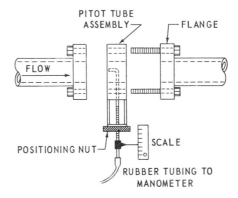

Figure 5
Pitot Tube Assembly Installed In Pipe

4. Place rubber gaskets between the pipe flanges and pitot tube assembly, then screw on the nuts and tighten them with a wrench. Care must be taken here to be sure that the rubber gaskets do not slip and cause a leaky joint.

Assembling and Checking the Equipment

1. Select a U-tube manometer with two foot legs and a scale between them. This size of manometer will permit a large range of flow rates. See Figure 6.

2. Fill the manometer legs with water until the top of the meniscus in both legs is at the zero mark of the scale.

3. Attach one end of the rubber tubing to a leg of the manometer, and the other end to the pitot tube assembly.

4. Select a flow rate and allow water to flow in the pipe.

5. Check the equipment for plugged lines and leakage by turning the screw clockwise to advance the pitot tube into the stream flow.

6. Check that the pointer indicates zero on the scale when the pitot tube is against the near side of the tube.

Taking the Readings

1. Observe and record the difference in levels of the liquid in the manometer when the pitot tube pointer indicates zero. There will be a slight difference in levels indicated since the pitot tube has some diameter and will project into the flow area.

2. Advance the pointer by turning the screw clockwise to 0.05 inches and read then record the difference in levels in the manometer liquid again.

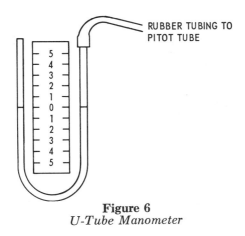

Figure 6
U-Tube Manometer

The difference will be greater since the velocity is greater in this part of the pipe.

3. Continue taking readings every 0.05 inches across the diameter of the pipe.

4. Repeat the process of determining the velocity by the difference of liquid levels in the manometer every 0.05 inches on the return trip through the pipe. This procedure will give two values at each point and their average will give a more representative value.

Plotting the Results

1. Determine the average of the pressure difference at corresponding points in the pipe.

2. Set up a Cartesian coordinate system with the difference in levels of the manometer ($\Delta\eta$) as abscissa and the distance from zero ($\Delta\chi$) as ordinates. See Figure 7.

3. Plot the points and connect them with a smooth curve. This curve represents the velocity profile of the water in the pipe. See Figure 7.

As may be seen in Figure 7, water does not flow at a constant speed across the diameter of a pipe. The velocity is zero at the inner edge of the pipe and maximum in the center of the pipe. This can be explained by the fact that there is relative motion between the water and the wall of the pipe. The water must therefore shear and the velocity profile shown indicates the symmetrical distribution of shear stress and velocity.[2]

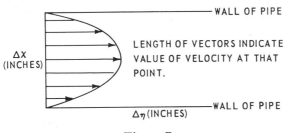

Figure 7

---

[2]A student paper by Paul Wergin, written to meet an assignment for writing an explanation of a process.

Here is an example from a commercial manual explaining the process of making quality concrete. The process, though not complex, involves many complicated details. The explanation is intended to instruct a "lay" individual — a farmer — in the fine craft of making quality concrete. Its details are very specific and precisely enumerated. Note how illustrations are integrated and used to promote clarity of explanation. This process involves serious expenditures in money, time, and effort on the part of the reader — the customer. The company putting out this "do-it-yourself" manual recognizes the importance of the technical writing to explain the process properly. Note the conformity to the specifications of this chapter in the example.

Making Quality Concrete for Farm Construction (10:3-11)

Concrete—a Versatile Building Material

Concrete has several unusual properties that make it a versatile and widely used building material. For example, it sets or hardens in the presence of water. This property is of great importance in the construction of buildings or parts of structures, such as footings and foundations, built on the earth or in wet locations.

Freshly mixed concrete can be formed into practically any shape. Many materials can be molded only when heated and they require elaborate casting facilities, but concrete can be molded at normal temperatures, which gives it tremendous flexibility.

Since concrete is composed of mineral materials, it is impervious to bacterial attack and the resultant decay. It also resists vermin, termites and rodents. Concrete is made of materials that cannot burn. Portland cement is manufactured at temperatures up to 2,700 deg. F. Aggregates are also incombustible. Thus concrete has a high fire resistance so important in rural areas where fire protection is often inadequate.

The basis ingredients of concrete — portland cement and aggregates — are available almost everywhere at reasonable prices. Hence, concrete is economical in all parts of the country.

These many properties combine to give concrete structures and improvements low initial costs, minimum maintenance requirements and long life.

What Is Concrete?

Concrete is a mixture of portland cement, water and inert materials called aggregates. Aggregates are commonly divided into two sizes, fine and coarse. Fine aggregate is sand, and coarse aggregate is usually gravel or crushed rock.

During mixing, the cement and water form a paste which coats the surface of every piece of aggregate. Usually within two to three hours the concrete starts to set due to hydration, a chemical reaction between the cement

Well-graded aggregate has particles of various sizes. This photo shows 1½-in. maximum size coarse aggregate. Pieces vary in size from ¼ to 1½ in.

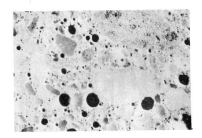

A greatly magnified section of air-entrained concrete. Black round areas are air bubbles formed by air-entraining agent. Pieces of aggregate visible are sand particles of various sizes.

and water. As hydration continues, the cement-water paste hardens much like glue and binds the aggregate together to form the hard, durable mass that is concrete.

Thus the quality of concrete is directly related to the binding qualities of the cement paste. Concrete of various degrees of strength and durability can be produced by changing the proportion of water to cement. As the amount of water per sack of cement is reduced, the strength of the past increases and the concrete is stronger and more durable.

THE INGREDIENTS OF QUALITY CONCRETE

Portland Cement — The Magic Powder

Portland cement was patented by Joseph Aspdin, an English mason, in 1824. He called his product portland cement because it produced a grey concrete that resembled a fine building stone quarried on the Isle of Portland. In 1872 the first cement plant was built in America at Coplay, Pa., to give birth to the portland cement industry in the United States.

Cement is sold in bags and in bulk. Each bag of cement holds 1 cu.ft. and weighs 94 lbs. Many large contracting firms, block manufacturers and ready-mixed concrete producers buy cement in bulk. Large quantities are delivered by rail or truck and are quickly unloaded by automatic equipment into silos or hoppers for storage.

White Portland Cement

Most portland cement is grey in color. However, white portland cement is manufactured from raw materials that have a pure white color. It is used for paints, decorative work and other special jobs.

Mixing Water

Water to be used for concrete should be clean and free of acids, alkalis, oils, sulfates and other harmful materials. Many specifications for construction work state that the water used should be fit to drink. This is a good rule for use on all farm jobs, too.

Aggregates

Aggregates are used in concrete to provide volume at low cost. They comprise 66 to 78 per cent of the volume of concrete and can be called a filler material. Aggregates must be clean and free of materials such as dirt, clay, coal or organic matter. They must be durable, hard, and have few long, sliverlike particles.

Fine Aggregate

Natural or manufactured sands are the most common fine aggregates used in farm work. The fine aggregate should have particles ranging in size from $\frac{1}{4}$ in. down to some that will pass through a sieve having 100 openings to the inch.

Coarse Aggregate

Gravel or crushed rock are widely used coarse aggregates. The aggregate should have pieces in all sizes from $\frac{1}{4}$ in. up to the maximum size used for the job. The common maximum sizes are $\frac{1}{2}$, $\frac{3}{4}$, 1, and $1\frac{1}{2}$ and 2 in.

Bank-Run and Commercially Recombined Aggregate

Bank-run material is aggregate that is used as taken from the quarry or gravel pit. The user should be certain that the material is clean and free from any ingredient that can harm the concrete. Many bank-run materials have a high percentage of sand, which makes it difficult to obtain a high yield of concrete per sack of cement and at the same time maintain the necessary strength. It is generally advisable to screen the bank-run material over a $\frac{1}{4}$-in. screen to separate it into fine and coarse aggregate. Some commercial firms also sell a mixed aggregate. Sand and gravel are separated and then recombined into the correct proportions for concrete.

Lightweight Aggregates

Lightweight aggregates are available in many parts of the country. They consist of cinders or expanded materials such as clay, shale or slag. They reduce the weight of concrete approximately one-third and improve the insulating qualities. Very lightweight aggregates, such as pumice or expanded mica, are also used for further weight reductions and greater insulation.

Size of Aggregate Is Important

Generally, the most economical mix is obtained by using the largest size coarse aggregate that is practical. This is usually considered to be about one-fifth of the thickness of vertically formed concrete and one-third of the thickness of flat concrete work such as floors or walks. When reinforcing steel is used, the largest size aggregate should not exceed three-fourths of the distance between bars. For example, if reinforcing steel is placed in a slab 1 in. from the bottom, the largest aggregate should be ¾ in.

Air — Another Important Ingredient

In the late 1930's it was discovered that air in the form of microscopic bubbles improves the durability of concrete. Concrete containing these air bubbles is called air-entrained concrete. Air entrainment virtually eliminates scaling due to freezing and thawing and salt action. Air-entrained concrete is more workable and cohesive, reduces segregation and bleeding and has improved sulfate resistance. Since it is more workable, it requires less mixing water — an added benefit.

Air-entrained concrete should be used where the hardened concrete is exposed to alternate freezing and thawing or to salt action. In northern areas it is recommended for practically all outside work. Because of its increased workability, it should be considered for many indoor jobs, too.

These tiny air bubbles in concrete are created by chemical compounds called air-entraining agents. These agents can be added at the time of mixing. Many cement manufacturers market portland cements that contain an air-entraining agent. Such cements are identified on the bag.

PREPARING TO MIX CONCRETE

The mixing site should be as close as practical to the point where the concrete is to be placed. The size of the mixer to use depends upon the size and type of the job and the manpower available. Concrete for large flat jobs is easy to place and a large mixer will help speed up the job. The mixer should be checked for capacity, which is usually given on an attached plate.

The aggregates should be placed so that they are convenient to the mixer operator. The cement must be stored in a dry location. If it is stored outside, it should be raised above the earth and covered with a tarpaulin, polyethylene or other suitable cover.

Ample water and containers to measure it accurately should be available.

HOW TO PROPORTION MATERIALS

The mix should be economical, and it must be workable so that it can be properly placed and finished. When the concrete has hardened, it must have strength to carry the loads, and it must be durable to resist elements to which it is exposed.

Select the Proper Water-Cement Ratio For Your Job

The strength and durability of concrete depend on the quality of the cement-water paste. To select the proper mix, refer to Table 1. First check the type of concrete work to be done to determine the gross amount of water to be used with each sack of cement. The first three columns of Table 1 give the net amount of water to use for sand of various degrees of wetness. Most sand as used on the job contains a surprising amount of water. Hence, the sand must be tested for water content and an allowance made.

TABLE 1. SUGGESTED MIXES MADE WITH SEPARATED AGGREGATES

Kind of work	Gal. of water added to each 1-sack batch if sand is:			Suggested mixture for 1-sack trial batches ††		
	Damp*	Wet** (average sand)	Very wet †	Cement, sacks (cu. ft.)	Fine, cu.ft.	Coarse, cu.ft.
5 gal. of water per sack of cement						
Concrete subjected to severe wear, weather, or weak acid and alkali solutions	With ¾-in. max. size aggregate					
	4½	4	3½	1	2	2¼
6 gal. of water per sack of cement						
Floors (such as home, basement, dairy barn), driveways, walks, septic tanks, storage tanks, structural beams, columns and slabs	With 1-in. max. size aggregate					
	5½	5	4¼	1	2¼	3
	With 1½-in. max. size aggregate					
	5½	5	4¼	1	2½	3½
7 gal. of water per sack of cement						
Foundation walls, footings, mass concrete, etc.	With 1½-in. max. size aggregate					
	6¼	5½	4¾	1	3	4

*Damp describes sand which will fall apart after being squeezed in the palm of the hand.
**Wet describes sand which will ball in the hand when squeezed but leaves no moisture on the palm.
†Very wet describes sand that has been subjected to a recent rain or recently pumped.
††Mix proportions will vary slightly depending on gradation of aggregates.

For small jobs it is possible to estimate the amount of water in the sand by squeezing it in the hand. The photographs on pages 245-46 show the appearance of damp, average wet and very wet sand.

After determining the relative moisture condition of the sand, refer to the proper column in Table 1 and find the net amount of water to use. For example, if a floor is being placed and the sand is wet (average), 5 gal. of water will be used with each sack of cement.

Mixing water must be carefully measured to obtain quality concrete. Here the mixer-operator marks correct water-line for a one-half sack batch.

Make a Trial Batch

After the net water-cement ratio has been determined, a trial batch of concrete should be made. Table 1 gives suggested proportions of cement, sand and coarse aggregate to try. These proportions are based on experience with typical aggregates. Note that the amount of water suggested is for a one-sack mix. Use proportionally smaller amounts if smaller mixers are used. Mix a batch using the correct amount of water and the proportions of other materials suggested.

Discharge a sample of the concrete from the mixer and examine it for stiffness and workability. If the mix is too wet, add a little more sand and coarse aggregate. If it is too stiff, the mix contains too much sand and coarse aggregate, and it will be necessary to reduce the amount of aggregate in subsequent batches. *Never add more water. Adjust the consistency of the mix by varying the sand and coarse aggregate.*

In some cases the mix may be too sandy or too stony. When this occurs, it is advisable to make a second trial batch and to vary the proportions of sand and coarse aggregate until the desired workability is obtained.

"Damp sand" falls apart after being squeezed in the hand.

Concrete should be mixed at least 1 minute and preferably for 3 minutes after all materials have been placed in the mixer. There is little advantage in mixing over 3 minutes.

"Wet sand," which describes most sands, will form a ball when squeezed in the hand. It leaves no noticeable moisture on the palm.

"Very wet" describes sand that has been exposed to a recent rain or washing. When squeezed in the hand, it forms a ball and leaves moisture on the palm.

PLACING CONCRETE

Concrete Requires Good Forms

The forms should be carefully set, accurately plumbed and leveled, and adequately braced. Fresh concrete exerts considerable pressure against the formwork. This pressure increases as the height of the form increases. The forms should be oiled with commercial form oil or clean motor oil before the concrete is placed so they may be easily removed.

Flat Concrete Work

The supporting soil for all flat concrete work should be adequately compacted to prevent unequal settling. Prior to placing the concrete, the earth or granular subbase should be dampened to prevent it from drawing water from the freshly placed concrete. Concrete should never be placed on frozen earth or earth that is flooded with water.

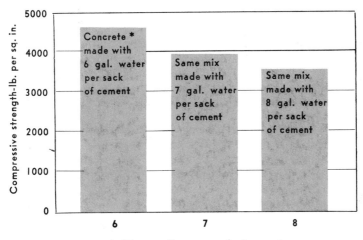

Figure 8
The amount of water used per sack of cement is the key to quality concrete. Note the effect of additional mixing water on the strength of the concrete when the proportions of cement, sand and coarse aggregate are held constant.

Placing Concrete in Walls

Concrete for walls should be placed in 12- to 18-in. layers around the entire wall. The concrete should be spaded with a flat scraper or other thin-bladed tool or mechanically vibrated. This is done to eliminate a condition called "honeycombing," which occurs when coarse aggregate collects at the face of the wall. In inaccessible areas the forms can be lightly tapped with a hammer to achieve the same result.

Vibrators Are a Great Aid

Spud vibrators are excellent tools to consolidate fresh concrete in walls and other formed work. The spud vibrator is a metal tube-like device which vibrates at several thousand cycles per minute. When inserted in the concrete for 5 to 15 seconds, the spud vibrator consolidates it and improves the surfaces next to the forms.

Spud vibrators are often used to consolidate vertically formed concrete to obtain a smoother surface.

FINISHING CONCRETE

How To Obtain the Texture That You Need

After the concrete has been placed, it is screeded or struck off with a straightedge by using a saw-like motion to level the concrete. The concrete is then sometimes floated with a long-handled float, called a bull float. This gives a more even finish than the screed. Further finishing should be delayed until the water sheen has disappeared from the surface of the concrete. Premature floating and steel troweling brings fine sand, cement and water to the surface and results in a surface with less wearability.

After the water sheen has disappeared, the concrete is floated with a wood, aluminum or magnesium float. Floating gives a gritty texture to the surface and is recommended for all areas where hogs and cattle walk. If a smoother surface is desired, the concrete should be steel troweled. Troweling is always

Newly placed concrete is first screeded or struck off with a straightedge to bring it to proper grade. One or two passes with the strike-off board is generally sufficient.

done following the floating operation. Feed troughs and bunks should be steel troweled. Sprinkling cement on the surface of fresh concrete to absorb the water is not recommended.

A rougher texture can be obtained by drawing a broom across the surface of the concrete after floating. Pressure on the broom and the texture of the bristles determine the degree of roughness.

Screeding is followed with floating. Here a bull float, which is a wood or light metal float with a long haindle, is being used to float the concrete.

Hand floats are often used for a final finish after screeding or bull-floating. A float finish has a gritty texture and is desirable for many farm jobs.

Brooming gives concrete a durable nonslip texture. On this job, water-tight kraft paper was spread over the new concrete as soon as brooming was done.

Where To Put Joints

Concrete expands and contracts slightly due to temperature differences. It may also shrink as it hardens. Joints should be put in the concrete to control expansion, contraction and shrinkage.

It is desirable to prevent slabs-on-ground, either inside or outside, from bonding to the building walls. Thus, the slab will be free to move with the earth. To prevent bonding, a continuous rigid waterproofed insulation strip, building paper, polyethylene or similar material, is placed next to the wall. These materials are also used next to other existing improvements, such as curbs, driveways and feeding floors. The continuous rigid, waterproof insulation strip acts as an expansion joint.

Wide areas, such as floors slabs and feeding floors, should be paved in 10- to 15-ft. wide alternate strips. A construction joint, also known as a key joint, is placed longitudinally along each side of the first strips paved. A construction joint is made by placing a beveled piece of wood on the side forms. This creates a groove in the slab edges. As the intermediate strips are paved, concrete fills this groove, and the two slabs are keyed together. This type joint keeps the slab surfaces even and transfers load from one slab to the other when equipment is driven on the slabs.

Contraction joints, often called dummy joints, are cut across each strip to control cracking. They do not extend completely through the slab, but are cut to a depth of one-fifth to one-fourth of the thickness of the slab, thus making the slab weaker at this point. If the concrete cracks due to shrinkage or thermal contraction, the crack usually occurs at this weakened section.

Contraction joints should be cut soon after the concrete has been placed in order to work the larger pieces of coarse aggregate away from the joint. A simple way to cut this type of joint is to lay a board across the fresh concrete and to cut the joint to the proper depth with a spade, axe or similar tool. A groover is then used to finish the joint. Contraction joints are generally placed 10 to 15 ft. apart on floor slabs, driveways and feeding floors. They are placed 4 to 5 ft. apart on sidewalks.

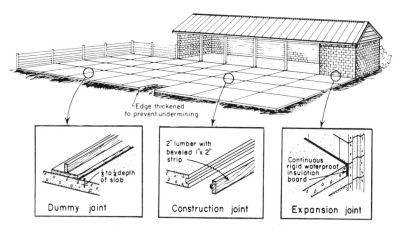

Figure 9
Joints are used to control contraction and expansion in concrete. They are easy to put in and should be used in all flat work such as floors, slabs, and walks.

Since the development of special blades that can saw concrete, the practice of sawing contraction joints is becoming common. The proper time to saw joints is generally 18 to 24 hours after the concrete has been placed. A sawed joint is clean and attractive and works very well when cut to the proper depth. On larger jobs a special mobile concrete saw is used. On smaller jobs, such as sidewalks and driveways, a portable electric hand saw has been used with success. The operator must be careful to keep the saw blade straight so that it will not shatter.

All open edges should be finished with an edger to round off the edge of the concrete slab to prevent spalling.

CURE CONCRETE FOR BEST RESULTS

Concrete should be cured for at least 5 days to prevent evaporation of the mixing water. If too much water escapes, there is not enough available to completely react with the cement, and the concrete does not gain the strength that it should.

Methods of Curing Concrete

Concrete can be cured by one of several methods. A common curing method is to cover the concrete with burlap, sand, straw or similar material. These materials are kept damp during the curing period. Another method is to cover the concrete with a vapor-sealing material, such as polyethylene

There are several ways to cure concrete. Polyethylene is being used here. It is moisture-tight and does an excellent job of preventing moisture from escaping from the concrete.

or water-resistant kraft papers, to seal in the water and prevent evaporation. Commercial curing compounds can also be used. These are sprayed on the fresh concrete to seal the surface. In vertically formed concrete, a simple way to prevent the concrete from drying is to leave the forms in place.

SPECIAL CONDITIONS

Cold Weather Concreting

Temperature has a considerable effect on the rate of hardening of concrete. The optimum temperature for placing concrete is assumed to be

70 deg. F. As the temperature drops, the rate of hardening slows. All new concrete must be protected from freezing. In buildings, heat is often supplied with an oil-fired stove called a salamander. The hot air from the salamander should not be allowed to come into direct contact with the concrete as it will dry it out.

When concrete is mixed in cold weather, the water and aggregates are often heated. The water should not be heated to more than 150 deg. F. as overheating is likely to cause a flash set in the concrete.

High-early-strength portland cement is frequently used in winter concrete work because it sets more rapidly than normal portland cement. When normal portland cement is used, calcium chloride can be added to the mix in cold weather. It is not an antifreeze material but an accelerator that speeds up the chemical reaction between the cement and water. The quantity of calcium chloride should not exceed 2 lb. per sack of cement. It should be dissolved in the mixing water, not added in powder form to the mix. Use of calcium chloride is not a substitute for normal cold weather precautions.

Hot Weather Concreting

As the temperature rises above 70 deg., the initial rate of hardening of concrete increases. Evaporation of water from the concrete is also more rapid in hot weather. A combination of wind, high temperature, and low humidity dries water from concrete rapidly.

In extremely hot weather it may be necessary to reduce the temperature of the freshly mixed concrete. Aggregates should be stockpiled in the shade, if possible, and the coarse aggregate sprinkled with water to reduce its temperature. It is often advisable to delay placing concrete until late in the afternoon to take advantage of lower temperatures. Curing must be applied promptly to prevent the evaporation of water.

ESTIMATING

How To Find the Amount of Concrete Needed

The unit of measure for concrete is the cubic yard, which contains 27 cu.ft. To determine the amount of concrete needed, find the volume in cubic feet of the area to be concreted and divide this figure by 27. The following formula can be used to determine the amount of concrete needed for any square or rectangular area:

$$\frac{\text{Width, ft.} \times \text{Length, ft.} \times \text{Thickness, ft.*}}{27} = \text{Cubic yards}$$

For example, a 4-in. thick floor for a 30x90-ft. building would require:

$$\frac{30 \times 90 \times 0.33}{27} = 33.33 \text{ cu.yd. of concrete}$$

---

*The thickness dimension must be changed to feet or parts of a foot. The decimal part of a foot was used in the example. However, the fractional part may be used instead. Table 3 gives both the fractional and decimal parts of a foot for several common thickness dimensions.

The amount of concrete determined by the above formula does not allow for waste or slight variations in concrete thickness. An additional 5 to 10 per cent will be needed to cover waste and other unforeseen factors.

How To Find the Amount of Materials To Order

Table 2 below gives the number of sacks of cement and the amount of aggregate (in cubic feet and in pounds) needed to produce a cubic yard of concrete for water-cement ratios of 5, 6, and 7 gal. of water per sack of cement. Since it is generally impossible to recover all of the aggregate, a 10 per cent allowance has been included to cover normal wastage.

TABLE 2. MATERIALS NEEDED PER CUBIC YARD OF CONCRETE MADE WITH SEPARATED AGGREGATES

Suggested mixture for 1-sack trial batches *			Materials per cu. yd. of concrete				
Cement, sacks (cu. ft.)	Aggregates		Cement, sacks	Aggregates			
	Fine, cu. ft.	Coarse, cu. ft.		Fine		Coarse	
				cu. ft.	lb.	cu. ft.	lb.
With ¾-in. maximum size aggregate							
1	2	2¼	7¾	17	1550	19½	1950
With 1-in. maximum size aggregate							
1	2¼	3	6¼	15.5	1400	21	2100
With 1½-in. maximum size aggregate							
1	2½	3½	6	16.5	1500	23	2300
With 1½-in. maximum size aggregate							
1	3	4	5	16.5	1500	22	2200

*Mix proportions will vary slightly depending on gradation of aggregate.

TABLE 3. HOW TO CHANGE THICKNESS IN INCHES TO FRACTIONS AND DECIMAL PARTS OF A FOOT FOR USE IN CALCULATING QUANTITIES OF CONCRETE.

Inches	Fractional part of foot	Decimal part of foot
4	4/12 or ⅓	0.33
5	5/12	0.42
6	6/12 or ½	0.50
7	7/12	0.58
8	8/12 or ⅔	0.67
10	10/12 or ⅚	0.83
12	1	1.00

The following excerpt deals with a biological process. In organization and structure the explanation of this process varies little in principle and explanatory techniques from the foregoing explanations of velocity profile and quality concrete.

CELL DIVISION

One of the most fundamental of all cellular processes and one in which the tissue culturist is particularly interested is that of cell division. In fact, the process to be described is nuclear division. The cytoplasm usually follows the nucleus in dividing (but this does not happen invariably).

Three types of nuclear division have been described. Amitosis is a simple division of a nucleus without formation of chromosomes. It is very uncommon and probably not a normal process. Consequently, it will not be discussed further.

Meiosis and mitosis are both characterised by the appearance of chromosomes which are segregated into two groups, one group going to each of the daughter cells. Meiosis is a special case, occurring in the formation of germ cells. During meiosis the normal (diploid) number of chromosomes is halved so that each daughter cell contains a haploid number. In mitosis, on the other hand, the cell at the time of division has material for a double set of chromosomes. On segregation of the chromosomes each daughter cell is left with the normal (diploid) number. This is the normal type of cell division occurring in somatic cells during growth and is almost the only type of cell division seen in tissue cultures. All subsequent discussion will be concerned only with mitotic division.

The early stages of mitosis are characterised by the appearance of the chromosomes. These structures carry the genes, which are mainly responsible for transferring information about hereditary characteristics to the daughter cells. Each cell has a double set of chromosomes and the number is characteristic of the species. The number of chromosome sets is referred to as the "ploidy" of the cells. Thus, normal cells with two sets are referred to as diploid, cells with three sets are triploid, with four sets tetraploid, and so on. Germ cells with half the diploid amount are referred to, somewhat inconsistently, as haploid.

The stages in cell division are as follows:

1. *The Intermitotic Phase* (Interphase). This is also, mistakenly, referred to as the resting phase. In growing cells most of the synthesis of new materials proceeds during this period. When the cell contents have been doubled the cell then enters prophase.

2. *Prophase.* This is, as a rule, the earliest recognisable stage of the cell division. Within the nucleus the chromatin condenses into a continuous skein and begins to break up into chromosomes. In the meantime, the centrosome divides and the two portions move in opposite directions. Between these two poles, the structure of the spindle (rather like visualised magnetic lines of force) appears and chromosomes arrange themselves on it. The nucleoli disappear and the nuclear membrane disintegrates. At the end of this phase, the cell has the typical appearance of metaphase.

3. *Metaphase.* The chromosomes are arranged in the equator of the cell. By time-lapse cinemicrography they can be seen to be highly active at this stage, often apparently rotating en masse. Quite suddenly the cell enters anaphase.

4. *Anaphase.* The chromosomes split longitudinally and the halves (chromatids) move in opposite directions toward the poles. When they are

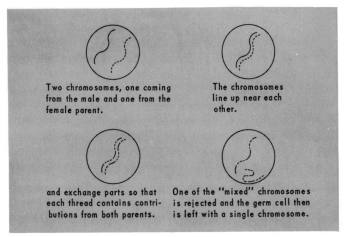

Two chromosomes, one coming from the male and one from the female parent.

The chromosomes line up near each other.

and exchange parts so that each thread contains contributions from both parents.

One of the "mixed" chromosomes is rejected and the germ cell then is left with a single chromosome.

Figure 10
The reduction process by which germ cells are formed.

completely separated, they are again regarded as chromosomes. As cytoplasmic division begins, the cell enters telophase.

5. *Telophase.* The chromosomes at each pole come together and begin to fade and disappear. Simultaneously a new nuclear membrane forms around each nuclear zone. A constriction appears in the cytoplasm between the two nuclei. It gradually deepens until the two new cells are separated. The cell surface "bubbles" very actively indeed at this stage and as the activity decreases the new daughter cells spread out over the surface on which they are growing.

The duration of cell division varies with different cells. With avian fibroblasts, it can be quite rapid and can be watched with the eye over a period of about twenty minutes (at $38°C$). The interphase period in rapidly growing cells is of the order of eighteen hours or less (11:15-17).

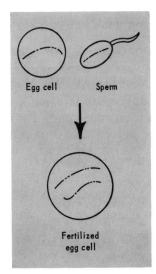

Egg cell   Sperm

Fertilized egg cell

Figure 11
Fertilization of the egg by the sperm restores two chromosomes of each kind to each cell (only one chromosome in each germ cell is shown.)

Methods of Analysis

Dictionaries define *analysis* as a systematic and logical process of separating or breaking up a whole into its parts, so as to determine their nature, proportion, function, or relationship. The word is derived from the Greek, *lyein*, meaning "to loosen," and the Greek prefix, *ana-*, meaning "up." To analyze, then, is to loosen or break a subject up into its logical entities.

The process of analysis is fundamental to all scientific and technical activity and to the reporting and communication of such activity. It is the process of resolving any problem into its component elements. In the science of chemistry, the process plays an important role in the separation of compounds and mixtures into their constituent substances to determine the nature, or the proportion, of the constituents and in the determination and nature of the proportion of one or more constituents of a substance, whether separated out or not. In mathematics, the process is used to aid in the solution of problems by means of equations and to examine the relation of variables, as in differential and integral calculus. In medicine, analysis of symptoms plays an important part in the proper diagnosis of a disease. In logic, analysis is used to trace things to their source and to resolve knowledge into its original principles. It aids clarity of thought by breaking down a complex whole into as many carefully distinguished parts as possible and helps determine how the parts are related within the whole. In technical writing, the process of analysis helps the writer to understand the subject under investigation; it enables him to see its component parts and aids him in identifying the relationship of those parts to each other and to the whole. Analysis helps the writer to distinguish and group together related things; it helps him to select data essential to his subject and problem; it helps him to select out of his mass of data the relevant material and to eliminate the irrelevant. In short, analysis is the process which helps the writer understand his material and organize it into a logical order for efficient communication.

The basic operational element in analysis is division. When a group is divided into its classes, the process is known as *classification*. When a whole is divided into its parts, the process is called *partition*.

Classification and partition are closely related to definition. Definition, as was seen in chapter 9, is actually a form of analysis which resolves a subject into its component parts by (1) identifying the class to which a subject belongs and (2) differentiating or distinguishing the characteristics by which the subject is set apart from other members of its class. To define is partly to classify and partly to partition. Analysis by classification examines one arm of definition — the genus or class. To *classify* is to determine the whole of which the subject is part. Analysis by partition examines the other arm of definition — the differentia or distinguishing elements. To *partition* is to determine the parts of which the subject is a whole. Classification defines a subject by revealing its essence through comparison; partition defines a subject by listing the details or parts of its essence.

Classification

All science starts with classification. Our universe is much too vast and complex to be examined as a whole; so scientists select manageable portions for observation and investigation. Thus, science is frequently defined as the branch of study concerned with the observation *and classification* of facts.

A group of things which have a defined characteristic in common is called a class. Examples are: compounds containing carbon, things made of iron, substances which are transparent, coordinate numbers, scientific theories. It should be noted that membership in a class does *not* imply being *exactly* like all members of the class, but being alike only with respect to the *specific* quality or characteristic on which the classification is based. One object can be a member of a large number of different classes, one for each of the qualities it possesses. For example, a dog is a member of the *Vertebrata*, the *Mammalia*, the *Carnivora;* it is fur-bearing, friendly to children, and vicious to intruders, and if the dog is young enough, it can be taught new tricks. Fogs may fit the classification of a colloid, a liquid, a gas, water, air, and, at times, ice crystals. Classifications can be structured in an infinite number of ways. The basis for the classification is always the practical consideration — the use to which it will be put.

Classification has become a fixture in the operation of some of the sciences, especially botany and zoology:

> Taxonomy, the science of classification, attempts to arrange organisms in relation to one another. The theory of organic evolution has greatly enriched this science, so that today the orderliness exhibited by living forms and recorded by the taxonomists has come to imply hereditary relations. In other words, classification of higher plants and animals is based on kinship and reveals lines of descent. It is thought of as a natural classification depicting the history of living forms through the ages. Although the theory of evolution has been the basic philosophy on which the science of taxonomy now rests, we must not overlook the fact that man's intellectual need for orderliness drives him to find a "natural" order that may later prove to be one of his own inventions. Any method of classification satisfies this human need and is valid in the light of knowledge at a given time: more knowledge often necessitates abandoning one system or orderliness and accepting another.
>
> In the case of higher organisms, knowledge of their embryological development has revealed a kinship that could not have been recognized from adult anatomy. Sometimes in order to place an individual organism in proper relation to other organisms, a knowledge of fossil records of related, pre-existing form, has been useful. The following classification of three well-known animals indicates the method of sub-grouping employed by biologists.
>
> A few comments on the . . . classification will help to show how the taxonomist attempts to distinguish between superficial and fundamental similarities. The fact that both birds and bats possess wings and fly is a similarity of no taxonomic importance. A study of the embryology and adult anatomy of these two animals reveals two similarities and significant differ-

CLASSIFICATION OF THREE WELL-KNOWN ANIMALS

Taxonomic Group	Blue Jay	Little Brown Bat	Man
Phylum	Chordata	Chordata	Chordata
Sub-Phylum	Vertebrata	Vertebrata	Vertebrata
Class	Aves	Mammalia	Mammalia
Order	Passiformes	Chiroptera	Primates
Family	Corvidae	Vespertiliomidae	Hominidae
Genus	Cyanoeitta	Myotis	Homo
Species	cristata	lucifugus	sapiens
Scientific Name	*Cyanoeitta cristata*	*Myotis lucifugus*	*Homo sapiens*

ences. Both birds and bats possess a segmented vertebral column and a dorsal hollow nervous system. These characteristics, together with others, place both these animals in the major phylum, Vertebrata. Feathers place birds in the class, Aves, and the possession of mammary glands and hair places bats within the class Mammalia, which also includes man. Bats possess claws with forearms modified for flight, which classifies them in the order Chiroptera. Man, with nails, is admitted to the order Primates.

A particular bat is identified by two names: one, the genus, to which his structure admits him, and the other, the species, which separates him from all other types of bats. A species common to eastern North America is a little brown bat, genus *Myotis*, and the species *lucifugus*....

Man and bat are both vertebrates. They are also mammals. At this point, however, their lines of descent diverge and we must arrange them in separate orders. If we study birds and bats in detail, we find that although they share a vertebrate ancestry they diverge from a common reptilian stock on the main trunk vertebrate line at this juncture. They must therefore be grouped directly into separate classes. In a word, bats and man are much closer relatives than birds and bats, in spite of the fact that birds and bats appear to have more in common with each other than either group seems to have with man. The similarities seen in birds and bats, however, are superficial compared to the closer relationship between man and bats revealed by their embryological development and their adult anatomy. The adult modification of forearm and skin flap used by the adult bat for flight is a species specialization just as is the erect posture seen in man.

The science of classification makes it possible for scientists working with bats to communicate with one another concerning a particular bat. An unknown bat may be identified by referring to the key of nomenclature, and if a new form is found the taxonomist is free to decide whether the new characters warrant establishing a new species or whether these new identifications merely indicate a new strain of a previously described species has been found. The formulation of rules for international nomenclature in taxonomy was an early step toward meeting the growing need for an easy and accurate exchange of essential information among all countries of the world (8:154-55).

In their book, *An Introduction to Logic and the Scientific Method*, Cohen and Nagel have this to say:

> There is a general feeling shared by many philosophers, that things belong to "natural" classes, that it is by the nature of things that fishes, for instance, belong to the class of vertebrates, just as vertebrates "naturally" belong to the class of animals. Those who hold this view sometimes regard other classifications as "artificial." Thus a division of animals into those that live in the air, on land, and in water would be regarded as artificial. This distinction involves a truth which is confusedly apprehended. Strictly speaking, the last division, or any division of animals according to some actual trait arbitrarily chosen, is perfectly natural. For in every classification we pick out some one trait which all the members of the class in fact possess and therefore we call it natural. All classification, however, may also be said to be artificial, in the sense that we select the traits upon the basis of which the classification is performed. For this reason, controversies as to what is the proper classification of the various sciences is interminable, since the various sciences may be classified in different ways, according to the objectives of such classification.
>
> Various classifications, however, may differ greatly in the logical or scientific utility, in the sense that the various traits selected as a basis of classification differ widely in their fruitfulness as principles of organizing our knowledge. Thus the old classification of living things into animals that live on land, birds that live in the air, and fish that live in water gives us very little basis for systematizing all that we know and can find out about these creatures. The habits and the structure of the porpoise or the whale may have more significant features in common with a hippopotamus or the horse than with a mackerel or a pickerel. The fact that the first two animals named have mammary glands and suckle their young, while all species of fish deposit their eggs to be fertilized, make a difference which is fundamental for the understanding of the whole life cycle. In the same way the fact that some animals have a vertebral column, or, to be more exact, a central nervous cord, is the key which enables us to see the significance of the various structures and enables us to understand the plan of their organization and functioning. Some traits, then, have a higher logical value than others in enabling us to attain systematic knowledge of science (5:233-34).

Principles of Classification

Classification is a useful technique in exposition. It enables the writer to systemize widely diverse facts for presentation in an orderly arrangement so that the reader can follow easily and understand the relationship of these facts. Classification may occupy an entire volume, as in a textbook on botany or on types of motors, or it may form a chapter, a section of a chapter, or a paragraph in a chapter. Whenever the writer must answer the question, "Where does this species or this idea belong?" he classifies. The analytical process of classification helps the writer in organizing his material for logical presentation. It enables the writer to examine his material in proper proportion. It enables him to differentiate the important from the unimportant, to give proper weights and values to items within his material, and to structure the presentation in parallel fashion or in subordinate fashion.

The recognition of similarities and differences among experiences through classification is a basic process in all scientific research. It offers important empirical techniques that make it possible to structure classes in such a way that mere membership in a class reveals probable attributes. In other words, certain characteristics appear to be associated with each other. Thus, a class of objects which have the properties of strength, susceptibility to corrosion, and capacity to be magnetized can also be assumed to be good conductors of electricity and will probably have the other properties of iron and steel. The process of classification, as we shall see later, helps the mind to interpret facts and data through inference.

Here is how the process of scientific classification enables the scientist inductively to identify an unknown — in this case, bacteria.

Let us suppose that we have collected a large number of observations concerning the structural and physiological characters of an unknown organism. The data that we have are listed below:

Shape:	Long rods; occur singly and in chains
Motility:	Actively motile
Gram Stain:	Gram-positive
Spore Formation:	Definite spore, central in position
Oxygen Requirement:	Grows best in presence of atmospheric oxygen; facultative anaerobic growth
Agar Colonies:	Spreading, grayish, irregular borders, adherent
Gelatin Stab:	Surface growth; liquefaction
Litmus Milk:	Alkaline; peptonized
Dextrose Phenol Red Agar Stab:	Acid; no gas
Saccharose Phenol Red Agar Stab:	Acid; no gas

Grows best at 37° C. Grows also, but less well, at room temperature. Organism was obtained by rubbing sterile, moist swab over new petri dish just unpacked from straw wrappings.

Here, then, are our data. We next consult *Bergey's Manual of Determinative Bacteriology.* ... To the experienced bacteriologist this organism at once suggests that it is a member of the family Baccillaceae, because it is a rod-shaped, spore-forming bacterium; and of the genus *Bacillus*, because it grows aerobically. By following out further details, we find that it corresponds to the detailed description of *Bacillus subtilis*. The inexperienced person who does not carry this information in his mind as part of his useful equipment can turn to the key and follow the logical procedure in tracing out the identification of an unknown organism... (8:205-206).

Rules for Classification

1. Know your subject.

2. Define the term (subject) to be classified. The purpose of classification is to bring clarification and order to a number of elements which are diverse and perhaps confused in the reader's mind. Grouping things together may be meaningless to a reader unless he understands what you are talking about in the first place. This requires careful identification and definition of the term. It may also mean sufficiently limiting the term so that it is useful to the purpose of the reader.

3. Select a basis for the classification which is useful and logical.

4. Keep the same basis for all of the items within a grouping. Changing the basis invites confusion. For example, a division of the cat family into lions, tigers, leopards, lynxes, pumas, jaguars, and cats-friendly-to-children does not follow a consistent principle of grouping.

5. Keep groupings, though related, distinct. Classes must not overlap. Classification of a technical report as examination, investigation, recommendation, research, or formal indicates overlapping because a single report could easily fit all the listed categories at the same time. Reduction of phenomena into a logical and orderly system offers frequent difficulties. Species do not always seem mutually exclusive. There are times when the subject seems to be neither fish nor fowl. The duckbill platypus is such an anomaly. This zoological curiosity has a beak, webfeet, and lays eggs. These factors would serve to classify it as a bird. However, it also has the reptilian characteristic of poisonous burrs, which are similar to the fangs of a snake. Because the furry duckbill platypus suckles its young, it has been classified as a mammal. This feature is significant in the evolutionary processes of animal forms, and from that approach its classification as a mammal is justified by zoologists.

6. Be sure to include all members in the classification. In other words, the classification should be complete. If incomplete, it does not serve as the "umbrella" to cover every member which logically belongs under it. For example, a classification of semiconductors would be incomplete if it did not contain thermisters.

7. Have at least two classes on each level of division. Since classification entails the breaking down of a grouping into its member parts on a logical basis, there must be at least two member parts in any such breakdown.

8. Arrange classifications in a logical order, convenient for the reader to follow. Two forms lend themselves to such a presentation — tabular and outline. The format of tables facilitates classification of data. The outline form readily enables the reader to see relationships and helps to reveal how subspecies are related to the parent genus.

Partition

Partition is a form of analysis. It is the division of a whole into its parts. It is exactly the process used by an automobile mechanic in taking a motor apart, by a chemist in breaking down a substance into its components, and by a biophysicist in breaking down the cell into its minute elements.

For example, an automobile may be partitioned into its principal assemblies:

1. Steel frame
2. Running gear — wheels, axles, springs, brakes, steering device
3. Propulsion — engine and clutch, change-speed gears, drive shaft
4. Body — including upholstery, glazing, doors, and interior fittings
5. Accessories and auxiliaries — radiator, fuel tank, defroster, cigarette lighter, lamps, bumpers, etc.

The vertebrate eye may be partitioned into the conjunctiva, aqueous humor, pupil, lens, iris, cornea, vitreous body, layer of rods, optic nerve, retina, pigment layer, choroids, and sclerotic.

A formal report has a cover, a title page, an abstract, a table of contents, an introduction, a body, a terminal section, a bibliography, and, frequently, appendixes.

Partition, although employing a logical division, is not the same as classification. Within the process of partition, a complete entity is broken up into its components. Within classification, a general group is logically divided into the various species which make up the class. For example, mollusks, anthropods, and vertebrates all have eyes or visual organs. The eyes are quite different types of sensory organs, but they do perform the seeing function for the animal of which they are a part. However, when an eye is partitioned, any of its parts — whether the pupil, lens, iris, retina, vitreous body, or optic nerve — is not, of itself, the eye.

Partition may be applied to abstract concepts, as well as to physical entities. In communication theory, for example, the abstract concept of the communication unit is divided into the source, transmitter, receiver, and destination. The practice of psychiatry might be divided into the prevention, diagnosis, treatment, and care of mental illness and mental defects. The study of color divides into shade, hue, and intensity.

The analytical process is basic to all fields of activities. The logician, Thomas Crumley, has observed:

> Partition is employed by the builder in laying out his work; it is indispensable to the playwright in fashioning his plot; it is an aid to the lawyer in drawing up his brief, to the orator in marshaling his argument, to the painter in balancing his composition, and to the musician in apportioning his theme (6:85-86).

Classification is a means of analyzing or explaining a plural subject. It examines relationships by determining similarities within the various species of its subject. Partition, on the other hand, analyzes and explains a single subject by examining the various components which make it up.

The analytical approach used in partitioning is similar to that used in classification.

The processes of classification and partition are obviously related to formal logic. Classification uses reasoning from the particular to the general (induction); partition uses reasoning from the general to the particular (deduction). The identity of a class can be established by examining and grouping particulars together on the basis of a common element. The entity of particulars can be established by showing that the particulars belong to a class whose identity has been established. Often deduction (partition) is possible only after the identity of a class has been established by a previous induction (classification). This interrelationship in practice is exemplified by a physician who diagnoses a disease by classifying the symptoms then detects others through the deduction of the diagnosis.

Rules for Partitioning

1. Be consistent in your approach or viewpoint in dividing your subject. Your viewpoint should be made clear to your reader. A logical approach to the partition might (as in the description of a mechanism) be from the viewpoint of the observer's eye, from the sequence by which the various parts are arranged, or from the way the parts function.

2. Separate and define clearly each part in the division. Parts should be mutually exclusive.

3. List all parts or explain any incompleteness. Within a generalized description — one representative of a class — partition may be limited to major, functional parts. In such a situation the partition must be preceded by a qualifying statement such as "the chief functional parts are. . . ."

4. Check your stated partition for unity. The mechanism, organism, thing, or concept must be properly introduced and defined. The point of view from which the partitioning is approached must be explained. The analysis follows with a logical listing, description, and explanation of parts and their relationship. The concluding element of analysis by partition will show how the various parts relate to each other and to the whole.

Partition as an expository technique may comprise an entire monograph or text; it may be included as a section or a chapter in a book, or a paragraph or two within a larger piece of writing. Maintenance manuals are characterized by their numerous uses of the partition explanatory technique.

Synthesis and Interpretation

Closely related to analysis are the two processes of synthesis and interpretation. Synthesis is the antonym of analysis; it is the putting together of parts or elements so as to form a whole. While opposed to analysis, the process of synthesis is frequently complementary to it. This is borne out graphically in the science of chemistry which utilizes both processes of taking apart and putting together. Analysis and synthesis aid in arriving at interpretation. Analysis breaks data down and arranges them into the logic demanded by the situation. The scientist or the technical writer may rearrange or recombine the various elements of the data in order to arrive at the meaning or explanation of the data through inductive and deductive inferences. Interpretation affords explanations, meaning, and conclusions regarding problems associated with the data. Analysis and synthesis provide a logical process for achieving appropriate conclusions and generalizations.

In his book *How We Think,* John Dewey shows how interpretation evolves through the operation of analytic and synthetic thought:

> Through judging [judgment], confused data are cleared up, and seemingly incoherent and disconnected facts are brought together. This clearing up is *analysis.* The bringing together, or unifying, is *synthesis....*
>
> As analysis is conceived to be a sort of picking to pieces, so synthesis is thought to be a sort of physical piecing together.... In fact, synthesis takes place wherever we grasp the bearing of facts on a conclusion or of a principle on facts. As analysis is *emphasis,* so synthesis is *placing*; the one causes the emphasized fact or property to stand out as significant: the other puts what is selected in its *context,* its connection with what is signified. It unites it with some other meaning to give both increased significance. When quicksilver was linked to iron, tin, etc., as a *metal,* all these objects obtained new intellectual value. Every judgment is analytic in so far as it involves discernment, discrimination, marking off the trivial from the important, the irrelevant from what points to a conclusion; and it is synthetic in so far as it leaves the mind with an inclusive situation within which selected facts are placed.
>
> . . . .
>
> The analysis that results in giving an idea the solidity and definiteness of a concept is simply emphasis upon that which gives a clew for dealing with some uncertainty. If a child identifies a dog seen at a distance by the way in which the animal wags its tail, then that particular trait, which may never have been *consciously* singled out before, becomes distinct — it is analyzed out of its vague submergence in the animal as a whole. The only difference between such a case and the analysis effected by a scientific inquirer in chemistry or botany is that the latter is alert for clews that will serve for the purpose of sure identification in the *wildest possible area of cases*; he wants to find the signs by which he can identify an object as one of a definite kind or class even should it present itself under very unusual circumstances and in an obscure and disguised form. The idea that the selected trait is already plain to the mind and then is merely isolated from other traits equally definite puts the cart before the horse. It is selection as evidence or as a clew that gives a trait distinctness it did not possess before.
>
> Synthesis is the operation that gives extension and generality to an idea, an analysis makes the meaning distinct. Synthesis is correlative to analysis. As soon as any quality is definitely discriminated and given a special mean-

ing of its own, the mind at once looks around for other cases to which that meaning may be applied. As it is applied, cases that were previously separated in meaning become assimilated, identified, in their significance. They now belong to the same *kind* of thing. Even a young child, as soon as he masters the meaning of a word, tries to find occasion to use it; if he gets the idea of a cylinder, he sees cylinders in stove pipes, logs, etc. In principle this is not different from Newton's procedure in the story about the origin in his mind of the concept of gravitation. Having the idea suggested by the falling of an apple, he at once extended it in imagination to the moon as something also tending to fall in relation to the sun, to the movement of the ocean in the tides, etc. In consequence of this application of an idea that was discriminated, made definite in some one case, to other events, a large number of phenomena that previously were believed to be disconnected from one another were integrated into a consistent system. In other words, there was a comprehensive synthesis.

It would be a great mistake, however, as just indicated, to confine the idea of synthesis to important cases like Newton's generalization. On the contrary, when any one carries over any meaning from one object to another object that had previously seemed to be of a different kind, synthesis occurs. It is synthesis when a lad associates the gurgling that takes place when water is poured into what he had thought was an empty bottle with the existence and pressure of air; when he learns to interpret the siphoning of water and the sailing of a boat in connection with the same fact. It is synthesis when things themselves as different as clouds, meadow, brook, and rocks are so brought together as to be composed into a picture. It is synthesis when iron, tin and mercury are conceived to be of the same kind in spite of individual differences (7:126; 129-30; 157-59).

Examples of Classification

FOG (11:682-83)

Condensation and consequent formation of water droplets (or ice crystals) in the air at the earth's surface will produce a fog. Fogs are classified in many ways. One of the simplest is the use of formation cause or processes as a basis for differentiation among the various types.

1. Advection fogs are fogs that owe their existence to the flow of air from one type of surface to another. Surface temperature contrast between two adjacent regions is necessary in causing the formation of advection fogs.

a. The usual type of advection fog is formed when relatively warm and moist air drifts over much colder land or water surfaces. Examples of this type are found over land when moist air drifts over snow-covered areas, or over water when moist warm air drifts over currents of very cold water. The latter happens with the southerly or easterly winds blowing from the gulf stream over the Labrador Current.

b. Coastal and lake advection fog forms when warm and moist air flows offshore onto cold water (summer) or when warm, moist air flows onshore over cold or snow covered land (winter).

c. Sea smoke, arctic fog, or steam fog form in very cold air when it flows over warm water.

2. Radiation fog is the type that develops in nocturnally cold air in contact with a cool surface. Radiation fog forms over land and not over water because water surfaces do not appreciably change their temperature during hours of darkness.

3. Upslope fog is caused by dynamic cooling and air flowing uphill. Upslope fog will form only in air that is convectively stable, never in air that is unstable because instability permits the formation of cumulus clouds and vertical currents.

4. Precipitation fog forms in layers of air which are cooler than the precipitation which is falling through them. The greater the temperature difference between a relatively warm rain (or snow) and the colder air layer, the more rapidly will the fog develop. Fogs associated with fronts are largely precipitation fogs.

Visibility in fogs varies from a few feet up to a mile. Often in fog blankets, all ranges of visibility are present.

COUPLINGS AND UNIVERSAL JOINTS (4:70-71)

couplings

The broad term "coupling" applies to any device that secures two parts together. Line shafts made up of several lengths of shafting may be held together by shaft couplings. There are several types of shaft couplings. We will check on a few of those commonly encountered.

PLAIN SLEEVE COUPLING

The plain coupling consists of a sleeve which receives the ends of the two shafts it is to join, and is secured by set screws or pins so the assembly can turn as one.

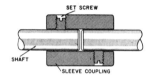

CLAMP TYPE SLEEVE COUPLING

Used to join two closely aligned shafts. It also offers an adjustment in shaft relationship. This coupling consists of a sleeve, the ends of which are slit so clamps can be applied to hold the two shafts firmly together in the sleeve and turn as one.

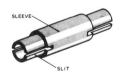
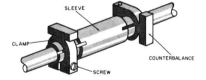

FLANGE COUPLING

The flange coupling finds use on shafts which tend to be pulled apart in operation. A pair of flanges, secured to the ends of the shafts by set screws, are pulled together by bolts and nuts. Splines or keys are resorted to in the case of heavy duty drives.

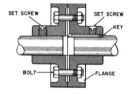

OLDHAM COUPLING

The important feature of the Oldham coupling is that it can be readily connected or disconnected. It is used to join shafts that do not require perfect alignment. Because of its spring loading it also finds use as an expansion joint in long shafts. The Oldham coupling consists of a pair of disks, one flat and the other hollow, pinned to the ends of the shafts. A third disk, with a pair of lugs projecting from each face, fits between the two shaft disks. The lugs fit into slots of the two end disks, and enable one shaft to drive through the disks to the other shaft.

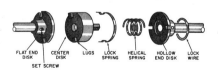

EXPLODED VIEW OF OLDHAM COUPLING

A helical spring, housed within the center and hollow end disks, forces the center disk against the flat disk. With the coupling assembled on the shaft ends, a flat lock spring is slipped into the space about the helical spring. The ends of this flat spring are formed so that when it is pushed into place the ends spring out and lock about the lugs. A lock wire is passed between holes drilled through the projecting lugs to guard the assembly. It is a simple matter to remove the lock wire, withdraw the lock spring, compress the helical spring by pushing and sliding back the center disk, and disconnecting the shafts. Reinstallation requires no measuring or setting as the assembly fits quickly and smoothly into place.

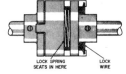

LOCK SPRING
SEATS IN HERE

LOCK WIRE

OLDHAM COUPLING ASSEMBLED

universal joints

A universal joint is a shaft coupling which allows shaft movement in any direction within a limited angle and conveys a positive motion to the driven shaft. Many universal joints will function with shafts up to 45 degrees out of alignment.

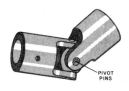

PIVOT PINS

CROSS PIN UNIVERSAL JOINT

The universal joint is widely used although the joint changes velocity of the driven shaft with two high points and two low points each complete revolution. This fluctuating velocity induces similar torque variance, causing vibration and wear throughout the associated mechanism. Because of the two pins, one shaft can drive another even though the angle between the two is as great as 25 degrees.

CONSTANT VELOCITY UNIVERSAL JOINT

A new development in universal joint design provides smooth torque even at unbalanced angles, resulting from ball bearings applying power in a plane that bisects both shaft axes. The torque developed is shown in the graph.

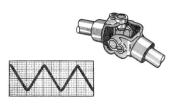

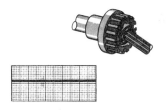

Examples of Partition

Examples cited in the previous pages on the description of mechanisms and organisms also illustrate partition.

The outer zones of the earth are partitioned in the following exposition:

As it presents itself to direct experience, the earth can be physically described as a ball of rock (the lithosphere), partly covered by water (the hydrosphere) and wrapped in an envelope of air (the atmosphere). To these three physical zones it is convenient to add a biological zone (the biosphere).

The *atmosphere* is the layer of gases and vapor which envelops the earth. It is essentially a mixture of nitrogen and oxygen with smaller quantities of water vapor, carbon dioxide and inert gases such as argon. Geologically, it is important as the medium of climate and weather, of wind, cloud, rain and snow.

The *hydrosphere* includes all the natural waters of the outer earth. Oceans, seas, lakes and rivers cover about three-quarters of the surface. But this is not all. Underground, for hundreds and even thousands of feet in some places, the pore spaces and fissures of the rocks are also filled with water. This ground water, as it is called, is tapped in springs and wells, and is sometimes encountered in disastrous quantities in mines. Thus there is a somewhat irregular but nearly continuous mantle of water around the earth, saturating the rocks, and over the enormous depressions of the ocean floors completely submerging them. If it were uniformly distributed over the earth's surface it would form an ocean about nine thousand feet deep.

The *biosphere*, the sphere of life, is probably a less familiar conception. But think of the great forests and prairies with their countless swarms of animals and insects. Think of the tangles of seaweed, of the widespread banks of mollusks, or reefs of coral and shoals of fishes. Add to these the inconceivable numbers of bacteria and other microscopic plants and animals. Myriads of these minute organisms are present in every cubic inch of air and water and soil. Taken altogether, the diverse forms of life constitute an intricate and ever-changing network, clothing the surface with a tapestry that is nearly continuous. Even high snows and desert sands fail to interrupt it completely, and lava fields fresh from the craters of volcanoes are quickly invaded by the pressure of life outside. Such is the sphere of life, both geologically and geographically it is of no less importance than the physical zones.

The *lithosphere* is the outer solid shell or crust of the earth. It is made of rocks in great variety, and on the lands it is commonly covered by a blanket of soil or other loose deposits, such as desert sands. The depth to which the lithosphere extends downward is a matter of definition: it depends on our conception of the crust and what lies beneath. It is usual to regard the crust as a heterogeneous shell, possibly about twenty to thirty miles thick, in which the rocks at any given level are not everywhere the same. Beneath the crust, in what may be called the *substratum*, or *mantle*, the material at any given level appears to be practically uniform, at least in those physical properties that can be tested (9:29-30).

The following excerpt from a maintenance manual is a fine example of the partition expository technique:

Projectile Ring Drive Assembly

The projectile ring drive assembly consists of the components listed below and described in detail in the following text:
Electric motor
Flexible coupling
Brake
Wormwheel and pinion bracket
Controller
Control stations
Electric motor. The projectile ring drive motor is a horizontally positioned, squirrel cage, induction type of commercial manufacture arranged with a flexible coupling at one end of the rotor shaft and a brake at the other end. Motor design and specification data are as follows:

ELECTRIC MOTOR DATA

Type	squirrel cage, induction
Design features	horizontal, direct drive fan cooled, waterproof enclosure
Horsepower	7.50

RPM, full load 1180
Rotation (shaft end) ....clockwise
Voltage 440
Amperes, full load 13.1
Amperes, locked rotor 106
Phases 3
Cycles 60
Temperature rise, C 75
Weight, including brake, lb 380
Manufacturer ....Reliance Electric and Engineering Co.
Manufacturer's Designation ....type AA, frame C-324
Drawing number 563103

Flexible coupling. A coupling of commercial manufacture provides direct drive from the electric motor rotor shaft to the pinion shaft of the worm-wheel and pinion bracket (gear reducer). The coupling is composed of two duplicate steel hubs and two identical steel sleeves which form the cover (Fig. 12). It will function properly with slight misalignment between connected units.

The hubs are broached to fit the six spline-rotor shafts and gear-reducer pinion shafts. The outer periphery of the hubs have spur gear teeth and the sleeves have internal gear-teeth which mesh with those on the hubs. A lubricant fills the space between the hubs and sleeves, and an oil plug is provided in one of the sleeves for filling and draining this lubricant.

Brake. The electric motor brake is a friction disc type brake of commercial manufacture. It is mechanically engaged when the motor circuit is de-energized, and electrically disengaged when the motor circuit is energized. The brake assembly (Fig. 13) is totally enclosed for protection against moisture, dust, and explosions.

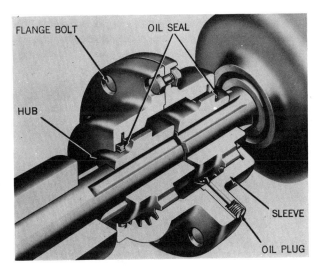

Figure 12
Projectile Ring Drive Motor Coupling Section.

This brake is mounted to the back end of the electric motor. The brake solenoid is connected to two phases of the motor circuit; therefore, when the electric motor circuit is energized, the brake solenoid is also energized.

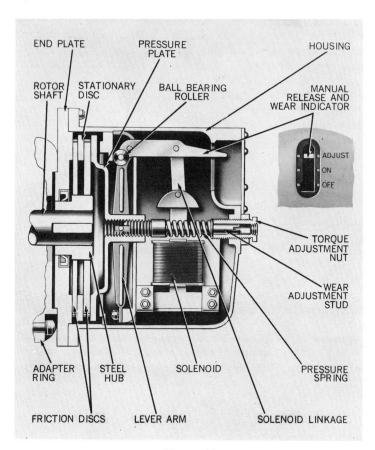

Figure 13
Section. Motor Brake.

This action retracts the spring-loaded pressure plate and releases the motor for operation. However, when the brake circuit is de-energized the brake solenoid is also de-energized. The spring-loaded pressure plate then bears against friction discs and locks the motor, thus preventing motor rotation. This type of brake is capable of applying a stopping torque of 35 pound-feet to the electric motor shaft.

A brake housing is secured to an adapter plate at one end of the motor. Rotating friction discs are keyed to the rotor shaft hub and straddle a stationary friction disc keyed to the inside of the housing. A spring-loaded circular pressure plate, mounted on an axial stud, bears against the end rotating friction disc. The pressure plate is connected to a push-pull solenoid through a series of linkages.

Two adjustments are provided on the brake; one for wear and the other for torque. The wear adjustment is accomplished by a stud which is connected to the pressure plate. Turning the stud counterclockwise brings the pressure plate closer to the rotating friction discs to compensate for wear on the discs. Need of performing this adjustment is indicated on the housing by a manual release and wear indicator lever which is connected to the solenoid. The torque adjustment is performed by a nut threaded over the wear adjustment stud and bearing against the pressure plate spring. By turning the nut, the pressure plate spring preload can be varied. Both adjustments are accessible from the back end of the brake housing, and their use is described in the Adjustments section of this chapter.

Wormwheel and pinion bracket. The electric motor drives the rotating projectile ring through a combination spur gear and worm and wormwheel reduction unit of the design arrangement (Fig. 14). This unit reduces the motor drive from 1180 to approximately 8.5 revolutions per minute; a reduction ratio of 140 to 1.

The worm and wormwheel are case enclosed and oil bath lubricated. Lubricant is supplied to the case through an oil filling pipe (Fig. 14) which extends above the projectile ring roller deck. Upper and lower oil level plugs are provided on the side of the case for checking the amount of oil present. The input pinion and gear, however, are provided with separate lubrication fittings. Drain plugs are located on the underside of the unit.

The worm is a type with a single right-hand thread cut on the worm shaft, and the wormwheel has 500 teeth cut on its edge; the ratio between the two elements is therefore, 50 to 1. The pinion is keyed to the wormwheel shaft by two square keys. A section in the upper casing is cut away to permit the pinion and training rack to mesh. The pinion has 15 teeth and a pitch diameter of 7.429 inches. Other gear data may be found in the training circle section.

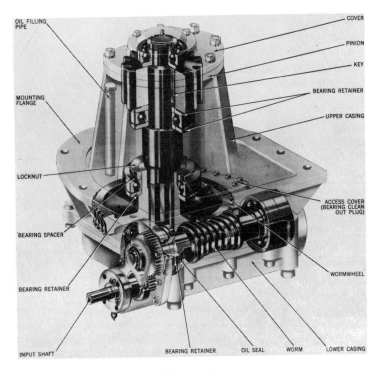

Figure 14
Wormwheel and Pinion Bracket. Phantom.

The reduction unit is split horizontally in two parts along the worm center line. The upper and lower casings are aligned by two taper pins and secured together with fitted bolts and nuts. The mounting flange of the reduction unit is horizontally machined and is cast integral with the upper casing.

Controller. The controller is an assembly of commercial design and manufacture. It provides control and protection for the projectile ring train drive motor described in the preceding paragraphs. The controller is mounted on the outer circular compartment bulkhead at the electric deck. The controller is equipped with an ON-OFF contactor located on the right-hand side of the unit and four push-button switches, two located at the upper projectile handling flat and two at the lower projectile handling flat. Controller design and specification data are as follows:

CONTROLLER DATA

Type .... semi-automatic, magnetic, across-the-line starter, remote push-button control, watertight enclosure
Ampere rating, full load 13.1
Protection:
    Overload, amp 33.3
        Adjustable range, amp ....14.13 to 17.27
    Normal setting, amp 15.7
Short circuit:
    Power circuit, main line, amp 40
    Brake circuit, amp 10
    Control circuit, amp 10
Undervoltage:
    Drop-out voltage, minimum 176
    Sealing voltage, maximum 352
Shock rating ....high impact
Weight, lb:
    Controller 120
    Pushbutton switches, 4 (16 lb each) 64
Manufacturer ....Cutler-Hammer
Drawing numbers ....563101 and 563102

Control stations. The rotating projectile ring drive motor can be regulated from four push-button control stations located at the upper and lower projectile handling flats. Each control station or push-button switch consists of five push buttons arranged in a single water-tight enclosure which is secured to the circular bulkhead on the right and left sides of the projectile hoists on the handling platform levels. The upper control stations are designated STATIONS No. 1 and 2; the lower STATIONS No. 3 and 4.

The pushbutton enclosure consists of a sheet metal cover and base clamped together by eight swinging eye bolts. The cover gasket is seated in a groove which rests over the lip of the base when the two parts are clamped, thus providing a watertight seal. Each switch aperture on the top face of the cover is sealed with a flexible rubber gasket. A cable conduit is threaded into the top and bottom ends of the base. The pushbutton assembly is 18.0 inches long and 6.625 inches wide; fully assembled it weighs 16 pounds. The push button is of a commercial manufacture.

The five push buttons are arranged in the enclosure as indicated below.
SIGNAL RESPONSE
START RIGHT
RESET-EMERG
START LEFT
STOP
The START RIGHT and START LEFT push buttons are 4-pole normally open switches. The SIGNAL RESPONSE push button is a 4-pole switch, with two normally open and two normally closed contacts. The RESET-EMERG push button is a 2-pole, normally open switch. The STOP push button is also a 2-pole normally closed switch.

The projectile ring drive is controlled by two handlers, either at the upper or lower positions. The stations at one level are interlocked for coordinated starting control and audible signal purposes. However, the drive can be stopped from any one of the four control stations. Likewise, emergency run operation can be accomplished by holding in the RESET-EMERG button at any station (1:17-20).

Discussion Problems and Exercises

1. Choose a simple mechanism, device, or organism with which you are familiar. Prepare an outline similar to that recommended in this text for a technical description. Before describing it or outlining it, make one or as many drawings of the subject as necessary to familiarize yourself with it. Then write the description; use headings for each major division and subdivision of your subject. Include whatever illustrations or drawings you need to supplement the description.

2. Write a detailed explanation of an experiment you have performed or of something you have built or made, such as a lamp, rock garden, bookcase, costume, mobile, etc. In writing your exposition, use the first person active voice, and answer the following questions:
   a. What was the experiment you performed or the object you made?
   b. Why did you perform the experiment or make the object?
   c. What tools or instruments did you use in performing the experiment, or in making the object?
   d. What materials did you use and where did you get them?
   e. How did you proceed in the process of performing the experiment or in making the object? What steps did you follow in the process of doing your work?
   f. What was the result of your work?

3. Rewrite the assignment in exercise 2 in the third person.

4. What is analysis? From your personal experience select an illustration of a form of analysis you have undertaken in the laboratory and one from a

simple analysis you have carried out in solving a personal problem. Present orally or in writing, as your instructor desires, the two problems with their respective backgrounds and point out the ways in which they are similar and dissimilar. What have the two problems in common? Synthesize the results.

5. Locate, read, and analyze examples of technical descriptions and processes appearing in literature, as for example, the making of blubber into oil in *Moby Dick,* by Herman Melville; the description of Dover Cliff, Act IV, Scene 6, in Shakespeare's *King Lear;* the description of the Waterloo battlefield in Victor Hugo's *Les Miserables.*

6. Write a description of a simple mechanism or device for: (a) a reader who would be interested only in knowing in a general way how the subject looks and functions. (b) Write the same description for a reader who needs to operate the device. (c) Write the same description for a reader who needs to fabricate the device.

7. Write a set of directions for a young or nontechnical reader on any of the following:
   a. How to fold a sheet of paper into a ship.
   b. How to start a fire without matches or lighter.
   c. How to do a trick of magic.
   d. How to mitre a board.
   e. How to remove and test a TV vacuum tube.
   f. How to milk by hand.
   g. How a bird flies, how a glider flies, or how a rocket flies.
   h. How to prepare a specimen for a microscope.
   i. How offset printing works.
   j. How color photography works.

8. Locate examples of handbooks of instruction or technical manuals. Study the elements of their composition. Identify and discuss which expository techniques explained in chapters 9 and 10 operate within them.

9. Locate and discuss examples of classification and partition in three of your texts covering different disciplines.

10. Show the relationship of definition to classification and partition.

11. If partition is the division of a whole into its parts, is there any validity to the statement that the whole is often equal to more than the sum of its parts?

12. Discuss the role of analysis and synthesis in interpretation according to John Dewey.

13. Partition a book. Classify books into a minimum of 5 possible bases of groupings.

14. How are classification and partition interrelated?

15. (a) How would you test the validity of a classification you made?
    (b) How would you test the validity of a partition you devised?

16. Write a classification of abstract concepts. Then write a partition of one of the members of the group. Follow the suggestions in this chapter.

References

1. "Ammunition Stowage," *6-Inch, 2-Gun Turrets, Main Armament for USS Worcester Class, Ammunition Stowage and Hoist Equipment, OP 760 (Vol. 3)*, Bureau of Ordnance, Department of Navy, Washington, D.C.: 9 July 1956.

2. Bardossi, Fulvio, *Multiple Sclerosis: Grounds for Hope*, Public Affairs Pamphlet No. 335A, New York: Public Affairs Pamphlets, 1971.

3. *Chevrolet Passenger Car Shop Manual*, Detroit, Michigan: General Motors Chevrolet Division, 1954.

4. Chief of Bureau of Naval Weapons, *Weapons Systems Fundamentals: Basic Weapons Systems Components, NAVWEPS OP3000, Volume 1*, Washington, D.C.: U.S. Government Printing Office, 1960.

5. Cohen, Morris R. and Ernest Nagel, *An Introduction to Logic and the Scientific Method*, New York: Harcourt, Brace and Company, 1934.

6. Crumley, Thomas, *Logic: Deductive and Inductive*, New York: The Macmillan Company, 1947.

7. Dewey, John, *How We Think*, New York: D. C. Heath and Company, 1923.

8. Grant, Madeleine Parker, *Microbiology and Human Progress*, New York: Rinehart and Company, 1953.

9. Holmes, Arthur, "Interpretations of Nature," *The Crust of the Earth*, (Edited by Samuel Rappaport and Helen Wright) New York: Mentor Books, 1955.

10. *Making Quality Concrete for Farm Construction*, Chicago: Portland Cement Association, 1960.

11. Paul, John, *Cell and Tissue Culture*, Edinburgh: E. and S. Livingstone, Ltd., 1959.

12. *Van Nostrand's Scientific Encyclopedia*, New York: D. Van Nostrand Co., Inc., 1958.

11

The Technical
Article and Paper

A giddy hostess once asked Professor Albert Einstein to explain his theory of relativity in "a few well-chosen words."

"Let me tell you a story instead," said the scientist. "I was once walking with a blind man. The day was hot. I remarked that I would like a glass of milk."

"What is milk?" asked my blind friend.

"A white liquid," I replied.

"Liquid I know, but what is white?"

"The color of swan's feathers."

"Feathers I know, but what is a swan?"

"A bird with a crooked neck."

"Neck I know, but what is crooked?"

"Thereupon, I lost patience," said Dr. Einstein. I seized his arm and straightened it. "That's straight," I said. And then I bent it at the elbow. "That's crooked."

"Ah!" said the blind man. "Now I know what you mean by milk!"

This little story illustrates the difficulties in communicating about the advances of science and technology.

Writings about science and technology can be classified into three major types:

1. Monographs and technical papers for fellow specialists in professional and trade publications
2. Review and semitechnical articles for other scientists
3. Popular pieces for the general public

Technical Papers

It is traditional for the scientist to write and publish the results of his work in the specialized journals of his field. Scientific periodicals are a forum where the work of the scientist is evaluated. Through publication, the scientist frequently establishes his reputation in the scientific world. It has become a practice for authors to obtain reprints of their papers and circulate them among other workers in their field. This practice has reached the point where it has been humorously said that eminent scientists obtain the bulk of their knowledge of other men's research from reprints that are sent to them.

Nevertheless, many scientists and engineers are loathe to write a paper, report, or article, claiming that they are too busy or, if the paper is to be read to a group, that they dislike the idea of standing up before an audience of learned men to read the results of their work. Scientists and engineers are reluctant to write for the same reason other professionals are — fear of being unable to express their thoughts in speech or on paper. They have a wealth of information and knowledge but not the writing aptitude, or, more accurately, the writing *experience*. Yet by necessity and tradition writing is part and parcel of the scientific pursuit and a requisite, "normal" activity.

Research is not complete until the data have been recorded and disseminated, usually by publication. It is surprising how much work is carried to completion with great expenditure of funds and time and then left to languish in notebooks until it is too late to be of any use. This state of affairs may have been permissible in the days when scientists financed their own research. It is hardly acceptable when the money comes from others. Publication should be prompt and complete.

Research and significant advances may be reported in a number of ways. They may be set forth in report form, not for publication but for use within a given organization or for an organization's clients. Thus, some scientific and engineering research organizations confine themselves to investigating and testing their own experiments. Their only tangible product is the report. In such an instance, the report may be circulated within the organization and/or to outside organizations that have a need for, or are entitled to, the research information developed. At other times, the research may be published as a short communication to an editor, as a full-length paper, as a longer article for one of the journals which specializes in the field of the research, and/or as a book.

Communications to the Editor

The Communication or Letter to the Editor is a form which permits quick publication and dissemination of short summaries of significant research on topics of high current interest. Some important work is never published in any other form; details of problem, technique, data, and skills are exchanged among a small group of interested investigators and are never widely disseminated. The communication should be written to contain only that information which other investigators require to carry on similar work; preparation of the full-length publication should then be pushed as vigorously as if the letter had not appeared.

Conventions of the format of the communication to the editor vary from publication to publication. The *Proceedings of the IEEE* carries the communication in a section called "Proceedings Letters"; *The Journal of Chromatography* in a section entitled "Short Communications"; *The Journal of Medicinal Chemistry* under a section of "Notes"; the *Journal of Experimental Psychology*, as "Short Reports"; *The Journal of Biological Chemistry*, as "Communication"; *The Journal of the Oil Chemists Society*, as "Letters to the Editor"; *Science* titles its section, "Reports"; and the American Physical Society, which publishes nine periodicals, receives communications in such great numbers that it also publishes a separate semimonthly periodical entitled *Physical Review Letters*. Since Communications to the Editor permit quick publication and dissemination of new information and results in the form of short summaries, this format has become popular with scientists who wish to establish priority of discoveries or wish to obtain feedback on their reported activities. A number of societies and commercial publishers have instituted periodicals devoted solely to this form. Among such periodicals are *Analytical Letters*; *Applied Physics Letters*; *Astrophysical Letters*; *Environmental Letters*; *JETP Letters* (a translation from a Soviet publication); *Information Processing Letters*; *Polymer Letters*; and *Lettre Al Nuovo Cimento della Societa Italiano di Fisica*, an Italian journal which also publishes within its pages letters written in English.

In their published form, these short communications to the editor bear little resemblance to a conventional letter. Though signed by the author at the end of the communication, they are actually short reports. The size of the communications varies from about two hundred words to almost two thousand. Their organizational structure comprises an introduction (which explains the problem and its means of investigation); body (which discusses the resultant data); and terminal section (which offers generalizations or conclusions and, if appropriate, recommendations). Some contain sectional headings, abstracts, and documentation; many have illustrations, tables, and charts to clarify significant points.

Here is an example of a typical communication appearing in the "Reports" section of *Science* (5:106-107):

Radiotelemetry of Physiological Responses
in the Laboratory Animal

(Introduction)   The measurement of physiological reactions in laboratory animals has long presented problems. Recording is
usually accomplished under conditions varying from
complete restraint to the partial restraint of extended
wires. The animal may never grow entirely accus-
tomed to the attachments, and this factor may constitute a source of stress. Restraint is reported in the
literature as being extensively used as a stressor, the
work of Selye on this topic being widely known. It is
apparent, therefore, that measurement of physiological
variables under conditions of even minimal restraint
may yield a distorted picture of such responses. In addition, the cues inherently present in attachments of
any kind, as well as the transfer of an animal to the
study environment, are major — if not, at times, fatal —
methodologic obstacles to classical conditioning. The
development of the transistor offers new possibilities for
classical conditioning.

(Method of   With this in mind we set out to develop a system
Experiment)   of radiotelemetry for use as an adjunct to studies in
classical conditioning. This methodological advance will
permit the monitoring and recording of selected physiological reactions in intact, unanesthetized laboratory
animals during their normal daily routines in a simulated normal environment uncontaminated by the
intervention of the experimenter and experimental procedures, except for planned changes in the controlled
environmental chambers. Transistorized, miniaturized,
battery-power packages are being developed in our laboratories and permanently implanted in laboratory
animals. The modulation of the radiofrequency carrier
with biological information permits short-distance
propagation of selected physiological activities, with the
possibility also of providing remote control of selected
stimuli (1). Continuous measurement of physical conditions within the environmental chambers — ambient
temperature, humidity, air ionization, barometric pressure, air velocity, light intensity, chemical composition
of the air within the chambers, and such other physical
parameters as may be shown to be significant — may be
recorded along with the physiological activity specific
to the animals. The behavior of the animals may be
observed by a remote visual system. Classical conditioning experiments with animals may be conducted over
periods of weeks and months, with observation periods

before, during, and after experimental manipulation. We feel this innovation, growing out of the technique of telemetering, to be important and an improvement over techniques more familiar to those engaged in classical conditioning investigation.

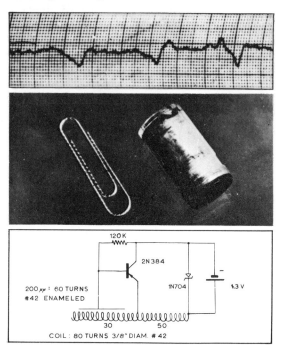

(Discussion)

Figure 1 (top) shows a signal output of respiration from a laboratory rat, obtained by means of the accelerometer principle incorporated into a small capsule (Fig. 1, center) very much like that reported by Mackay (2). A brass pellet mounted on a rubber diaphragm near the oscillator coil modulates the radiofrequency carrier. We have found that we can obtain a radiofrequency signal of 6.8 Mcy/sec and of about 250 μV at the antenna terminals of the receiver, with the circuit shown in Fig. 1 and battery current of 200 μa. This by no means represents a lower limit to power requirements, but an arbitrary stopping point for the moment. The characteristics of the transistor are such that an increase or decrease in power is affected by changing the collector voltage and adjusting the emitter bias to increase or decrease the collector current. There is no reason that the current cannot be of the order of 50 μa, if this is permitted by the radiofrequency noise level within the environmental chamber. A transistorized, miniature, radiofrequency amplifier can be constructed within the antenna probe and supplied with low-voltage d-c power through the coaxial cable. We have found that, by exciting the capsule from an external source of radiofrequency power of about 3.9 Mcy/sec from a 100-watt exciter, about 2 ma of reverse current can be realized in the battery circuit within the capsule, in which we have included the Zener diode rectifier shown in the circuit diagram (Fig. 1). A rechargeable dry cell is being tested for the planned indefinite implantation of the capsule. After the transmitter has been assembled and dipped in toughened paraffin, it is inserted by surgical procedures into the abdominal cavity of the rat, where ballistic movements are sensed by the accelerometer. The signal is picked up by a "ferri-loop-stick" antenna mounted within the environmental chamber and conducted through a coaxial cable to a communications receiver. The incoming frequency-modulated signal is mixed with a beat-frequency oscillator within the receiver, or with another radiofrequency signal from a frequency meter. The audio output of the receiving system is recorded on magnetic tape for subsequent playback and data analysis.

(Results) While the base line of the trace (Fig. 1, top) is an accurate (0.2° per 1000 cy/sec of audio output) and reproducible (±0.5 percent) measure of the core temperature of the animal, a transistor alone will not detect rapid changes of temperature. The response time constant of the radiosonde used at this writing is of the order of 100 seconds. It is therefore necessary to introduce a thermistor transducer into the circuit to effect this measurement.

(Conclusion and Recommendations) The improvement in techniques inherent in micro-miniaturization and telemetry permit the coupling of classical conditioning experiments into an on-line digital computer to form a closed-loop systems approach to experimentation. This, in conjunction with automatic data processing and reduction, would seem to lead to qualitatively different testing of old and new hypotheses with multiple independent and dependent variables, under conditions of experimental control previously impossible.

Samuel J. M. England
Benjamin Pasamanick
Columbus Psychiatric Institute and Hospital, Columbus, Ohio.

References:
1. M. Verzeano, R. C. Webb, Jr., M. Kelly, *Science* 128, 1003 (1958).
2. R. S. Mackay, *IRE Trans. on Med. Electronics* 6, 100 (1959).

Papers for Professional and Technical Journals

The major form of scientific publication is the full-length paper appearing in a specialized journal. Despite special interests, scientists want to know—must know — what their colleagues elsewhere in the country and in the world are doing, and they find out by reading the professional journals and the technical or trade publications of their special fields.[1]

The chief difference between the professional journal and the technical journal is usually the source of publication. A professional journal, as the term implies, is published by a professional organization for its members and the profession at large. The technical journal is a periodical published by a commercial concern dealing with a specialized technical area for workers in the field. Thus, the *Proceedings of the IEEE* is a professional journal for radiofrequency engineers published by the Institute of Radio Engineers. On the other hand, *Electronics* is a publication of the McGraw-Hill Publishing Company, dealing with the same or similar general area of electronics.[2] However, the *Proceedings of the IEEE* is more scholarly in its approach and esoteric in the handling of its subject matter. Another important distinction is the proportion of material devoted to basic as against developmental research. *Electronics* deals on the whole with developmental and applied research and other practical matters which have direct bearing on how the line engineer-reader can do his job better. Some professional organizations publish both a journal on the state of the art of its profession and a technical periodical for the industry of the activity served. Thus, the American Society of Mechanical Engineers publishes its *Transactions* series to cover the former and publishes *Mechanical Engineering* to cover the latter.

The Technical Article for the Professional Journal

The article reports to the journal's readers a subject researched or experienced for the purpose of exchanging information, facts, and ideas current and important to the field. The subject matter may be general in nature

---

[1]In the present context technical and trade publications are considered synonymous since the technical articles in each are by "experts" for "experts." It is true that distinction can be and is made frequently. Some associate the term "horizontal" with the technical journal and "vertical" with the trade journal. The implication is that the latter covers a very narrow industry or technical specialty and the former covers a more general area or a group of related industries.

[2]There are scholarly technical publications following the very formal and erudite approach of the professional journals, serving a profession but not formally published by a professional society. Among a few notable ones are *The Journal of Chromatography, Journal of Colloid and Interface Science,* and *American Journal of Science,* the first scientific journal in the United States. Most have academic associations. There are others which have associations with respected government or private research organizations. *The Journal of Research* of the National Bureau of Standards, *IBM Journal of Research and Development,* and the *Bell Systems Technical Journal* are such examples. Similarly, many companies publish technical house organs equivalent to the technical journal. Among these are the *Westinghouse ENGINEER,* the *Western Electric Engineer, Diagnostica* of the Miles Laboratories, and the *International Nickel Magazine.*

or a new theoretical concept arrived at, a new commercial application, a new discovery—theory, device, technique, process—or a change in existing conditions within the field.

In content, organization, and structure, the technical article resembles the report. The basic difference between the technical report and the article is not in the subject-matter and organizational structure, but rather in motive. The technical article is written to share information; the report is written to obtain action. The article is written for a reader who may or may not choose to read it; the report has a "captive" audience. The writer of the technical article has the burden of capturing the attention of his audience in his beginning and keeping it throughout the article. Elements of style play an important part in keeping the reader's interest.

The most frequently included elements of the technical article in the professional journal are:

1. An abstract
2. An introduction stating the purpose or thesis and background information or background theory
3. A discussion of results
4. An analysis of results and conclusions
5. Generalizations about the significance of the subject

Publications serving different fields and disciplines have their own peculiar characteristics based on individual requirements and situations. In general, however, professional papers within various journals bear many similarities, varying usually in detail only. Table 1, see p. 284, compares a representative group of professional journals serving representative technical fields. Elements of the comparative analysis include content, structure, style, and mechanical details.

Each professional publication has its own requirements both as to content and style. Before beginning to write, you might survey for the publication most suited for your subject and purpose. Check with the editor for his interest in the content of your article and in your approach and for his stylistic and mechanical requirements. Most publications and societies have style manuals or style sheets which are helpful to contributors. If a publication does not have a style sheet, the editor will be happy to indicate by letter his requirements. A study of the periodicals is the best way for a writer to learn first-hand of the content material most suitable, structural aspects, and stylistic characteristics. An editor looks more favorably upon an article which is written specifically to his requirements. The less work he has to do on an article to bring it up to his publication requirements, the more receptive he will be to the contributed piece. A helpful guide is provided on the following pages.

In general, readership, as in any writing, guides the approach in the article for the professional journal. Readers are scientists and technical men in the same area of specialty; they share similar interests, training, knowledge, and experience. Consequently, they will have an advance interest in the subject-matter; few attention-getting devices for catching reader

Title and Society of Professional Journal	Title of Article	Abstract	Intro.	Theory	Method or Procedure and Materials
Journal of Experimental Psychology **Published by Am. Psychological Assn.**	Descriptive, Technical	100-250 words	Informal; explains purpose	If any, briefly mentioned in Intro.	*Always* Formal Section follows Intro.
Journal of Forestry **Published by Soc. of Am. Foresters**	Descriptive, usually non-technical	None; Summary sometimes included at end of article	Brief, informal; explains problem or purpose	None	Freq. under formal section of method or materials
Microchemical Journal **Published by the Metropolitan Microchemical Soc. under the auspices of Academic Press**	Descriptive, Technical	Summary at end of article	Formal; explains background	Usually under section entitled "Experimental"	Formal section follows Theory (Experimental)
Science **Published by Am. Assn. for the Advancement of Science**	Descriptive, freq. short and non-technical	Long subtitle	Brief, informal; explains background quickly	If any, not under formal section	Not under formal Head, but may be extensively treated
The Journal of Biological Chemistry **Published by the Am. Soc. of Biological Chemists**	Long, descr., scholarly limited to two printed lines by instructions to author	Summary 100-350 words	Informal untitled section Includes purpose and review of literature	Within procedures only	In experimental procedures or materials and methods
Mechanical Engineering **Published by The American Society of Mechanical Engineers**	Short, from three to six words usually	Brief, annotated summary below title, prepared by editor	Not formally labeled. Lead paragraph styled to intrigue reader interest	None	Generalized into text
Journal of Applied Physics **Published by the American Institute of Physics**	Long descrip. and scholarly	Short, descriptive	Formal Introductory section	Follows Intro. Sometimes as separate section with word "Theory" in section title, but more often than not emphasis on theory	Usually in Section of Experimental Technique
Proceedings of the IEEE **Published by the Institute of Electrical and Electronics Engineers, Inc.**	Technically descriptive	50-200 words	First section usually so titled; presents problem	Usually in second section often most of article discusses theory	If applicable, under Section titled "Experiment"

1

Representative Professional Journals

Title and Society of Professional Journal	Discussion	Results	Analysis	Conclusion	Layout
Journal of Experimental Psychology **Published by Am. Psychological Assn.**	*Always* Follows Results or Method section	*Usually* follows Method section	In Discussion section	At closing of Discussion section	2-col. spread. Simple book of 6¾" x 10"
Journal of Forestry **Published by Soc. of American Foresters**	Follows Procedure or Materials	Frequently under formal section or under Discussion	Under Discussion	Frequently under Discussion	Three col. book: 8½" x 11"
Microchemical Journal **Published by the Metropolitan Microchemical Soc. under the auspices of Academic Press**	Usually under one heading "Results and Discussion"	Usually under one heading "Results and Discussion"	Under Discussion. Sometimes under own heading	Sometimes under separate section or under Discussion	One column, simple book: 6" x 9"
Science **Published by Am. Assn. for the Advancement of Science**	Not under formal section but under descriptive heads	Not under formal head but Discussion and Analysis may be together under several descriptive heads		Sometimes under formal head at end of article but may be under descriptive head at end of article	Three (3) columns. Illustrations frequently enter design layout of page. Book: 8⅜" x 11⅛"
The Journal of Biological Chemistry **Published by the Am. Soc. of Biological Chemists**	Significance indicated	Section so labeled includes charts	In Discussion section	In Discussion section	Two-column simple spread book: 8½" x 11"
Mechanical Engineering **Published by The American Society of Mechanical Engineers**	See General Comments	See General Comments	See General Comments	See General Comments	Attempts at artistic layout of pages. Book: 8¼" x 11¼"
Journal of Applied Physics **Published by the American Institute of Physics**	Usually Discussion and Results in same section	Usually Discussion and Results in same section	In Discussion section	Sometimes in separate section and sometimes with Discussion	Two columns, not justified on right. Book: 8¼" x 11¼"
Proceedings of the IEEE **Published by the Institute of Electrical and Electronics Engineers, Inc.**	Most of article	Section usually so titled near end of article, if applicable	Frequently follows Result section	At end of article usually so titled	Two column simple spread book: 8⅜" x 11"

Table 1—
Comparative Analysis of

Title and Society of Professional Journal	Foot-notes	Bibliog-raphy	Length	Types of Illustrations	Specialized Style Require-ments	Use of Formulae and/or Special Symbols
Journal of Experimental Psychology **Published by Am. Psychological Assn.**	Yes	References at end of article	Up to 20 printed pages	Line diagrams and charts	Formal, scholarly	Statistical
Journal of Forestry **Published by Soc. of Am. Foresters**	Yes	Under "Literature Cited"	1000-3500 words	Photos, charts, line drawings, maps	Less formal	Some; statistical symbols
Microchemical Journal **Published by the Metropolitan Microchemical Society under the auspices of Academic Press**	Occa-sion-ally	Under "Refer-ences"	1200-6000+ words	Line drawings, diagrams, charts, photographs	Formal, scholarly	Chemical, mathe-matical
Science **Published by Am. Assn. for the Advancement of Science**	Yes, also under notes with refer-ences	Extensive list of references at end	1000-5000 words	Photos, line diagrams, charts	Less formal; aimed at general science edu-cated reader	Mathe-matical, chemical, statistical
The Journal of Biological Chemistry **Published by the Am. Soc. of Biological Chemists**	Yes	References	2000 upward words	Diagrams, charts, mostly; few photos	Instructions to authors available. Formal and scholarly.	Many chemical formulas. Uses standard chemical abbrevia-tions
Mechanical Engineering **Published by The American Society of Mechanical Engineers**	Few	List of references for back-ground materials	200-2000 words	Illustr. with photos and drawings; many articles depend on illustrations to comm. info.	More in-formal than formal. Suc-cinct and breezy. See General Comments	Rare
Journal of Applied Physics **Published by the American Institute of Physics**	Yes	None: Footnotes provide the references	1500 to 6000	Line drawings, diagrams, charts. Photos are rare.	Formal, scholarly. Follows style man-ual of AIP	Profuse math. formulas and symbols
Proceedings of the IEEE **Published by the Institute of Electrical and Electronics Engineers, Inc.**	Yes	References	1500 to 10,000+ words	Diagrams, line draw-ings, charts, photos	Formal, scholarly. Follows IEEE style sheet	Extensive use of formulas and symbols.

Continued
Representative Professional Journals

Title and Society of Professional Journal	Frequency of Issuance	Advertisements	Acknowledgments	Tables	Comments
Journal of Experimental Psychology **Published by Am. Psychological Assn.**	Monthly	Some, back of book, separated from text	In footnote on title page	Occasionally	Reports on original experiments
Journal of Forestry **Published by Soc. of Am. Foresters**	Monthly	Some articles continued on pages with ads	Not usual	Frequent	Article frequently of a generalized nature
Microchemical Journal **Published by the Metropolitan Microchemical Society under the auspices of Academic Press**	Bimonthly	In back of book except for inside cover, separated from text	Occasionally	Frequent	Significant state of the art articles
Science **Published by Am. Assn. for the Advancement of Science**	Weekly	Most text separated from ads. Some continued pieces on page with ads	Rare	Many	Work of specialists aimed at general science educated reader. Many generalized articles.
The Journal of Biological Chemistry **Published by the Am. Soc. of Biological Chemists**	Semi-monthly	Limited to separate sections	At end of article	Many	Basic and significant research reported
Mechanical Engineering **Published by The American Society of Mechanical Engineers**	Monthly	Ads separated from articles		Occasional, graphically attractive	Styled to compete with commercial trade periodicals. Articles of a generalized nature
Journal of Applied Physics **Published by the American Institute of Physics**	Monthly	None	At end of article	Many	Basic, theoretical articles which have practical applicability
Proceedings of the IEEE **Published by the Institute of Electrical and Electronics Engineers, Inc.**	Monthly	Ads separated from article pages	Frequently at end of article	Frequent use	Basic, theoretical significant applied analytical articles; contains frequent tutorial, invited papers.

interest are necessary. Technical terms, mathematical data, and theoretical concepts can be used without explanation or clarification. The reader will be especially interested in illustrations, diagrams, and curves which pinpoint the data. The article for the professional journal usually includes a bibliography and footnotes.

It would be well for you to review chapters 4, 5, 6, 8, 9, 10, and 12 if you are writing a technical paper.

Technical Periodicals

Technical or trade periodicals are geared for readers who are not esoteric specialists but workers in the field. They may be professionals, administrators, or high level technicians. Each field of activity has its own publications servicing that field. The subject-matter will be of an applied nature. Since these are commercial publications, their financial support comes from advertisers within that area of activity. Consequently, articles are interspersed within many pages of advertising. The advertising itself is of great interest to the readers. Many new products, processes, and services are of professional interest to them. New knowledge of the field is frequently disseminated in the form of announcements of new products, instruments, and services. Readers of the technical publications read the advertising with as much interest as they read articles.

Whereas articles in the professional magazines are more fundamental in their subject-matter and more scholarly in their approach, articles in the technical publications are streamlined because their readers can give them only a minimum amount of time. Language is less formal than in the professional magazine. Long subtitles frequently are used to present a synopsis of the article. Paragraphs are shorter, as are sentences and words. Many details have been deleted to conserve the reader's time. There is little use of footnotes. Illustrations in the form of diagrams, charts, and photographs play an important part in telling the story. There is greater use of headings. Editors are quite concerned with attractive page layout. Where a number of the professional magazines use uncoated stock and worry not at all about the attractive appearance of a page, the technical publication must compete for the reader's time. The whole approach in a technical journal is to make the article easy to read and easy to understand. The writing, the layout, and the illustrations are determined by this end.

The reason for the tendency to streamline the article in technical journals has been expressed by a former managing editor of *Product Engineering:*

> Some significant changes in industry during recent years are having a wide influence on methods of communication, particularly in technical areas. These changes are increasing the complexity of subject matter while at the same time calling for faster means of communication. . . . Publications must reflect this increased pace by more frequent publication; where technical magazines were once monthly or quarterly, now an increasing

number come out weekly. Thus, in the face of increased complexity of content, we now have much less time in which to prepare material for publication.

The impact of these changes on publishing has brought a new element in our competition for readership. Busy readers have much more reading they ought to do, and much less time available for reading. They have to become selective in their reading. The objective of each publication, therefore, is to be the one selected. . . .

Publications are responding to these changing times in a variety of ways. There is quite generally a sharpening definition of readers' needs, functions and interests. Publications are more carefully examining their readership — concentrating on more important areas and dropping overlapping groups of lesser importance. Editorial content is being more precisely fashioned for the specific interest of the central group. . . .

By careful planning and selection of article topics, editors are reaching for a better balance of subject matter in each issue. The objective is to represent all major interests of the reader group in every issue of the publication so each reader can expect to find at least one contribution important to him in every copy. . . .

Articles must be edited for faster communication. Readers must be able to get the message with less reading time — with no sacrifice in content. There is good reason for the current interest in reading improvement. . . . *PRODUCT ENGINEERING* today makes more use of graphical material, which often means converting text into pictorial or graphic form. We find the mathematics often can be converted into English, delivering the message faster to a wider group.

But perhaps our most effective step toward faster communication has been through what we call "tight editing." This is a sweaty operation involving repeated revisions and rewriting by the editor in the interest of eliminating every unnecessary paragraph, sentence or word — which sometimes does apparent damage to a conventional technical style. It also requires a heading, subhead and lead illustration that clearly show the reader whether the article hits a topic of current interest to him. Where once our intention was to try to interest the reader in every article, we now try to tell him at a glance whether he should pass this one up and turn to the next one. . . .

Reviews and Semitechnical Articles

Scientists write articles for the purpose of exchanging the knowledge of their fields with scientists in related and other fields. Such articles commonly present a review of a field or the broad implications of a new theory, process, instrumentation, or trend. A review summarizes and interprets the state of knowledge about a field or the current status of research on a significant problem. In the semitechnical article, particular processes, techniques, and descriptions of apparatus and tests are generalized; conclusions and significance rather than methods are emphasized. Both the review and the semitechnical article use less specialized language, fewer mathematical data, and fewer symbols since their readers are not always familiar with the technical diction of the specialized field. Typical media[3] are the *Scientific American, American Scientist, Psychology Today,* and the Science Section of the *Saturday Review/World.* Readership is intelligent, alert, and has a variety of interests.

The structure and organization of the semitechnical article are more varied than in the papers appearing in professional journals. Stylistically, the semitechnical article may be written from an objective (third person) or from a very personal point of view. The objective style appears more frequently. The basic content is that of the technical report—purpose, results, analysis, conclusion—though not in that order. Generally, the writer will not begin with a statement of purpose since he must first catch his reader's attention. Beginning with significance may accomplish this. Review articles often resort to this mode of beginning. Writers of semitechnical articles find effective means for starting their piece through pertinent rhetorical devices—an interesting example, an extraordinary statistic, a startling statement, a striking figure of speech, an analogy, a comparison, a definition, or an appeal to personal interest.

For example, in his article "The Buoyancy of Marine Animals," appearing in the *Scientific American,* Erik Denton opens with an interesting comparison:

> Many fishes, like many people, must keep swimming just to keep from sinking. Bone and muscle are denser than sea water and so tend to drag the animal to the bottom. It is obvious that the ability to float can be advantageous for animals that live in the sea. Accordingly, fishes are equipped with swim bladders which give them neutral buoyancy — that is, an average density equal to that of sea water — and save them the labor of continuous swimming. Two other animals — the cuttlefish and cranschid squid — have developed quite different kinds of flotation organs. They anticipated man in using the working principles of the submarine and the bathyscaph, the one endowing the cuttlefish with active control of its buoyant, the other permitting the squid to live at greater depths (3).

The use of an extraordinary statistic to capture reader attention is illustrated by G. J. Levenbach's opening to his article, "Systems Reliability and Engineering Statistical Aspects," appearing in the *American Scientist:*

> Around the year 1800, the French mathematician Pierre Simon, Marquis de Laplace, once calculated the probability that the sun would rise the next day, utilizing the observation that it had done so each morning for the last 1,826,230 days since creation... (8).

An example of a startling statement is the one used by W. P. Monroe in his article "Cathodes Can Protect Your Projects," appearing in the *Consulting Engineer:*

---

[3]*Science,* the organ of the American Association for the Advancement of Science and *Physics Today,* a publication of the American Institute of Physics, carry many articles falling into the review and semitechnical article category.

There must be something in the make-up of some mechanical and civil engineers — those who lay out piping systems and other underground structures — that prevents their having faith in any electrical equipment that does not move like a motor (9).

John S. Edwards in his article, "Insect Assassins," which appeared in the *Scientific American*, makes good use of a literary quotation—an exhortatory doggerel couplet—to begin his article on insects that prey and live on other insects:

> *Catch 'em Alive! Gentlemen, I've*
> *Here such a dose as no fly can survive.*

With this bit of doggerel a Victorian poet described the perfect way to administer a perfect insecticide. The procedure is clearly impractical in human efforts to control insects, but it is used effectively by the insects themselves. The insect order Hemiptera includes a family of some of the most efficient predators in the animal kingdom. They are the Reduviidae, or assassin bugs, which literally catch their victims alive and administer a dose they cannot survive (4).

Wendell R. Garner in his article "Good Patterns Have Few Alternatives," published in the *American Scientist*, begins with a definition:

> Over half a century has passed since a school of psychology was founded and named for the German word *Gestalt*, a word which has been carried over into English because there is no translation of it which seems quite to carry all the connotations of the German word itself. In general terms, a gestalt is a form, a figure, a configuration, or a pattern. But gestalt is also the quality that forms, figures, and patterns have. Thus gestalt is both form and form-ness, pattern and pattern-ness. The school of psychology was given this name because of its emphasis on studying the form and pattern characteristics of stimuli, rather than on studying the elements which make a stimulus but which do not in and of themselves constitute the pattern (6).

A most interesting narrative beginning was employed by Douglas W. Schwartz in his article "Prehistoric Man in Mammoth Cave," which appeared in the *Scientific American*:

> One day some 2,220 years ago, a man wearing sandals and a loincloth woven of fibers from the papaw tree descended the steep slope leading to the entrance of the great cavern in Kentucky that is now known as Mammoth Cave. He was going to mine gypsum from the walls of the inner passages of the cave. As he entered the cave, he lit a torch of cane reeds from a slow burning wick of twisted bark. Then he walked into the darkness... (10).

These examples of beginnings illustrate only a few of the rhetorical devices which can be used by the writer of the semitechnical article. The purpose of the beginning is to entice the reader into the depth of the article. This emphasis on literary devices, however, should not mislead you into thinking that the primary purpose of the semitechnical article is to entertain. Its purpose is to *inform*—bring to the reader definite information of practical or intellectual value. Because the reader of the semitechnical piece is not a specialist in the field written about, the specialized character of that discipline or the details and terminology may either bore him or frighten him off before he reaches the matter of interest. Hence, the interest-creating devices are called for.

The Review Article[4]

The review article in the semitechnical magazine may also begin with a rhetorical device for the purpose of gaining the reader's immediate attention. A beginning more in keeping with the purpose of the review—to inform scientists of what is being accomplished outside their own narrow fields of specialization—will be one of emphasis on the significant developments within a field; presentation of historical background information; or a definition if the subject area is esoteric or not well known.

A typical opening paragraph of a review article is the following one on plasma physics which utilizes all three indicated literary devices:

> In the past ten years a new field of physics has become important and popular. This is plasma physics, which deals with the motion of ionized media such as gases in the presence of external forces, generally magnetic fields. Perhaps the surprising thing is that plasma physics was not initiated earlier since it involves only ideas and laws known to Maxwell a hundred years ago. This late developement is perhaps due to the fact that only recently have applications of plasma physics been recognized (7).

The structure of this article is typical of reviews. The quoted first paragraph is part of this introduction, which provides additional details developing the significance, definition, and historical background of the subject reviewed. The introduction terminates with an elucidation of various applications of the process and the potentials plasma physics could hold for us after we have learned its secrets.

This leads into the body which begins to examine the theoretical basis of a plasma; reviewed are the two most important properties of a plasma— its tendency in the presence of a magnetic field to be constrained to move along lines of force, and its abundance of charge which quickly shields out any electric field in the absence of a magnetic field or shields out any components of electric field along the lines of a magnetic field.

Next to be considered in the theoretical examination of this phenomenon is its behavior capabilities in the form of three types of energy densities.

Following the theoretical background considerations, the article examines some of the significant roles plasma physics may play in astronomy. Reviewed last and analytically are practical design applications of plasma in terrestrial phenomena.

---

[4]Review articles frequently appear in professional and technical journals for the same purpose that they appear in the semitechnical publications.

The terminal section is a short one, pointing out the problems posed by our present limited state of the art in applying this phenomenon to the design of a practical fusion reactor. A note on the challenging potentials of plasma physics brings the review to a close.

The preceding synopsis illustrates the summary and interpretive function of the review article.

Structuring the Semitechnical Article

Today, with many publications and other demands competing for a reader's time, many articles receive only a glance or a quick scanning. Mere recital of facts, interesting in themselves, will not hold the reader because his reading set is "Why should I read this article? What can I do with this knowledge?" Purpose or value to the reader is primary in the semitechnical article.[5] In planning the article, you, the writer, must ask yourself, Why should the reader of the publication be interested in this information? How can I present it so that its significance reaches him quickly and easily? Many editors refer to articles as "stories." Each article tells a "story." Like a story, the article's narrative or sequence needs plotting—an organized structure. The structure or sequence of information is influenced by the purpose and the thesis. The basic objective in the semitechnical piece is communication of useful information. Interest is maintained through emphasis on the significance of the information, through relation of the subject-matter to the reader's experience, and through illustrations. Photography, flow charts, diagrams, and color also are used to advantage.

In some semitechnical articles interest may be created through reader identification with the writer who may unfold a series of obstacles or developments and in the writing tell how they were overcome or achieved. The semitechnical article, unlike the professional or technical article, will make use of the narrative structure which will lead to a climax or to a high point—the reaching of the objective in which the reader has an interest. The objective is the satisfaction the reader achieves through his desire to learn about the subject.

The Popular Science Article

In a market survey, *Author and Journalist* found that the best opportunities for writers were primarily in nonfiction; it foresaw no end in a decline of markets for fiction. Many prominent magazines exclude fiction entirely from their pages. The magazines publishing both fiction and nonfiction were devoting more pages to the latter. A significant point mark-

---

[5]This factor is even more important in the technical article in the trade or professional journal because the reader there also asks: "How can I adapt this information to help solve my problem? Can I go a step further and extrapolate from the author's results?"

ing the trend was the fact that confession magazines were beginning to carry articles.

The reason for the situation, *Author and Journal* said, was that the changing world we live in is so absorbing that to most readers articles explaining the changes carry much more interest than fiction. The Apollo flights have made the layman aware that he is on the threshold of a great age of scientific and technological progress. Science fiction can no longer compete with the events of our space age. The layman wants to be ready for Buck Rogers' age, which he feels is but a door away. He reads avariciously and wants to understand.

There is a great gulf separating the scientist from the layman. Less than one hundred years ago there was no such gulf. A contradiction of our times is that we can be cultivated in the humanities and still be ignoramuses. The cultured, educated man of one hundred years ago was up not only on the Spenserian stanza but also on the theory of evolution, the laws of thermodynamics, and the principles of telegraphy. Knowledge since has grown to the extent that no one man can encompass its boundaries.[6]

The basic content ingredients of a popular science article are those of the technical report. The writer treats the material more generally. He soft-pedals or eliminates all reference to complex data or theory. He simplifies methods and gives only general or metaphoric descriptions of equipment; for example, he might describe an oscilloscope as a TV-type gizmo no bigger than a bread box.

The organization varies in the popular article even more than in the semitechnical piece, depending on the readership and the technical knowledge level of the audience of the particular magazine or newspaper. A major problem in writing about science is that the simplest technology is incomprehensible to the literate and educated layman not trained in the symbols, language, methods, and customary thought processes of the specialist in the field.

The technical writer writing for the general reader should remember Pope's admonition, "The proper study of mankind is man." The more the subject-matter deviates from the reader's interests and experiences, the greater the skill the writer must exercise to gain and hold his reader's attention.

Popular articles on science and technology differ in the degrees to which they must be popularized because magazines and newspapers vary in the average educational level of their readers. Newspapers with mass circulations require articles written in the simplest terms. Yet, some newspapers, for example the *New York Times* and the *Christian Science Monitor*, have a readership level above that of the *Cosmopolitan* or *Reader's Digest*.

The following publications are ranked in order of increasing difficulty:

---

[6]This is true not only in the sciences but in the humanities. Witness the number of Ph.D. dissertations on such esoterics as "Non-conforming Tercet Rhymes in the Terza Rima."

Parade
Popular Science Monthly
Cosmopolitan
Reader's Digest
Playboy
Psychology Today
Atlantic Monthly
Harper's Magazine
Science
Other professional journals such as *Journal of Experimental Psychology*, *Proceedings of the IEEE*, *Journal of Research of National Bureau of Standards*, etc.

Establishing contact with the reader at the very start is all-important. If the reader's interest is lost after a few sentences, it is never regained. The reader has gone to another article, another magazine, or another diversion.

The types of beginnings noted in our previous examination of the semi-technical article are even more appropriate for the popular piece. These approaches are based on human psychology. They include appealing to the reader's own self-interest and curiosity, his interest in other people, and his interest in the concrete or tangible matters in his everyday experience. Some examples of types of openings based on these major appeals not previously mentioned are:

Appeal to a basic interest
Direct question
Anecdote
Historical background
Direct statement of thesis
Narration
Reference to a specific occasion

Steps in Writing the Popular Science Article

The first step is selection of the subject. Choose one with which you are intimately familiar or have researched enough to be technically competent to know what you want to say about it and to know who would want to read about the subject.[7] Then survey which magazine readers primarily compose the audience you want to reach. If more than one magazine has

---

[7]Writing about science and technology for the general reader is one of the most difficult challenges for the free-lance writer. There are free lancers whose specialty is writing the popular science piece. Though some may command from $500 to $5,000 per article, researching the piece costs money out of their own pocket and can involve considerable personal travel and time. Most magazine editors have their favorite science writers (who have proved themselves) whom they commission to write on subjects of current interest. Some professional technical writers, engineers, scientists, and teachers do occasional free-lance writing, utilizing their knowledge of specialized technical subjects. They are usually most successful, however, in contributing to trade publications.

the audience you want, aim for the one with the highest circulation (you are making a market survey).

Obtain a number of issues of the periodical over a period of about two years. Study style, structure, and layout of their feature articles. Locate issues with articles parallel in scope to the one you have in mind. Analyze and reduce several to outline form to give you a better idea of structure and presentation. In studying your target periodical, analyze the literary style and physical format of the articles. Are they written in the third person or first person? How are difficult, technical words handled? What devices are used to explain theory or abstract concepts? To what extent are illustrations used? How are they used as expository devices? How long is the average paragraph, sentence, word? What types of openings are used in the article? What rhetorical devices? How are heads used? Do heads aid in explanation? Do heads spell out the organization of articles?

Study the ads carefully. Advertisers are the major support of the general magazine. Illustrations and text of ads should be revealing about the periodical's audience and its level of education, intelligence, and technical understanding.

Following such a study, you are ready to rough out an outline for your article. List topics, areas, and ideas you will cover. Set down the approach or point of view from which you will present the information.

Then, if necessary, obtain permission to write about the subject from the person, company, or organization involved.

Next, contact the periodical selected to publish the article, inquiring about its interest in the subject-matter. A brief outline accompanies or is included in your letter of inquiry. Your letter should explain why the information of your proposed article would be of interest and value to the magazine's readers. Indicate, also, the extent and type of illustrations you are planning to have.

The Query Letter

Here is what an editor of a general magazine has recommended as the ingredients of a good query letter:

> First of all, it should reflect some knowledge on the part of the writer of the general editorial content of the [magazine]. It should also indicate whether the topic being suggested has been covered elsewhere previously, and, if so, when and where. The author's sources of information for this story material should also be given.
>
> An outline of the story line of the proposed article should be included (this need not be more than 150-200 words). Perhaps most important, the specific angle, or peg for the piece should be described. The editor should be told why the writer feels the particular topic would be pertinent and interesting for the current issue of the magazine. Some indication should also be given as to how many words the writer feels he would need to develop the subject properly (our single-page stories run between 850-1000 words, and our double-page stories between 1500-2000). The writer trying

us for the first time might also list some of his previous magazine and/or newspaper credits.

We are impressed by succinctness in queries. I might also say that we prefer to have writers confine themselves to not more than two of three suggestions at one time. More than that is a little too much to digest at one sitting (11).

An experienced editor and publisher considers the following example as a particularly fine query letter:

Dear Sir:

The second largest telescope in the world is now nearing completion at Lick Observatory, Mt. Hamilton, California. Under the direction of Mr. Donald Shane, this 120-inch mirror has been specifically designed for study of outer stellar galaxies and mapping of the entire universe.

In the 20 year since the building of the 200-inch telescope, Palomar, many advances in electronics and optics made the most advanced and — according to the engineers — the most accurate telescope in the world.

I am co-ordinating public relations for Judson Pacific-Murphy Corporation, builder of the telescope, and the University of California Public Information Office to present, at last, the story of this fantastic astronomical instrument to the public.

The story, I believe, will interest your readers; it deals with the building of the telescope. Here a firm known for building bridges — with tolerances in inches — successfully completes this 145-ton instrument with tolerances of one-two thousandth of an inch or one-twentieth the thinness of the paper this is written on and using only bay area firms as subcontractors rather than European or Eastern specialists as has always been the case in the past.

Complete records, progress photos, engineering data, and the co-operation of the astronomer at Lick Observatory are available to me for a feature story geared to the format you require on this number one astronomical event in this International Geophysical Year (11).

Writing the Article

After you have received word from the editor of his interest in the proposal, outline the article more fully (preferably in sentence form) before you begin the actual writing.

Keep the Reader in Mind. Start the rough draft. The opening is important. It must capture the reader's attention, interest him to read further, and lead him to the point or thesis of the article.

To maintain the reader's interest, the writing must remain within the range of the reader's intelligence and experience, structured within an organizational plan which will maintain the interest. Useful expository techniques to bring very complex or technical aspects of a subject within the realm of the reader's experience are such rhetorical devices as anecdotes, analogies, and literary allusions. The anecdote is one of the most frequently used devices in popular writing because it illustrates points more effectively than will long explanations. References to prominent people in history, to episodes, or to situations in literature can sometimes aid in establishing a common bond between the writer and reader and make the point the writer wants to make more graphically. Sigmund Freud (a superior writer of science) incisively used literary allusions to Greek myths and the characters in Shakespeare's plays.

A widely used and effective device for bringing unfamiliar ideas within the reader's experience is the figurative analogy. The analogy is a comparison between two unlike things which nevertheless have certain essentials in common. Among commonplace analogies are the comparison of the earth to a ball, the heart to a pump, the camera to the eye, or the computer to a super-brain. There is danger in the use of analogies in that readers often understand only the analogical counterpart but not the matter compared.

Handling Scientific Terms. Science is constantly at work exploring new fields or modifying old techniques; it keeps reaching out for new words and phrases to describe its innovations. These descriptive terms are usually based on or coined from Latin and Greek, though occasionally borrowed from other languages—German in the case of nuclear and rocket research. Many new terms are created by the addition of prefixes or suffixes to word roots. *Transistor* is an example (transfer and resistor). These new, strange, and sometimes unpronounceable words, together with complicated sentence structure, create difficulty for the reader—lay or otherwise.

When it is necessary to use new terms, immediate definition or explanation should follow without interrupting the article. This can best be accomplished by defining or explaining the term in an appositional phrase set off by commas. Definitions should start with what the reader knows and then add no more than he is able to comprehend. Never define a hard word by a harder word. For example do not define "calorie" to the general reader as do some dictionaries by explaining it as the quantity of heat necessary to effect a rise in temperature of 1° C. of a cube of water. The general reader would understand calories better if you say that "100 calories of energy can be derived from three cubes of sugar or from a small pat of butter" or that "man uses up 100 calories an hour to keep his body running and 160 calories when he is working hard."

Robert F. Acker and S. E. Hartsel, in their article "Fleming's Lysozyme" in the *Scientific American*, very effectively explain difficult words and terms by offering the more common meaning first and then following it by the more difficult term, as exemplified in the following paragraph:

> Bacterial anatomists are indebted to the late Sir Alexander Fleming for a sensitive chemical tool with which they have been studying bacteria, dissolving away the cell wall, and exposing the cell *body* or *cytoplasm* within. In 1922 at St. Mary's Hospital in London, six years before his epochal discovery of penicillin, Fleming found "a substance present in the

tissues and secretions of the body, which is capable of rapidly dissolving certain bacteria." Because of its resemblance to enzymes and its capacity to *dissolve,* or *lyse,* the cells, he called it "lysozyme" (1).

The writers, instead of using first the difficult terms "cytoplasm" and "lyse," used first the more commonly known terms "cell body" and "dissolve." This very effective device gives the reader confidence when he goes from the common term to the more uncommon term. He has no difficulty understanding what the writers have written. Psychologically this is a very effective device in that it builds confidence in the reader, making him feel that he understands the more difficult terms because he has first been exposed to their meaning.

Some Final Words on Structure and Style. Use personal pronouns; they create empathy between writer and reader and add vitality and interest to the article. Use clearly worded sentences, expressing one idea at a time. Cement the sentences of your paragraph with a topic sentence. Organize paragraphs into divisions clearly identified by heads that point to the direction in which your article is going.

A plan of organization is imperative to keep the reader's interest. The plan should be simple enough for the reader to grasp and follow. Wherever drama, conflict, and suspense are inherent in the subject, the narrative elements delineating these will hold the reader. However, not all matters in science and technology are suitable to the narrative form. Cause and effect, analysis of contributing factors, evidence and conclusions, predictions, and other methods may be more appropriate.

The ending of the popular science article is as important as the beginning. It must clinch the reader's understanding of the thesis. Sometimes a summary is appropriate, sometimes a typical application of the technical points developed in the body, and sometimes a final astonishing fact. The ending should demonstrate the significance of the article so that the reader will leave it feeling it has been worth his while to read.

A STUDENT'S COMPARATIVE ANALYSIS OF FOUR DIFFERENT FORMS OF TECHNICAL WRITING

The approach to any piece of writing depends upon its reading audience. Different audiences require a different style, organization, and language.

In comparing a report, a professional paper, a semitechnical article, and a popular article, all on the same subject, many differences as well as some similarities become apparent.

The subject of monomolecular films for preventing water evaporation from reservoirs is relatively new. Means for making these films practical has initiated much research.

An article appearing in the *Denver Post* on Sunday, November 9, 1958, entitled "Chemical Film May Save Western Slope Water" by Roscoe Fleming was used as an example of a popular article.

"Reducing Evaporation from Small Reservoirs" by Nedavia Bethlahmy was selected as an example of a semitechnical article. This article appeared in *Northwest Science,* Vol. 33, No. 3, 1959.

A reprint from the *Australian Journal of Applied Science*, Vol. 10, No. 1, pp. 65-84, 1959, entitled "The Influence of Monolayers on Evaporation from Water Storages" by W.W. Mansfield was used as an example of a professional paper.

Geological Survey Professional Paper 269, entitled "Water-Loss Investigations: Lake Hefner Studies, Technical Report," published in 1954, was used as an example of a technical report.

It may be generalized from the titles alone, that as the material becomes more technical, the title tends to become more specific concerning the material contained and therefore often less appealing to the lay reader. As the writing becomes more technical, the terminology used in the title also tends to become more technical.

The differences in length of the various types of writings are very noticeable. The popular article was a little shorter than the semitechnical article, which was four pages long, about 1600 words. The professional paper was nineteen pages long, while the report was 158 pages long. It is interesting to note that the popular article was printed in a Sunday edition of the newspaper, when more space is available.

The organization differed greatly among the various pieces of writing. The popular article first created interest and aroused the curiosity of the reader. Then the concept of the "chemical films" was explained, some of the experiments being carried on were cited, and the potential importance of these films was explained. Then after tracing the history of the interest in the films, the article explained how the chemical films function. Simple terminology as well as analogy was used. The last three paragraphs mentioned some of the organizations which were working on the experimental projects. No references were cited. There did not appear to be any superimposed pattern of organization for the article. Rather, its text was organized for maintaining reader interest.

The semitechnical article also began with a section designed to create interest. Some of the historical aspects of the research on chemical, evaporation-reducing films were mentioned, and some of the details concerning what the films are and how they work were explained. An experiment carried out in Oregon was described. A section discussing the practical applications of the chemical films concluded the article. A bibliography of eight items followed the body of the article. In general, the semitechnical article was designed to be of informative value to an educated reader who lacked technical knowledge on this particular subject. It provided some background on the subject, cited some experiments being carried out, and discussed the usefulness and practicality of the hexadecanol.

The professional paper presented a sharp contrast to the popular and semitechnical articles. The paper was divided into two sections; the first was concerned with evaporation and seepage from water storages, and the second with the action of wind, waves, and dust upon monolayers. Each of the two sections was similarly organized and written. A short summary of the entire article preceded the introduction, which went immediately into the subject without any devices for creating reader interest. The various aspects of the problems involved were each systematically discussed; experiments were cited and results given. A discussion section followed; then came a section of acknowledgments, and finally, a list of references. The organizational pattern of the paper was definitely evident, designed to communicate specific information to those who might be involved in the same problem, or to those who are technically competent to receive this specialized information.

The technical report began with preliminary sections: a foreword, a preface, a table of contents, an abstract; then there followed a section stating the conclusions and recommendations. The formal introduction included an historical review; the statement of the problem; a listing of personnel and their qualifications; this was followed by a section giving the reason for the choice of Lake Hefner for the experiment.

The body of the report contained a number of individual sections, each of which dealt with a certain aspect or component of the study. Each of these sections contained a long description of the instrumentation and methods used, followed by a discussion of the results obtained, conclusions, and then a list of references. The report ended with a summary and recommendations section, appendix, and an index. The entire report was carefully, systematically organized. Each of the individual sections in the middle was written by a different person, but all of the sections were organized and presented in the same organizational structure. The report's data were systematically and concisely presented, aimed toward a technically trained reader.

The writing style of the different types varied greatly. In the popular article, the author was concerned in making his information as interesting as possible. He used devices to pique curiosity and he succeeded with the devices to maintain reader interest. Though the writer succeeded in sticking to facts for the most part, he tended at times to generalize and make sweeping statements, designed to create interest, but the statements were not always entirely accurate. The author thereby reduced highly technical information to the level of understanding of the general reader. The article was created by involving the reader in the matters explained.

The semitechnical article made an attempt to gain reader interest in the opening, but did not follow through with interest creating devices in the rest of the paper, whereas the popular article made successful use of interest creating devices throughout its entirety. Many technical concepts were explained in simple terms. There were many statistics offered, with references. Facts were related in simple language. A short summary of the present state of knowledge concerning evaporation-retarding films was directly and effectively presented.

In the professional paper, there was no attempt to gain or hold reader interest. The introduction presented the background of the problem; the rest of the paper was devoted strictly to analyzing the problem and sharing its present state of knowledge with interested, technically trained persons. The analysis progressed in a systematic, scientific manner. Highly technical concepts were used without explanation.

In the report, the background material necessary for the full understanding of the investigation was provided. There was no attempt to gain or hold the interest of the reader. The report related directly the investigations and experiments being carried out, explained why and how this was being done, and what the results were. Many technical terms and concepts were used without definition or explanation. The methods and instruments used were described in detail. Both positive and negative results were reported. A difference in style of writing was discernible in each of the authors who contributed the various sections of the report.

As the technical level of publications increased, the language used became more technical:

In the popular article, only three technical terms were used, and two were defined. The term "molecule" was not defined. The term "hexadecanol" was defined, but its abbreviated, popular form, "hexy" was thereafter used.

In the semitechnical report, five technical terms were used; only two of these were defined. The language was relatively simple; technical terminology was used only when necessary.

There were thirty-seven technical terms in the professional paper. None of these were defined or explained; exceptions were factors in equations which were defined by the use of another equation.

The report had innumerable technical terms included within it, none of which were defined or explained. As in the professional paper, factors in equations were defined by the use of another equation.

In the popular article, analogy was used three times to explain concepts and terms; comparison was used once, example was used once, and a simple explanation was used once.

Simple explanation was the only method used to explain technical concepts or terms in the semitechnical article.

In both the professional paper and the report, no definitions or explanations were given. There were no analogies, anecdotes, or other special rhetorical devices used.

No special technical background in monomolecular films is needed to understand the popular article. Technical background is not necessary, although it would be of help to understand the semitechnical article. A background in advanced mathematics and physics is especially necessary for the professional paper and the technical report.

As the level of writing became more technical, the more precise the information became; graphs, tables, and equations communicated many important data in the writings.

The popular article had no graphs, table, or equations. There was one photograph of a simple, but interesting process. This picture was fully explained in the text and in the caption.

The semitechnical article employed one minor table, used for comparison. There were no graphs, equations, or pictures.

There were fifty-one equations, eight graphs, and one table appearing in the professional paper. Also used were many statistics which, along with all the equations, rendered this paper extremely technical.

The technical report contained twenty pictures, forty-four tables, three maps, eighty-nine graphs, and one hundred eight equations. Through the use of the graphs, tables, and equations, very precise information was easily recorded and compared.

As the writing became more technical, more references were cited. There were no direct references cited in the popular article. The semitechnical article cited eight references. The professional paper cited twenty, and the technical report cited a total of one hundred thirty-four. It is interesting to note that many of the references cited in the technical report were written in foreign languages, primarily German, indicating a deeper study of the literature in the technical report than in any of the other writings.

In summary, it might be generalized that the similarities in the four types of writing here compared were few. Other than a common subject, each was consciously directed to a specific audience. Each had an informative purpose. Elements to evoke emotion or controversy were deliberately avoided.

On the basis of the present comparison, one could easily plot a linear curve to show that the greater the intended reading audience, the less technical the information becomes, the less formal the organizational structure, the less objective the style of writing, the less precise the language use, the less specific the data and details, and the more the use of rhetorical elements and literary devices, and the more emphasis on personal elements to involve the reader.

Conversely, as the intended reading audience becomes more specialized, the more technical and specialized is the information communicated, the more formal the organizational structure, the more objective the style, the more specific the details and data, the more precise the use of language, entailing more and more graphic presentational devices (use of diagrams, drawings, charts, tables, etc.) and the greater the use of specialized terminology, symbols, and processes.

Roger M. Hoffer

Discussion Problems and Exercises

1. Locate a report, a professional paper, a semitechnical article, and a popular article — all dealing with the same subject. Discuss the organization, style of writing, language, opening, and viewpoint of the four types of writing. Comment on the writer's success in transmitting his information to his intended readers of each of these types.

2. Choose a technical journal and a popular magazine, each carrying advertisements. Compare copy, illustrations, and layouts of the ads in each periodical. Are there more differences than similarities? How much specialized background does the reader need to have to follow the text or copy of the ads in the technical journal? What inferences can be drawn about writing articles for a particular publication from studying their ads?

3. Choose a technical concept which you feel sufficiently informed about. Structure an analogy which will make this concept more intelligible to the uninformed reader.

4. Establishing contact with the reader at the very beginning of an article is particularly important. How do you do this? Does capturing your reader's interest at the beginning guarantee success of your article? How do you continue to keep the reader's interest throughout the article?

5. Choose a technical subject with which you are familiar (for example, how an automobile carburetor works, how a plant grows from seed to flower, how a rock is formed, how immunization works, how a TV tube works) and write two explanations: one to a friend with a scientific background and one to a ten-year-old child.

6. Go to the *Reader's Guide and Index to Periodical Literature* and other literature guides; look up a subject of your own special technological field of interest. Find an article in one of the professional journals and one in a

popular magazine such as *McCall's Magazine* or a Sunday supplement such as *Parade*. Study and compare the two articles. Note particularly:
Opening paragraphs
How many technical terms used and how many defined by each?
What literary devices were used to help explanation of concepts — analogies, contrast, repetition, anecdotes, humor, etc?
What devices were used — formulas, tabulations, illustrations?
Is there a marked difference in level of specialized knowledge necessary to understand these special devices? In the number of illustrations?

7. Refer again to the subject you are looking for in the *Reader's Guide and Index to Periodical Literature* and the other guides. Can you tell from the titles of the articles as to the technological level of the periodicals in which the articles appear?

8. Read a number of articles listed in the guides about the particular subject of your interest. Study them as to organization, style, language, illustrations, viewpoint, literary devices. Compare them according to reading levels and note their special ingredients.

9. Why do you suppose professional and technical periodicals carry abstracts or summaries of their articles?

10. How and where does one get ideas for technical articles? How would a free-lance writer get his material for an article on a highly technical matter?

References

1. Acker, Robert F. and S. E. Hartsel, "Fleming's Lysozyme," *Scientific American,* June 1960.

2. Carson, Robert W., "Hidden Treasure—The Technical Article," *1959 Proceedings, Institute in Technical and Industrial Communications,* Herman M. Weisman and Roy C. Nelson, Editors, Fort Collins: Colorado State University, 1960.

3. Denton, Erik, "The Buoyancy of Marine Animals," *Scientific American,* July 1960.

4. Edwards, John S., "Insect Assassins," *Scientific American,* June 1960.

5. England, Samuel J. M. and Benjamin Passaminick, "Radiotelemetry of Physiological Responses in Laboratory Animals," *Science,* January 13, 1961.

6. Garner, Wendell R., "Good Patterns Have Few Alternatives," *American Scientist,* January-February 1970.

7. Kulsrud, Russell M., "Plasma Physics," *American Scientist,* December 1960.

8. Levenbach, G. J., "Systems Reliability and Engineering Statistical Aspects," *American Scientist,* September 1965.

9. Monroe, W. P. "Cathodes Can Protect Your Projects," *Consulting Engineer,* February 1958.

10. Schwartz, Douglas W., "Prehistoric Man in Mammoth Cave," *Scientific American,* July 1960.

11. *Writer's Digest,* March 1958.

PART THREE

12

Technical Style

Technical English is not a substandard form of expression. Technical writing meets the conventional standards of grammar, punctuation, and syntax. Technical style may have its peculiarities — in vocabulary and at times in mechanics — but it is, nevertheless, capable of effective expression and graceful use of language. The purpose of this chapter is to examine the characteristics of desirable technical style.

Jonathan Swift said that the "proper words in the proper places" make for style. Seneca and Lord Chesterfield said that style is the "dress of thoughts." Expressed straightforwardly, style is the way a person puts words together into sentences, arranges sentences into paragraphs, and groups paragraphs to make a piece of writing express his thoughts clearly. Technical style, then, is the way you write when you deal with a technical or scientific subject.

Technical style can be described by the richness or poverty of vocabulary, by the syllabic lengths of words, by the relative frequency of sentences of various lengths and types, and by grammatical structure. The style varies with the writer and subject-matter. Nevertheless, it has certain basic characteristics.

Qualities of Technical Style

By tradition technical style is plain, impersonal, and factual. Its fundamentals and ideals were enunciated very early in the modern scientific movement by Thomas Sprat in his *History of the Royal Society*, 1667. The members of the Society tried

> . . . to reject all of the amplifications, digressions, and swellings of style: to return back to the primitive purity, and shortness, when men delivered so many *things*, almost in an equal number of *words*. They have exacted from all their members a close, naked, natural way of speaking; positive expressions; clear sentences; a native easiness: bringing all things as near the mathematical plainness as they can: and preferring the language of Artizans, Countrymen, and Merchants, before that, of Wits or Scholars (11:13).

The tradition of plainness and directness has come down to the present time, but the scientist's technical vocabulary has extended far beyond that of "Artizans, Countrymen, and Merchants" and has made plainness a difficult achievement. Scientists have discovered so much and named qualities and things in such numbers that the average person has come to identify "big, strange, and unpronounceable words" as the most conspicuous trait of scientific writing.

Technical style is characterized by a calm, restrained tone, an absence of any attempt to arouse emotion, the use of specialized terminology, the use of abbreviations and symbols, and the integrated uses of illustrations, tables, charts, and diagrams to help exposition. Technical writing is characterized by exactness rather than grace or variety of expression. Its main purpose is to be informative and functional rather than entertaining. Thus, the most important qualities of technical style are clarity, precision, conciseness, and objectivity.

Clarity, Precision

Clarity and precision are frequently interdependent. *Clarity* is achieved when the writer has communicated meaningfully to the reader. *Precision* occurs when the writer attains exact correspondence between the matter to be communicated and its written expression.

Faults in clarity and precision result when:

1. The writer is not familiar enough with the subject-matter to write about it.
2. The writer, though generally familiar with his subject, cannot distinguish the important from the unimportant. (The essence of the problem has escaped him.)
3. The writer has, perhaps, a thorough mastery of his subject-matter but is deficient in communication techniques.

4. The writer is unfamiliar with his reader and has not directed his communication to the desired level of audience understanding.

Conciseness and Directness

Concise writing saves the reader time and energy because meaning is expressed in the fewest possible words and readability is thereby enhanced. Directness also increases readability; it eliminates circumlocutions (round-about expressions involving unnecessary words) and awkward inversions. Take this example:

> An important factor to be cognizant of in relation to proper procedures along the lines necessary is to consider first and foremost in connection with the nature of the experiment that effectuation is dependent on a fully and complete darkened interior enclosure.

The following sentence says the same thing more concisely, directly, and clearly.

> The experiment requires a completely darkened room.

Circumlocutions and awkward inversions so plague technical and scientific writing that many company manuals on report writing invent humorous and exaggerated examples to make the point:

> A 72-gram brown Rhode Island Red country-fresh candled egg was secured and washed free of feathers, etc. Held between thumb and index finger, about three feet more or less from an electric fan (General Electric No. MC-2404, Serial No. JC23023, non-oscillating, rotating on "high" speed at approximately 1052.23 ±0.02 rpm), the egg was suspended on a string (pendulum) so it arrived at the fan with essentially zero velocity normal to the fan rotation plan. The product adhered strongly to the walls and ceiling and was difficult to recover; however, using puttyknives a total of 13 grams was obtained and put in a skillet with 11.2 grams of hickory smoked Armour's Old Style bacon and heated over a low bunsen flame for 7 min. 32 sec. What there was of it was excellent scrambled eggs (2:8).

A more direct and concise rewrite was:

> Very good scrambling was produced by throwing an egg into an electric fan. The product was difficult to recover from the walls and ceiling, but the small amount that was recovered made an excellent omelet. A shrouded fan was designed to improve the yield, in preparation for additional experiments.

Objectivity — The Scientific Attitude

Scientific style is characterized by objectivity. Personal feelings are excluded; attention is concentrated on facts.

The use of the passive voice and the third person point of view is a long established tradition in technical writing. The feeling is that the exclusion of personal pronouns produces a style consistent with objectivity and the use of the passive voice places emphasis on the subject matter. As a result, technical writing is much more impersonal and much drier than it has to be. True, directions for assembling a window fan are quite different from explanations of basic research. It is also true that much "literary" writing is as guilty of mechanical, lifeless expression as is much technical writing. Still, a major current complaint is that basic research is reported in such a cut-and-dried manner as to drive all interest and excitement from it. The personalness, directness, and thoroughly human qualities that emerge clearly from the explanations of Harvey, Newton, Huxley, and others make for fascinating and comprehensible reading; whereas much current scientific research is reported in such an impersonal manner as to be as dull and lifeless as a book of recipes.

Opinion differs as to whether impersonality in scientific and technical writing requires the third person and passive voice. Most texts today place emphasis on reader requirements as determining the use of these devices. The informal situation involving writing for the general or less knowledgeable reader requires the frequent use of devices to inveigle interest. The first person point of view and the active voice aid in the creation of interest. Informative and functional technical writing can be objective and still include personal pronouns. Many successful technical writers can establish effective rapport with readers by playing an active role in their narration as if they were the principal actors involved. Readers frequently will allow themselves to be involved in exposition if it is made interesting for them by the actual participation of a human being in elaborating data which otherwise could appear as foreign and dry to them as a lobster's tail. This is exactly what Thomas H. Huxley did, and it was Huxley who did more to popularize science than any other person at a period when scientific activity was beginning to expand and was in need of public understanding and backing.

Two excerpts from Huxley's writing follow. Both exemplify excellent scientific style and effective communication. Each is aimed at a different audience. The first example is aimed at an intelligent but uneducated and unsophisticated audience.

THE LOBSTER'S TAIL

I have before me a lobster. When I examine it, what appears to be the most striking character it presents? Why, I observe that this part which we call the tail of the lobster, is made up of six distinct hard rings and a seventh terminal piece. If I separate one of the middle rings, say the third, I find it carries upon its under surface a pair of limbs or appendages, each of which consists of a stalk and two terminal pieces. So that I can represent a transverse section of the ring and its appendages upon the diagram board in this way (See Fig. 36).

If I now take the fourth ring, I find it has the same structure, and so have the fifth and the second; so that, in each of these divisions of the tail, I find parts which correspond with one another, a ring and two appendages and in each appendage a stalk and two end pieces. These corresponding parts are called, in the technical language of anatomy "homologous parts." The ring of the third division is the "homologue" of the ring of the fifth, the appendage of the former is the homologue of the appendage of the latter. And, as each division exhibits corresponding parts in corresponding places, we say that all the divisions are constructed upon the same plan. But let us consider the sixth division. It is similar to, and yet different from, the others. The ring is essentially the same as in the other divisions; but the appendages look at first as if they were very different; and yet when we regard them closely, what do we find? A stalk and two terminal divisions, exactly as in the others, but the stalk is very short and very thick, the two terminal divisions are very broad and flat, and one of them is divided into two pieces.

I may say, therefore, that the sixth segment is like the others in plan, but that it is modified in its details (4:21).

The other example from Huxley's technical writing is an excerpt from a textbook intended for advanced students of zoology. Note that the first person is entirely missing and that most of the sentences are structured in the passive voice. Recall the use of nontechnical terms in the first passage and compare the use of technical terminology in the one below, in which the details are more complete, the approach is more formal, and the sentences are much longer.

THE ABDOMEN OF THE ENGLISH CRAYFISH

The body of the crayfish is obviously separable into three regions — the *cephalon* or head, the *thorax*, and the *abdomen*. The last is at once distinguished by the size and the mobility of its segments. And each of its seven movable segments, except the telson, represents a sort of morphological unit, the repetition of which makes up the whole fabric of the body.

The fifth segment can be studied apart. It constitutes what is called a *metamere*; in which are distinguishable a central part termed the *somite*, and two *appendages*.

In the exoskeleton of the somites of the abdomen several regions have already been distinguished; and although they constitute one continuous whole, it will be convenient to speak of the *sternum* (Fig. 36, st. XIX, [These are callouts in the drawing of a transverse section through the fifth abdominal somite.] the *tergum* (t. XIX) and the *pleura* (pl. XIX), as if they were separate parts, and to distinguish that portion of the sternal region, which lies between the articulation of the appendage and the pleuron, on each side, as the *epimeron* (ep. XIX). Adopting this nomenclature, it may be said of the fifth somite of the abdomen, that it consists of a segment of the exoskeleton, divisible into tergum, pleura, epimeron, and sternum, with which two appendages are articulated; that it contains a double ganglion (gn. 12), a section of the flexor (f.m.) and extensor (e.m.) muscles, and of the alimentary (h.g.) and vascular (s.a.a., i.a.a.) systems (5:141).

Using Active and Passive Voices in Technical Writing

In grammar, *voice* is the term used for one form of a verb to show connection between the subject and the verb. The voice of a verb tells the reader whether the subject performs an action (active voice) or receives an action (passive voice). In active voice construction the subject is the doer of the action or is the condition varied by the verb:

> The canal starts at Whalen Dam.
> Man bites dog.
> Vertebrates learn mazes readily.
> The variable resistor controlled the circuit.

When the subject of a verb receives the action, the verb is in the passive voice:

> The canal at Whalen was built many years ago.
> The dog was bitten by a man.
> Mazes are learned readily by vertebrates.
> The traveling wave tube is to be used in the circuit.

The passive voice form is used in all tenses. It usually consists of a form of the verb *to be*, but the past tense of the verbs *to get* and *to become* is sometimes also used to form a passive:

> The component became overheated.
> The traveling wave tube is to be used in the circuit.
> During mitosis, the heterochromatic segments get stained more strongly than the euchromatic regions.

Much technical writing is concerned with the description of work so objective that the reader does not care who did it. He is interested solely in the work itself and is not at all interested in the agency or agent involved. The conventional, impersonal passive construction is suitable for this kind of subject. Compare the first two sentences below with the two that follow:

> The other end of the tie-beam was connected to an anchor pile by a bolt of 2½-in. diameter, inserted through an opening in the pile.
>
> As the experiment progressed, additional water was added to equal the amount lost from evaporation.
>
> I connected the other end of the tie-beam to an anchor pile, using a bolt of 2½-in. diameter which I inserted through an opening in the pile.
>
> As I went on with the experiment, I decided to add water equal to the amount I determined was lost through evaporation.

A comparison shows the third person passive construction to be more objective and efficient than the first person active example.

But compare:

> It is desired to ascertain how the success was achieved in increasing the yield of Russian wild rye.

with:

> We want to know how you increased the yield of Russian wild rye.

and:

> The agglutination is caused by substances analogous to antibodies which are present in the serum.

with:

> Substances analogous to antibodies present in the serum caused the agglutination.

or:

> It is believed that the city should increase its reserve water supply.

with:

> We believe the city should increase its reserve water supply.

Active verbs are more lively than passive verbs and call for simpler sentence structure. Usually, therefore, they are more efficient and are to be preferred in usage. There are instances, however, when the passive voice is to be preferred:

1. When the doer of an action is not known to the writer or when the writer does not want to reveal his identity:
   a. In 1917, when this site was selected, the water table was low and pump irrigation was necessary in many areas of the valley.
   b. The decision was made against the employee.
2. When the writer desires to place the emphasis on the action or on the doer at the end:
   a. The mineral is mined in Wyoming.
   b. The research was carried out by the director.

In normal usage, however, passive constructions are considered weak because they have actionless verbs, invert the natural word order, and require additional phrasing.

Some publications and organizations have a specific policy forbidding the use of "I" or "we." In writing for such a publication or organization, live with their rules until you have achieved the position and skill of a Thomas Huxley. Then you can change their rules.

Elements of Style

Thus far I have been discussing qualities of style. Qualities are overall impressions or characteristics. Qualities result from the writer's characteristic or individual use of the elements of style: grammatical construction, diction, phrasing, sentence length, and figures of speech. Sentence structure and diction (choice of words) are two of the major concerns of the technical writer.

Sentence Structure

The technical writer should understand the capacity of the sentence for expressing simple and complex relationships. This understanding is valuable to the technical writer in formulating observations, generalizations, and conclusions.

For stating an uncomplicated, unqualified observation a *simple sentence* is used:

> Lavoisier, the great French chemist, named the new gas oxygen.
> Honesty is the best policy.
> Fill the test tube with water.
> The man kicked the dog.
> The light is absorbed in the retina by a pigment called visual purple or rhodopsin.

The *compound sentence* expresses coordinate ideas in balance or in contrast:

> And there was evening, and there was morning, one day. (Genesis 1:5.— RSV)
>
> The flow is dark brown and blocky on the surface, but it continues to steam from hot viscous lava beneath.
>
> Science is nothing but trained and organized common sense, differing from the latter only as a veteran may differ from a raw recruit; and its methods differ from those of common sense only as far as the guardsmen's cut and thrust differ from the way in which a savage wields his club. (Thomas H. Huxley)

In the *complex sentence*, dependent clauses are used to express ideas subordinate to the thought of the main clause. The word *complex* is derived from Latin. It does not mean difficult, but literally "woven into." The dependent clause (an incomplete thought having a subject and predicate which modify or support in some way a word or sentence element of the whole sentence) is woven into the design:

> A substitute for wood, which would really take its place, has not been developed.
>
> We wondered when you would come.
>
> Any sentence can be analyzed if you use sufficient patience and don't go into unnecessary detail.
>
> When all else fails, read the instructions.

A *complex-compound* sentence contains two or more independent clauses (expressing coordinate ideas in balance or contrast) and at least one dependent clause expressing thought(s) subordinate to the main clause it modifies.

> If the revised process is to succeed, the first stage may be completed under the careful supervision of the shop personnel, but the second stage must be directed by high-level engineers.

> Uniformity in report writing undoubtedly has some advantages, but when the uniformity results in dull and difficult reading, I seriously question its value.

Professional writers today use about 50 percent complex sentences, 35 percent simple sentences, and only about 15 percent compound sentences, compound-complex sentences, and fragmentary sentences. Out of twenty sentences, three are compound; seventeen, or 85 percent, are either simple or complex. The most frequently written sentence in current English is a complex sentence from twelve to thirty-six words long.

Which type of sentence to use depends mainly on the thought to be communicated. Other factors bearing on or pertinent to the communication are: the purpose of the communication and the reader of the message. Efficient technical writing depends on clarity, precision, conciseness, and objectivity in the composition. For the sake of example and analysis, let us examine the sentence below. At first appearance it seems to express an uncomplicated thought concisely and clearly.

> A heavy object will sink readily.

On closer examination the thought of the sentence is not as clear as it might be. The meaning of the modifiers *heavy* and *readily* is vague. Heavy and readily are subjective judgments. The sentence, though a short one, does not express a clear, precise, objective thought. Shortness is not the equivalent of conciseness. Conciseness requires succinctness. Let us try rewriting the sentence to communicate the thought more efficiently.

> An object of great density will sink very quickly.

The new attempt, also a simple sentence, is an improvement. Though the word *density* carries a certain factual significance in the rephrased thought, the adjective *great* reflects subjectivity, as does the intensive *very*. *Quickly*, while having an aspect of subjectivity, is a more appropriate word than *readily* because it suggests an inherent capability for speed of action; whereas, *readily* suggests speed in compliance or response. Let us try again. This time let us aim for adequate expression no matter how many words or clauses are to be used in the sentence.

> A heavy object is defined as one whose density is such that it will not displace its own weight when placed in water, and, therefore, it will sink quickly.

This 29-word complex-compound sentence has achieved objectivity. Its heavy-handed structure is frequently seen in current technical writing. It can be improved:

An object which does not displace its weight in water will sink quickly.

Both the critical qualities of the weight factor which causes the object to sink and the manner of sinking are clearly, objectively, precisely, and concisely expressed in the complex sentence. The complex sentence is effective in technical writing because the dependent clause has the capacity of being a modifier of exacting precision and clarity.

Readability

Research has been going on since the early 1920s to find a formula for making writing more readable. One of the more prominent researchers is Rudolf Flesch, who has been a consultant for the Associated Press, a number of newspapers, and many corporations in an endeavor to make their printed communications, annual reports, and other writing more readable. According to Flesch, writing will have "reading ease" if sentences are short—an average of not more than 19 words each; if sentences have short words—not more than 150 syllables per 100 words; and if the writing has liberal use of words and sentences possessing human interest—that is, liberal use of personal pronouns (3).

Any parent who has glanced at the primer his child brings home knows that a page made up of short, simple sentences—even with personal pronouns—is dreadfully monotonous. Variety in sentence length—change of pace—helps sustain reader interest. How short is a "short" sentence? What does "readable" mean? Shortness and readability are, of course, relative terms. The intelligent person's concept of sentence length changes with age, education, and reading experience. One study has shown that sentences written by school children increase in length from an average of 11.1 words in the fourth grade, to 17.3 in the first grade of high school, and 21.5 in the upper college years (9:180).

A sentence that looks long to the fourth grader may look short to the college junior. The man whose only readings are Li'l Abner, tabloid headlines, and TV captions will find a sentence long that looks short to the student of literature.

A story for children has sentences averaging less than fifteen words; whereas, writing for educated adult readers may average between twenty and thirty words. A higher or lower average indicates that a writer should examine his sentences critically, but *it does not necessarily mean* that anything is wrong. Variety, type, construction, and length are on the whole more important than the average number of words.

A long sentence is suited to grouping a number of related details clearly and economically. Thus, the writer of a newspaper story answers in his first sentence the six essential questions: Who? What? Where? When? Why? How? Causes and reasons, lists, results, characteristics, and minor details may all be expressed tersely and clearly in long sentences. One of the most effective sentences in the English language is eighty-two words long:

> It is rather for us to be here dedicated to the great task remaining before us—that from these honored dead we take increased devotion to that cause for which they gave the last full measure of devotion—that we here highly resolve that these dead shall not have died in vain—that this nation, under God, shall have a new birth of freedom—and that government of the people, by the people, for the people, shall not perish from the earth.

Readability formulas may be used sometimes as diagnostic instruments. They are not a quick and easy answer to making your writings more readable. The chief ingredient for readability is within the definition of communication: the establishment of a common interest between writer and reader. Readability formulas will not guarantee understandability. Some of Gertrude Stein's writing of the ilk of "a rose is a rose is a rose . . ." or that of the enigmatic Franz Kafka would score higher than Lincoln's Gettysburg Address. Readability formulas measure neither content and format nor matters of organization since these cannot be measured quantitatively; nor will readability formulas improve the writer's style in a written communication. But formulas do point out certain qualities in written communication pertaining to understandability. Therefore, they may be helpful if discriminatingly used.

Summary of Factors Improving Readability

A number of factors will aid readability of written communications:

1. A thorough knowledge of the matter to be communicated
2. A knowledge of the reader
3. Organization of the text to lead reader to the thesis sentence
4. Use of format devices like display heads, enumerations, uncrowded spacing and margins
5. Use of illustrations, tables, and other graphic devices to supplement the text
6. Use of topic sentences in paragraphs
7. Use of structural paragraphs, sentences, and words and phrases
8. Variety in sentence type and length
9. A clear, concise, direct, and objective style of writing
10. Relegation of long derivations, computations, and highly technical matter to the appendix

The Paragraph — The Basic Unit in Writing

The function of the paragraph is to break the text into readable units. The paragraph helps the reader by grouping sentences around a central idea— a topic sentence. The paragraph is therefore the basic unit in the development of a subject. The paragraph has two principal uses: It holds together thoughts or statements that are closely related; it keeps apart thoughts and statements that belong to different parts of the subject.

Paragraphs clothe the skeleton of the outline. The sentence outline, as we have discussed earlier (see chapter 5), provides the topic sentences for paragraphs. The topic sentence expresses the main idea developed by the paragraph. The writing of paragraphs has been compared to the sorting of material into labeled baskets: the material is the sentences conveying ideas or information; the label on the material is the topic sentence. The topic sentence serves as a guide to both the writer and the reader. Like an efficient sentry, it excludes material that does not belong within the paragraph. It controls the unity, coherence, emphasis, and proportion of the paragraph. The topic sentence is usually placed at the beginning of the paragraph, but not always. It may appear at the end, in the middle, or in any position in the paragraph.

The inexperienced technical writer would do well to begin his paragraph with a topic sentence. It is a useful technique in helping to set the flow of necessary information on paper. The development of the paragraph works in much the same way as the development of an outline. Just as a thesis sentence requires substantiation, so the statement of a topic sentence demands a certain kind of development of or substantiation. A good procedure for developing paragraphs is to ask yourself two questions:

1. What is the main point of this paragraph?
2. What must I tell my reader to support, explain, or clarify it?

Expository devices similar to those used in developing definitions (see chapter 9) are used to develop the topic sentence — illustration, explanation, comparison or contrast, analogy, definition, detail, reasons or facts, diagram, cause to effect or effect to cause, and analysis.

Let me illustrate by a few examples how the reader's expectations are satisfied by the development of the topic sentence.

> It is a commonplace fact that scientific discoveries are a function of methods used.

Here is the reader's natural question: Why is it a commonplace fact? The writer has an obligation to give the reasons in the rest of the paragraph.

> Several occurrences during the experiment confirmed this opinion.

The reader logically expects specific instances to confirm the opinion of the writer.

> In certain other aspects, especially its spatio-temporal aspects as revealed by the theory of relativity, nature is like a rainbow (7:3).

The reader is interested in knowing how the analogy applies. The writer (in this case, Sir James Jeans) elaborates an effective explanation of his analogy.

The rattle is the most characteristic feature of the rattlesnake and is one of the most remarkable structures in nature.

The reader expects an explanation to answer his question, why?

Operation of an autopilot may be seen from the diagram in Fig. 12.

The rest of the paragraph becomes supplemental to the diagram in the explanation of how an autopilot operates. Graphic illustrations are used with great effect in technical writing to aid the development of the topic sentence of a paragraph. Frequently they are the most important substantiating details in such development.

Test your paragraphs by these criteria:

1. What is the central idea?
2. What must the reader know to support it or explain it?
3. Is there anything in it not related to that idea?
4. Are the sentences organized in a sequence which is sufficiently logical to support or explain the topic sentence clearly?

Structural Paragraphs

There is considerable similarity between the topic sentence of a paragraph and the first paragraph of a section. Both are necessary to summarize, preview, and connect — to orient the reader and give him proper perspective. Furthermore, closing paragraphs of sections often summarize the contents of that section and show their significance to the whole.

Structural paragraphs are commonly placed between the major heading and the first subheading of the section. Such paragraphs usually point both backward and forward. Their primary purpose is to introduce the subject to be discussed. Like a road map, their function is to help the reader as he moves along from one phase of the presentation to the next. This type of paragraph prods him to look back to where he has been and to look ahead to where he is going. The technique is simple: the writer tells the reader briefly the major points he has covered, and then indicates what he is going to do next.

An example of such a paragraph is the one which begins Chapter Six of John Dewey's book, *How We Think:*

> We have in previous chapters given an outline account of the nature of reflective thinking. We have stated some reasons why it is necessary to use educational means to secure its development and have considered the intrinsic resources, the difficulties, and ulterior purpose of its educational training—the formation of disciplined logical ability to think. We come now to some descriptions of simple genuine cases of thinking, selected from the class papers of students (1:91).

A well-organized paper has structural paragraphs which have duties of introduction, transition, review, and summarization. Like headings, they keep the reader informed of the design of the whole composition; in addition, they connect the larger parts of the report.

Paragraph Summarized

The paragraph is the basic unit in the writing process in the development of a subject. It must have unity, coherence, and emphasis. If you are to meet your reader's needs efficiently, you must talk about one thing at a time. Keep things together that *belong* together. Make clear relationships among the ideas of your paragraph with reference words, repetition, connectives and relative words, phrases and clauses. Indicate important ideas by properly subordinating the unimportant ones. Finally, help your reader to make transitions from one idea to the next, one subject to the next, one section to the next by using linking and bridging words, sentences, and paragraphs.

Discussion Problems and Exercises

1. Here are two paragraphs saying just about the same thing. Which has better technical style? Why?

   a.) . . . it is the opinion of the writer that it is the appropriate moment to reexamine the style of writing which might most effectively be used by members of the engineering profession. It is also the writer's belief that a long-lasting tradition about the inappropriateness of the active voice and the personal pronoun for technical writing has been making for a great deal of inefficiency. This kind of writing has been exemplified in the past by numerous national publications. It would appear that an application of the principles of engineering to the problem would be beneficial and it would seem the result might be that such style would be eliminated (10:664).

   b.) . . . I think it's about time we stop insisting on impersonal style for engineers. I think that our national publications could set a good example in breaking this strait jacket. After all, the engineer wants efficiency and a "common-sense" approach in his professional work. Why not encourage him to apply this practical method to his writing? If he does, he'll save time, money, material and his reader's temper (10:664).

2. (a) Lawyers are the worst offenders in the use of the very specialized language of a profession in their writings. The writing of most lawyers is shackled by the gobbledygook of legal terminology, the use of the third person and passive voice. Below are examples from an article by a lawyer writing in an engineering magazine. How would you rewrite the excerpts to make sense for a reader? Keep in mind that the qualities of technical style are clarity, precision, conciseness, and objectivity.

   The argument has been made that the contractor is a third party beneficiary of the promise of the engineer or architect to the owner that he, the architect, will act in a certain way with respect to the contractor, such as, for example, timely approval of plans. . . .

   Another interesting possibility presents itself with respect to the plaintiff contractor's problem of avoiding the exculpatory and notice provisions of a construction contract. These are the "fine print" clauses that attempt to shift onto the contractor all responsibility for everything that does or does not happen. Such provisions, either by their terms or because a plaintiff contractor fails to comply with them, often effectively deprive a plaintiff contractor of the fruits of what otherwise could have been a succueessful suit against the owner. It has been suggested that when the actions of the engineer or architect have given rise to the cause of action, and the owner has protected himself as noted, there may exist enforceable legal liability against the engineer or architect independently of the owner under conditions where the architect will not receive the benefit of the protection afforded the owner (6:49-50).

(b) Check your revision, if at all possible, with a lawyer to see whether your rewriting is accurate and conforms with the legal nuances in the original. Obtain his opinion about the specialized terminology of the legal profession. Does he see anything wrong in the original quotation? What is his opinion of legal writing—that is, writing by lawyers for laymen, as well as the writing of lawyers for lawyers?

(c) Locate a Supreme Court decision and compare its writing style and clarity with that in the quoted material. An alternative reference might be the decision of U. S. District Court Judge John M. Woolsey lifting the ban on the book *Ulysses* by James Joyce (8:ix-xiv).

3. Obtain a copy of the Patent Office *Gazette* in your library. Adapt one of the patent descriptions into an expository article for a nontechnical reader. Underline the topic sentence in each paragraph. Encircle the transitional words, phrases, and sentences.

4. Prepare documentation references and a bibliography for the report you are writing under problem 2, chapter 6, in accordance with the procedures indicated in the present chapter.

5. On an average, how many words per sentence and how many syllables per hundred words do you have in your report?

6. Are half of the sentences in your report complex sentences? Simple sentences comprise what percentage of your paper? Compound? Other types of sentences?

7. On your rough draft, underline the topic sentences in your report. Encircle transition devices and structural paragraphs.

8. Test each of your paragraphs by these criteria:
    What is the central idea (topic sentence)?
    What must the reader know to support it or explain it?
    Is there anything in it not related to the topic sentence?
    Are the sentences organized in a sequence which is sufficiently logical to support or explain the topic sentence clearly?

9. Check your report for the common grammatical errors noted in this chapter.

10. Check your report for the format considerations noted in this chapter.

11. Does your report have sufficient variety in sentence length and structure to promote interest and avoid monotony?

12. Do subjects and verbs of your sentences have grammatical agreement?

13. Does each "it," "who," "which," and "that" clause refer to a definite word?

14. Does the punctuation help your reader reach the exact meaning you want him to reach?

15. Are your wordings accurate and precise?

16. Is all spelling accurate?

References

1. Dewey, John, *How We Think,* D. C. Heath and Company, 1923.

2. E. I. Dupont de Nemours and Company, Explosive Department, Atomic Energy Division, Technical Division, Savannah River Laboratory, *The D(ratted) P(rogress) Report,* July 12, 1954.

3. Flesch, Rudolf, *The Art of Readable Writing,* New York: Harper and Brothers, 1949.

4. Huxley, T. H., *Essays,* New York: The Macmillan Company, 1929.

5. Huxley, T. H., *An Introduction to the Study of Zoology,* New York: D. Appleton and Company, 1893.

6. Jarvis, Robert B., "The Engineer and Architect — As Defendants," *Civil Engineering,* April 1961.

7. Jeans, Sir James, *The New Background of Science,* Cambridge: Cambridge University Press, 1933.

8. Joyce, James, *Ulysses,* New York: The Modern Library, 1940.

9. Perrin, Porter G., *Writer's Guide and Index to English,* Third Edition, New York: Scott, Foresman and Company, 1959.

10. Shurter, Robert L., "Let's Take the Strait Jacket Off Technical Style," *Mechanical Engineering,* August 1952.

11. Sprat, Thomas, "The Vanity of Fine Speaking," *A Science Reader,* Lawrence V. Ryun, Editor, New York: Rinehart and Company, Inc., 1959.

13

Reference Index and Guide to Grammar, Punctuation, Style, and Usage

There are some writers who take refuge in the popular notion that grammar is not important. However, grammatical writing is easier to read and understand than ungrammatical writing. Language is a code. Unless people signal to each other in the same code, they cannot exchange intelligible messages. Their agreement to use the same code is the basis of correctness or grammar. Incorrect or ungrammatical means "not customary to the accepted code." A writer may not change the customary code without risking the possibility of his message being garbled.

This Reference Index is not meant to be a text on grammar and usage. Its purpose is to provide a ready and convenient guide to some of the more common problems the technical writer will meet in his writing requirements. Fuller treatment of some of these problems will be found in the Selected Bibliography within the General References subsection at the end of this book.

*　　　　　　　　*　　　　　　　　*

a, an. *a* is used before words beginning with a consonant (except *h* when it is silent), before words beginning with *eu* and *u*, pronounced *yu*, and before *o*, pronounced as in *one*. *An* is used before all words beginning with a vowel sound and words beginning with a silent *h*.

a	*a*	*a*	*an*	*an*
beacon	haft	eugenicist	alloy	herb
drift	helix	one-hour rating	eel	*H*
shunt circuit	hydrate	unit	*N*	homage
tentacle	hysteresis	uropod	impulse	hour

abbreviations. Abbreviations are used frequently for convenience in technical writing. However, they are acceptable only if they convey meaning to the reader. Nonstandard abbreviations should be explained fully when introduced in a report. Abbreviations are appropriate in compilations, tables, graphs, and illustrations where space is limited. Their use in the text should be limited except when preceded by numerals, as in giving dimensions or ratings. Unfortunately, the spelling of abbreviations is not universal. The American National Standards Institute has a list which has been approved by many scientific and engineering societies. However, other professional societies, as for example the Institute of Electric and Electronic Engineers, and government agencies have their own abbreviations. Dictionaries give frequently used abbreviations in their main listing of words; some also provide a special list in a back section. It might be well to include a glossary of abbreviations used if there are a goodly number in a report. When the technical writer is in doubt as to the proper abbreviation, he should write the word out. Periods are used with abbreviations with the exception of acronyms, for instance: NBS, NSF, DOD, AEC, UNESCO, NATO, IEEE, PTA.

activate, actuate. *Activate* is an old English word that was once considered obsolete. It has been reintroduced as a special term in chemistry and physics to denote the process of making active or more active as: to make molecules reactive or more reactive; it is used especially in the sense of promoting the growth of bacteria in sewage and of making substances radioactive. The military have adopted *activate* to mean: to set up or formally institute a military unit with the necessary personnel and equipment. *Activate,* however, should not replace the word *actuate* which means to set a machine in motion or to prompt a person to action.

above. Grammatically, this word occurs most frequently as an adverb (the equipment identified *above* [above modifies the verb *identified*]) or a preposition (*above* the lithosphere). *Above* is sometimes used as an adjective (the *above* diagram) or as a noun (the *above* is an equation of merit). Many competent writers avoid using *above* except as a preposition because the sentence has a stilted construction and the thought transmitted tends to be ambiguous since the reference of above, especially as a noun, is vague. Above as an adjective or adverb, in the context previously indicated, should be used with discretion and specificity. I would avoid its use as a noun entirely.

accept, except. *Accept* (verb) means to take when offered, to receive with favor, to agree to. *Except* (verb) means to exclude or omit; *except* (preposition) means with the exclusion of, but. The customer *accepted* delivery of four of the equipments; of the five equipments built, only one was *excepted* from delivery. All *except* model 5 were delivered.

access, excess. Access means approach, admittance, admission (to gain access to the laboratory). *Excess* means that which exceeds what is usual, proper, or specified (there was an *excess* of ten liters of liquid in the tank.)

adapt, adopt. These two words are often confused, even though their meanings are entirely different. *Adopt* means to take by choice into some sort of relationship (he *adopted* the older engineer's design). *Adapt* means to modify, to adjust, or to change for a special purpose (he *adapted* the cam to provide a slower movement).

adjective. An *adjective* is the part of speech that modifies or limits nouns and pronouns. Its purpose is to clarify for the reader the meaning of the word it modifies. The adjectives in the expressions *electrostatic* reaction, *inverted-V* antenna, *igneous* magna, *alpha* particle, *standstill* torque, and *heat-resisting* steel provide more meaning to the reader than the nouns by themselves. Adjectives are classified as demonstrative, descriptive, limiting, and proper. A demonstrative adjective points out the word it modifies; *this* power supply; *these* cells; *that* alloy; and *those* theories. A descriptive adjective denotes a quality or condition of the word it modifies: an *automatic* titration; a *progressive* disease; a *bent* pipe; and a *slower-flow* rate. A limiting adjective designates the number or amount of the word it modifies: a 5 percent potassium chloride solution; a *600,000 candela* intensity; *5* grams of ascorbic acid; and a *10 millimeter-per-second* flow. The articles *a, an,* and *the* are limiting adjectives. A proper adjective is one derived originally from a proper name: *Josephson* effect; *Schick* test; *Gibb's* function; *Californian* jade; and *Brownian* movement. Proper adjectives that have lost their sense of origin are written in lower case: *india* ink; *macadamized* road; and *angstrom* unit.

adverbs. An *adverb* is a word, phrase, or clause that modifies a verb (the experiment went *smoothly*); an adjective (the *very* slow titration took longer than I expected); another adverb (*almost* immediately the change occurred); or an entire sentence (*Fortunately*, the power had already been turned off). Most adverbs are adjectives plus the ending *ly*: quick*ly*, complete*ly*, correct*ly*, simp*ly*, final*ly*. Some adverbs, derived from old English, have no special adverbial sign: *now, quite, since, below, much,* and *soon.* Some adverbs have the same form as adjectives: *best, early, fast, slow, straight, well,* and *wrong.* Some of these adverbs also have an *ly* form: *slowly, wrongly.*

adviser, advisor. Both spellings are acceptable. What adds to the confusion is that adviser is used more frequently, but the spelling of *advisory* is the correct one.

affect, effect. These words are frequently confused because many people pronounce them to sound alike. *Affect* is almost always a verb, meaning to influence or to make a show of: The new economic measures *affected* the industry's recovery; he *affected* an English accent. *Affect* as a noun is used as a term in psychology, pertaining to feeling, emotion, and desire as factors in determining thought and conduct. *Effect* is used infrequently as a verb. As a noun it means result or consequence: The *effect* of cooling on the system increased its efficiency; the *effect* of the design change was a saving in time and money. *Effect* as a verb means to bring about: The change in design *effected* a great saving.

agree to, agree with, agree on. These are idiomatic expressions. You *agree to* things, *agree with* people, and people *agree on* something by mutual consent: He *agreed to* the contract; he *agreed with* the client; they *agreed on* the details of the contract.

alternate(ly), alternative(ly). As adjectives and adverbs, these two words are frequently confused. *Alternate* means by turn; *alternative* means offering a choice: The problem had two *alternative* solutions. The interstate highway is the shortest route to the city, but since it becomes crowded very early, there are several *alternative* routes to take. The weather *alternated* between rain and sunshine. The wall had *alternating* layers of brick and stone. *Alternate* as a noun means one that takes the place of or alternates with another: If you cannot attend, send an *alternate*.

although, though. Both words often are used interchangeably to connect an adverbial clause with the main clause of a sentence to provide a statement in opposition to the main statement, but one that does not contradict it. *Although* is preferable to introduce a clause that precedes the main clause; *though* for a clause that follows the main clause: *Although* we were short of test equipment, we managed to check out the system. We managed to check out the system, *though* we did not have all the test equipment.

among, between. *Among* denotes a mingling of more than two objects or persons; *between*, derived from an old English word meaning "by two," denotes a mingling of two objects or persons: *Among* several designs, his was the most practical. The instrument's accuracy was *between* ±1 and ±2 percent. *Among*, however, expresses a collective relationship of things and *between* seems to be the only word available to express the relation of a thing to many and surrounding things, severally and individually: disagreement *between* bidders; to choose *between* courses; the space lying *between* three points.

amount, number. *Amount* refers to things or substances considered in bulk; *number* refers to countable items as individual units: the amount of gas in the container (but the number of containers of gas); the amount of corn in the silo (but the number of bushels of corn).

and/or. This expression, usually frowned upon in literary writing, is commonly used in agreements, contracts, in business and legal writing, and in technical writing to show there are three possibilities to be considered: He offered his house and/or automobile as security for the loan. (He offered his house, or his automobile, or both as collateral.)

ampersand. *Ampersand* is the word for the symbol &. Its primary use is to save space. Its use as a substitute for the word *and* is frowned upon in formal writing. Many firms use the *ampersand* as a formal part of their official or incorporated name as J.C. Mason & Co., Inc. In addressing firms, use the form indicated in their formal letterhead.

apostrophe. The most common use of the *apostrophe* is to show possession and ownership: Dr. *Stanton's* equation; the *agency's* director; the *formula's* deviation; the *book's* index. The *apostrophe* is used to indicate omission of one or more letters in a contracted word or figure: *'80* for 1980; *didn't* for did not; *they're* for they are. The *apostrophe* is used in the plural form of numbers, letters, and words: The *1980's*; *ABC's*; Programming is spelled with two *m's*; The second of the two *which's* is not necessary in the sentence. There is some tendency to omit the apostrophe in plurals of numbers and letters, such as: 1920s, four Ws, second of three thats. The *apostrophe* is not used with the possessive pronouns: his, hers, ours, theirs, yours, its.

appendix. The plural most commonly used is *appendixes.* Purists prefer *appendices.*

appositives and their antecedents. An *appositive* is a term which modifies a noun or other expression by placing immediately after it an equivalent expression that repeats its meaning: the resistor, V101; good judgment, the characteristic few of us have; our director, Dr. Roberts. *Appositives* should have clear antecedents. Vague appositive reference can be corrected by placing the appositive immediately after the word or phrase it modifies or by rewording the sentence.

> Faulty: Maxwell formulated a new theory while experimenting with Faraday's concept, a major contribution to the study of electricity.
>
> Revised: While experimenting with Faraday's concept, Maxwell formulated a new theory, a major contribution to the study of electricity.
>
> or: Maxwell contributed greatly to the study of electricity when he formulated a new theory based on Faraday's concept.

articles. There are three articles in the English language: *a, an,* and *the.* *A* and *an* are indefinite articles: *a* dog; *an* apple; dog and apple refer to typical or unidentifiable things. *The* dog, *the* apple refer to a particular or identifiable dog and apple.

assay, essay. These two words are sometimes confused. *Assay* means to test; *essay* means to attempt. *Assay* also has the meaning of analyzing or appraising critically: We shall *assay* the sample to determine its properties. Having obtained a more precise instrument, the scientist again *essayed* the experiment.

awhile, a while. *Awhile* is an adverb; *a while* is a noun phrase, often used as the object of a preposition: The odor lasts *awhile*. For *a while*, the odor remained in the room.

because. *Because* is used to introduce a subordinate phrase or clause, providing the reason for the statement in the main clause: *Because* of its low boiling point, the substance could not be used in space applications. When a sentence begins with *The reason is* or *The reason why . . . is*, the clause containing the reason should not begin with *because* but with the word *that*. *Because* in this usage is an adverb; *that* is a pronoun. The linking verb *is* requires a noun rather than an adverbial clause.

beside, besides. *Beside* means near, close by, by the side of. *Besides* means in addition to, moreover, also, aside from: The replacement assembly was placed *beside* the faulty instrument. *Besides* corrosion resistance, the new alloy had many other desirable properties.

brackets []. *Brackets* are used whenever it is necessary to insert parenthetical material. *Brackets* are also used to make corrections or explanations within quoted material. In quoting material *sic* in brackets [*sic*] is used to indicate that an error was in the original quoted material: He lives in New Haven, Conneticut [*sic*].

but. *But* is the coordinating conjunction used to connect two contrasting statements of equal grammatical rank. It is less formal than the conjunction *however* or *yet* and is more emphatic than *although*: Not an atom *but* a molecule. The signal was short *but* distinct. The crew worked industriously under his supervision, *but* the minute he turned his back, it sloughed off.

can, may. *Can* expresses the power (physical or mental) to act; *may* expresses permission or sanction to act. In informal or colloquial usage, *can* is frequently substituted for *may* in the sense of permission, but this substitution is frowned upon in formal writing. *Could* and *might* are the original past tenses of *can* and *may*. They are now used to convey a shade of doubt or a smaller degree of possibility: We *can* meet your stringent specifications. Under the new FDA regulations products with 3 percent hexachlorophene solutions *may* be sold by prescription only. The absorbed hydrogen *might* be removed by pumping out the system. The use of *could* suggests doubt or a qualified possibility: The possibility is remote that under those conditions the contractor *could* meet his delivery schedule.

cannot, can not.   Both forms are used. *Can not* is more formal, but *cannot* is used more often.

capitalization.   The rules for capitalization in technical writing are the same as in other formal writing. Use capitals for:

1. Titles, geographical places, and trade names:
   Origin of Species
   Atlantic Ocean, Antarctica, Ohio
   Freon, Polaroid, Teflon

2. Proper names and adjectives derived from proper names, but not words used with them:
   Dundee sandstone, Pliocene period, Gaussian, Coulombic, Boyle's law, Huntington's chorea, Einstein's theory of relativity

Words derived from proper nouns and which, through long periods of usage, have achieved an identity within themselves are no longer capitalized, for example:
   bunsen burner, galvanic cell, ohm, ohmic drop, petri dish

3. Scientific names of phyla, orders, classes, families, genera:
   Decapoda, Megaloptera, Colymbiformes, Urochorda

4. Heads, subheads, and legends require capitalization.

catalog, catalogue.   If you prefer the second spelling, you are a traditionalist fighting a losing battle.

center about, center around, center on, center upon.   In formal writing, the idiom is *center on* or *center upon*; in less formal writing, *center around* or *about* is the idiom found in use.

circumlocution.   A long word meaning wordiness, excessive verbiage. Technical writers are fond of overwriting and padding. Perhaps they have a subconscious belief that they are paid according to the number of words they write. Excess verbiage spreads ideas thin.

Wordy: Caution must be observed . . .
Revised: Be careful, don't . . .

Wordy: We are about to enter an area of activity.
Revised: We are ready to begin . . .

Wordy: It is not desirable to leave filters in a system after resistance has increased to the point where there is a substantial decrease in the flow of air.
Revised: Filters should not be left in a system after resistance has caused a substantial decrease in the flow of air.

Wordy: On the basis of the foregoing discussion it is apparent that . . .
Revised: This discussion shows . . .

Wordy: The pursuit and capture of winged, air-breathing anthropods is more easily effected when a sweet, as opposed to sour, substance is, for purposes of beguilement, made use of:

For example, the viscid fluid derived from the saccharin secretion of a plant and produced by hymenopterous insects of the super family *Apoidea* has proved to be more successful in this endeavor than has dilute and impure acetic acid.

Revised: More flies are caught with honey than with vinegar.

clause (subordinate). A *subordinate or dependent clause* is an element of a complex or compound sentence (see pages 313-315); it usually has a subject and a verb and is grammatically equivalent to a noun, adjective, or adverb. The dependent clause is introduced by a subordinating conjunction, e.g.: *as, because, since, when,* or by a relative pronoun, e.g.: *that, who, which.*

That the waveguide overheated was not a surprise to him (subject or noun clause).

Because the expert knows where to look for data on free-radical reactions in solution, he seems to have little sympathy for the nonkineticist who needs such data (adjective clause).

The administrator who received the report will not read it (adjective clause).

The endogenous RNA was removed when the preparation was subjected to alkaline hydrolosis (adverbial clause).

Few experiments have been carried out because nematodes have little or no capacity for regeneration (adverbial clause).

collective noun A *collective noun* is one whose singular form carries the idea of more than one person, act, or object: army, class, crowd, dozen, flock, group, majority, personnel, public, remainder, and team. When the collective noun refers to the group as a whole, the verb and pronoun used with it should be singular. When the individuals of the group are intended, the noun takes a plural verb or pronoun: The committee *is* here; The committee *were* unanimous in disapproval; The corporation *has given* proof of its intention. The plural of a collective noun signifies different groups: Herds of deer graze in the upper valley.

colon(:) The *colon* is a punctuation mark of anticipation to indicate that the material that follows the mark will supplement what preceded it. Use the colon:

1. After the salutation of a letter:
   Dear Dr. Reimann:
   Gentlemen:

2. In memoranda following TO, FROM, and SUBJECT lines:
   TO:
   FROM:
   SUBJECT:

3. To introduce enumerations, usually *as follows, for example,* the *following,* etc.
   The components are *as follows:*

4. To introduce quotations:
The statement read:
We the undersigned members of the corporation believe:

5. To separate hours and minutes:
The meeting will begin at 10:30 a.m.

comma. The modern tendency is to avoid using *commas* except where they are needed for clarity of meaning. Use commas:

1. To set off nonrestrictive modifiers or clauses. A word, phrase, or clause which follows the word it modifies and which restricts the meaning of that word is called a restrictive modifier or restrictive clause. Restrictive clauses are *not* set off by commas. When such a modifier merely adds a descriptive detail, gives further information, it is called *nonrestrictive* and is set off by commas. If the modifier is omitted, will the sentence still tell the truth or offer the meaning which you intend? If it does, then the clause or modifier is *nonrestrictive* and should be set off by commas. If it does not tell the truth and if it does not give the meaning intended, you must *not* use commas. Example: Our chief chemist, who got his Ph.D. degree from Stanford University, is now in Europe. *Who got his degree from Stanford University* is not necessary to the sense intended, so that clause is set off by commas. However: Our chief chemist who checked the computations of the experiment does not agree with the conclusions reached. The clause *who checked the computations* of the experiment is necessary to the sense of the sentence. It is restrictive and therefore should not be set off by commas.

2. With explanatory words and phrases or those used in apposition: To meet the deadline, Mr. Simon, our president, supervised the experiment.

3. With introductory and parenthetic words or phrases, such as *therefore, however, of course,* and *as we see it:* As you are aware, the legislation did not pass; This action, nevertheless, is necessary.

4. In inverted construction where a word phrase, or clause is out of its natural order: That the result was unexpected, it was soon apparent.

5. To avoid confusion by separating two words or figures that might otherwise be misread: By 1980, 30,000 students are expected to enroll.

6. To avoid confusion by separating words in a series: The client manufactures furniture, pottery items, electric fixtures, and steel garden implements.

7. To make the meaning clear when a verb has been omitted: I covered the door, John covered the hallway, Bill covered the windows, and Tom, the remaining exit.

8. To separate two independent long clauses joined by a coordinating conjunction *(and, but, for, yet, neither, therefore, or, so)*: Analog recorders of .05 percent accuracy are available, but frequent maintenance is often required to hold this tolerance. (A comma is not used when the coordinate clauses are short and are closely related in meaning: A panel of experts might be chosen from particular areas of specialization and their report would be published).

9. To separate a dependent clause or a long phrase from its independent clause: For complex mixtures of acids such as those found in physiological fluids, a five-chamber concave gradient is generated. Although good recoveries of both methyl pyruvate and methyl lactate could be obtained at column temperatures below 100° Celsius, there was no indication that such recoveries could be achieved when either of the acids were in excess.

10. To separate the day of the month from the year:
December 21, 1916
Usage is divided when the day of the month is not given, but usual practice today is to omit the comma: December 1916. In constructions in which the day precedes the month, no comma is used: 21 December 1916.

11. To separate town from state or country when they are written on the same line: Bethesda, Maryland; Washington, D.C.; Stockholm, Sweden; Fort Collins, Larimer County, Colorado.

12. In figures to separate thousands, millions: 528,121; 10,894,082.

compare, contrast. There is some confusion in the application of these two words because *compare* is used in two senses: (1) to point out similarities (used with the preposition *to*); and (2) to examine two or more items or persons, to find likenesses or differences (used with the preposition *with*). *Contrast* always points out differences.

compound predicate. A compound predicate consists of two or more verbs having the same subject. It is often used to avoid the awkward effect of repetition of the subject or the writing of another sentence: The Emperor Van de Graaff accelerator is precisely controllable. Another feature is that it is easily variable. It is also continuous. A more economical and smoother way of saying the previous sentences is to combine them with a compound predicate: The Emperor Van de Graaff accelerator is precisely *controllable, easily variable,* and *continuous.*

compound subject. Two or more elements that serve as the subject of one verb are called a *compound subject. Compound subjects* require the plural verb form: The *deposition time* and *amount of carbon available in the gases determine* (not determines) the resistance value in the film resistors.

council, counsel, counsul. These three words are sometimes confused. A *council* is an advisory group; *counsel* means advice and, in law, it means one who gives advice; a *counsul* is an official representing his government in a foreign country.

conjunction.   Conjunctions introduce and tie clauses together and join series of words and phrases. Types of conjunctions include:

   coordinating: *and, but, for*
   correlative: *either . . . or, not only . . . but . . .*
   conjunctive adverbs: *however, therefore, consequently*
   subordinating: *as, because, since, so that, when*

contractions.   *Contractions* are words from which an unstressed syllable is dropped in speaking. Their use is acceptable in information writing but not in formal writing.

   can't: use *cannot*
   didn't: use *did not*
   I'll: use *I shall, I will*

dangling modifiers.   The most prevalent fault in technical writing is the *dangling modifier.* The *dangling modifier* is usually a verb form (often a participle) that is not supplied with a subject to modify and which seems to claim a wrong word as its subject. It is said to dangle because it has no word to which it logically can be attached. Infinitive and prepositional phrases may also be dangling.

Wrong: Calibrating the thermistor through the temperature range of 17° to 19° Celsius, a value of 4.0 ± 01°C.1 molar was obtained.

Revised: Calibrating the thermistor through the range of 17° to 19° Celsius, we obtained a value of 4.0 ± 0.1°C.1 molar.

Wrong: After adjusting the valves, the engine developed more power.

Revised: The engine developed more power after the valves were adjusted.

Or: After adjusting the valves, the mechanic found the engine developed more power.

Wrong: To write a program for our computer, it helps to know Fortran.

Revised: To write a program for our computer you would find it helpful to know Fortran.

Better: A knowledge of Fortran would be helpful in writing a pro-program for our computer.

Wrong: Near Kamchatka, Alaska, Figures 11 through 17 show the typical appearances in both the television pictures and the infrared data, of several stages in the life cycle of the cloud vortices of a cyclonic storm.

Revised: Figures 11 and 17 show typical appearances in both television pictures and infrared data of several stages in the life cycle of the cloud vortices of a cyclonic storm near Kamchatka, Alaska.

dash (—). The *dash* is used to indicate a change of thought or a change in sentence structure. It is also used for emphasis and to set off repetition or explanation. The *dash* may be used in place of parentheses when greater prominence to the subordinate expression is desired. Dashes should not be used with numbers because they might be mistaken for minus signs.

The rectifier — that was the guilty component.

I am suspicious of prognosticators — but I would not be surprised if tangible evidence of extraterrestrial, intelligent life will be found before the end of the present century.

A traveling wave tube has been made with a helix fifteen-thousandths of an inch in inside diameter — about three times the diameter of a human hair.

These molecules — formic acid, acetic acid, succinic acid, and glycine — are the very ones from which living things are constructed.

data. The word *data* is the plural of the Latin word *datum.* The singular form, datum is rarely used in English. The word *data* is often defined as raw facts or observations. In science and technology, *data* are characterized by their tendency toward numerics or quantification. Because the singular form is rarely used, the word *data* is becoming more acceptable in all but the most formal English as either singular or plural. Historically, Latin purals sometimes become singular English words (e.g., *agenda, stamina*). Some writers treat *data* as a collective noun and use the singular verb with the word. For instance: The experiment's *data* was made available to the three laboratories. However, there is a loud and armed camp of data pluralists who will split an infinitive with alacrity but will scream with outrage and scorn at any usage of a singular English verb with the Latin plural form. My fatherly advice is that it is always safe to use the plural unless you can afford a battle — and can afford to lose it, because you may well do so.

disinterested, uninterested. *Disinterested* means fair, impartial, unbiased; *uninterested* means lacking interest or without curiosity. Although in recent years, the two words have been used interchangeably in informal speech, your use of disinterest to mean lacking interest may raise some readers' eyebrows.

division of words. Break words only between syllables. When in doubt about syllabication, look up the word in the dictionary. Avoid breaking up a compound word that requires a hyphen in its spelling. Do not divide words of two syllables if the division comes after a single vowel: among, along, atom, enough.

ditto marks ("). Ditto marks are a convenience in tabulations and lists that repeat words from one line to the next, but ditto marks should not be used in formal writing.

editorial "we." The substitution of *we* when *I* obviously is intended is considered by many as affected and pompous. In those circumstances where there are several authors of a report, the use of *we* is certainly called for. However, a writer may want to take the reader with him over intellectual territory, as for example:

> We have seen the effect of X on variable Y in our experiment; we will therefore proceed now to add Z into the mixture to see whether thus and thus will then occur.

In correspondence, the use of *we* may be desirable when the writer is expressing his organization's policy or desires.

either.   *Either* means *one or the other of two*. It can be either an adjective or pronoun: *Either* alternative is not sound. Nevertheless, I shall have to take *either* (one). To use *either* to refer to three or more objects is slovenly:

> Poor: *either* of three choices
> Better: *any* of three choices

e.g.   *e.g.* is an abbreviation of the Latin *exempli gratia,* meaning "for example." It is used to introduce parenthetical examples.

ellipsis (. . .).   A punctuation mark of three dots indicates that something has been omitted within quoted material. Four dots are used to indicate omission has occurred at the end of a sentence. The *ellipsis* also is used to indicate that a statement has an unfinished quality.

etc.   *etc.* is an abbreviation for the Latin *et cetera* which means "and so forth." It is sometimes used at the end of a list of items. Unless there is some reason for saving space, *etc.* should be avoided. An effective way to avoid this use is to introduce the list with *such as* or *for example*. The use of *and* with *etc.* is redundant.

every, everybody, everyone.   *Every* is an adjective; *everybody* and *everyone* are pronouns. *Every* and its compounds are grammatically singular:

> Every instrument in the bench setup was working.
> Everybody is here.
> Everyone who saw the phenomenon took back his own impressions.

exclamation mark (!).   An *exclamation mark (or point)* is used after an emphatic interjection or forceful command. It may follow a complete sentence, phrase, or individual word. It should be used with the greatest discrimination in technical writing.

farther, further.   In informal speech, there is little distinction between the two words. In formal writing, *farther* applies to physical distances; *further* refers to degree or quantity.

> We drove on sixty miles *farther.*
> He questioned me *further.*
> The more he read about the matter, the *further* confused he became.

good, well.   *Good* is an adjective; *well* is both an adverb and adjective. One can say:

I feel *good* (adjective).

I feel *well* (adjective).

However, the meanings are different. *Good,* in the first sentence implies actual body sensations. *Well,* in the second sentence, refers to a condition of health — being "not ill." In nonstandard spoken English, *good* is sometimes substituted for *well.* For example:

The centrifuge runs *good.*

Most educated persons would say:

The centrifuge runs *well.*

however. As a conjunction, *however* is a useful connective between sentences to show the relation of a succeeding thought to a previous one in the sense of "on the other hand" or "in spite of." Inexperienced writers tend to overuse *however. But* is often more appropriate because it is a more direct connective. Compare the following uses of *however:*

1. In an attempt to make certain that fragments of only the minus strand were synthesized, the recommended precautions were followed; *however,* when the reactions were carried out, the results indicated a vast excess of the plus strand.

2. In an attempt to make certain that fragments of only the minus strand were synthesized, the recommended precautions were followed, *but* when the reactions were carried out, the results indicated a vast excess of the plus strand.

3. In an attempt to make certain that fragments of only the minus strand were synthesized, the recommended precautions were followed. When the reactions were carried out, the results indicated a vast excess of the plus strand, *however.*

The more experienced writer will select use 2 or 3 over the usage in 1 above. *However* is also an adverb; it may modify an adjective or another adverb:

However great the difficulty was, he still did his best.

However hard he tried, he could not finish in time.

Hyphen (-), Hyphens are a controversial point in style. Modern tendency is to eliminate their use. Consult a good dictionary or the style manual of your organization. A *hyphen* is a symbol conveying the meaning that the end of a line has separated a word at an appropriate syllable and that the syllables composed at the end of one line and the beginning of the next are one word. The *hyphen* is also used to convey the meaning that two or more words are made into one. The union may be ad hoc — for that single occasion — or permanent. Light-yellow flame has a different meaning than light, yellow flame. *Hyphens* are, therefore, used to form compound adjectival descriptive phrases preceding a noun: beta-ray spectrum, cell-like globule, 21-cm radiation, less-developed countries. Conventionally, *hyphens* are used to join parts of fractional and whole numbers written as words:

Thirty-three
One-fourth

Hyphens are used to set off prefixes and suffixes in differentiating between words spelled alike but having different meanings:

Recover his composure, and re-cover his losses.
Recount a story, re-count the proceeds.
Fruitless endeavor, fruit-less meal.

Hyphens are used between a prefix and a proper name.

pre-Sputnik, ex-professor
pro-relativistic quantum theory

Hyphens are used between a prefix ending in a vowel and a root word beginning with the same vowel:

re-elected
re-enter

Hyphenation has become more of a publisher's worry than a writer's. Most organizations set a style policy in hyphenation. Some follow the Style Manual of the University of Chicago or the Style Manual of the U.S. Government Printing Office. Some professional societies have style manuals for the writing done in their fields (e.g., the American Institute of Physics and the Conference of Biological Editors).

i.e.   The use of *i.e.,* the abbreviation of the Latin *id est* (meaning *that is*), often saves time and space, but the English *that is* is preferable.

if, whether.   *If* is used to introduce a condition; *whether* is used in expressions of doubt.

If the weather holds good, the space launch will be made.
I wondered *whether* the space launch would be made.
I asked *whether* the space launch would be made.

imply, infer.   These two words are often confused. *Imply* means to suggest by word or manner. *Infer* means to draw a conclusion about the unknown on the basis of known facts.

in, into, in to.   *In* shows location; *into* shows direction.

He remained *in* the laboratory.
He came *into* the laboratory.

The construction *in to* is that of an adverb followed by a preposition.

He went *in to* eat.

its, it's.   *Its* is a possessive form and means belonging to it. *It's* is the contraction for *it is.*

lay, lie.   These verbs are often confused. *Lay* (principal parts: *lay, laid, have laid*) is a transitive verb meaning to put something down. *Lie* (principal parts: *lie, lay, have lain*) is an intransitive verb meaning to rest in a reclining position.

I *laid* the tool on the bench.
The old man *lay* resting on the sofa.

leave, let.   *Leave* means to go away or to part with; *let* means to allow or permit.

like, as.   The preposition *like* is used correctly when it is followed by a noun or pronoun without a verb. *Like* should not be used as a conjunction; *as* should be used instead.

He writes *like* an ignoramus.
He looks *like* me.
Winston tastes good *as* a cigarette should.
Watch and do *as* I do.

may be, maybe.   These two forms are often confused. *May be* is a compound verb; *maybe* is an adverb meaning perhaps.

misplaced modifiers.   Modifiers should be placed near the words they modify.

Wrong: Throw the horse over the fence some hay.
Revised: Throw some hay over the fence to the horse.

The accurate placing of the word *only* is critical to precise meaning in technical writing. Notice the change in meaning brought about by shifting the word *only* in the following sentences:

1. *Only* the physicist calculated the value of X in the equation.
2. The physicist *only* calculated the value of X in the equation.
3. The physicist calculated *only* the value of X in the equation.
4. The physicist calculated the *only* value of X in the equation.
5. The physicist calculated the value of *only* X in the equation.
6. The physicist calculated the value of X *only* in the equation.
7. The physicist calculated the value of X in the *only* equation.

money.   Exact sums of money should be written in figures:
25¢; $1.98; $10.00; $437.41; $83,755.00
Round amounts are written in words:
one hundred dollars; three thousand dollars; a million dollars

none, no one.   *None* is commonly used to refer to things (but not always); *no one* is used to refer to people. *None* may take either a singular or plural verb, depending on whether a singular or plural meaning is intended. *No one* always takes a singular verb.

number.   Number is the singular and plural aspect of nouns and verbs:

1. A verb agrees in number with its subject; a pronoun or pronominal adjective (my, our, your, his, her, its, their) agrees in number with its antecedent. Examples:

Right: Our greatest need *is* (not *are*) modernized plant designs.

Right: The remaining two tubes of the circuit are V-101A and V-101B. *They* (not *it*) control the power input.

Right: The closing sentence of both paragraphs and sections often *summarizes* (not *summarize*) the contents and *shows* (not *show*) their significance to the whole.

Right: Determination of the choice and of the placement of punctuation marks *is* (not *are*) governed by the author's intention of meaning to be conveyed.

2. A compound subject coordinated by *and* requires a plural verb, regardless of the individual number of the member subjects.

Right: A hammer and a saw *are* on the bench.

Exception: When the compound subject refers to a unity, a singular verb is used.

Right: A brace and bit *is* on the bench. (However, a brace and a bit *are* on the bench.)

Right: Johnson and Sons *is* a reliable distributor.

3. Nouns or pronouns appearing between a subject and a verb have no effect upon the number of the verb. The number of the verb is determined solely by the number of its subject.

Right: The director of engineering, in addition to two senior engineers and an engineering aid, *was* in the laboratory.

Right: He, as well as I, *is* to be assigned to the project.

4. Collective nouns (e.g., committee, crowd, majority, number) may be either singular or plural, depending on whether the whole or the individual membership is emphasized.

Right: The committee *is* holding its first meeting.

But: The committee *are* in violent disagreement among themselves.

5. Expressions of aggregate quantity, even though plural in form, generally are construed as singular.

Right: Two times two *is* four (some people say two and two are four).

Right: Two-thirds of the corporation's income *has* been embezzled.

Right: Forty kilowatt-hours *was* registered by the meter.

6. When the expressions of quantity not only specify an aggregate amount but also stress the units composing the aggregate, they are construed as plural.

Right: Two hundred bags of Philippine Copra *were* piled on the dock.

But: More than a billion pounds of Copra *was* purchased in the Philippines.

7. A relative pronoun (*who, which, that*) should not be taken as a singular when its antecedent is a plural object of the preposition *of* following the word *one*.

Right: He was one of the ablest hydrographic engineers who *have* devoted *their* skill to the Coast and Geodetic Survey.

numerals.   Practice varies in determining which numbers should be spelled out and which should be written as figures. Authorities frequently state rules that seem reasonable or preferable to them. A current trend is to use figures for all units of measure, such as meter, gram, liter, volt, hectare, and kelvin. Aggregate numbers are those resulting from the addition or enumeration of items. The tendency to differentiate between the two is rapidly disappearing. Most authorities recommend the spelling out of numbers under ten. Round numbers above a million are frequently written as a combination of figures and words:

a 56 million dollar appropriation
an employment force of 85 million

Most authorities recommend writing out a number at the beginning of the sentence. When two numbers are adjacent to each other, the first is spelled out:

six 120-watt lamps

The practice is to spell out small fractions when they are not part of a mathematical expression or when they are not combined with the unit of measure:

three-fourths of the area
one-third of the laboratory
one-half of the test population

However:

½ mile
¾ inch pipe
1/50 horsepower
¼ ton

Decimals are always written as figures. When a number begins with a decimal point, precede the decimal point with a zero as:

an axis of 0.52

Hours and minutes are written out except when A.M. or P.M. follow:

8:30 A.M. to 5:00 P.M.
six o'clock; nine-thirty; half past two
ten minutes for coffee breaks

Street numbers always appear in numerals:

1600 Pennsylvania Avenue, N.W.

When several figures appear in a sentence or paragraph they are written as numerals.

In Experiment 2, there was a total of 258 plants. These yielded 8,023 seeds — 6,022 yellow, and 2,001 green.

Numbers indicating order — first, second, third, fourth, etc. — are called *ordinal* numbers. (Numbers used in counting — one, two, three, etc. — are *cardinal* numbers.) First, second, third, etc. can be both adjectives or adverbs. Therefore, the *ly* forms — firstly, secondly — are unnecessary and are now rarely used.

oral, verbal. These two words should not be confused. *Oral* refers to spoken communication; *verbal* has the general meaning of communication — written or spoken — which uses words.

or. *Or* is a coordinating conjunction; it should connect words, phrases, or clauses of equal value.

parallel structure. Parallelism promotes balance, consistency, and understanding. Operationally, it means using similar grammatical structure in writing clauses, phrases, or words to express ideas or facts of equal value. Adjectives should be paralleled by adjectives, nouns, by nouns; a specific verb form should be continued in a similar structure; active or passive voice should be kept consistently in a sentence. Shifting from one construction to another confuses the reader and destroys the sense of the meaning. A failure to maintain parallelism results in incomplete thoughts and/or illogical comparisons.

1. Faulty: Assembly lines poorly planned and which are not scheduled properly are inefficient.

   Revised: Poorly planned and improperly scheduled assembly lines are inefficient.

   Faulty: This is a group with technical training and acquainted with procedures.

   Revised: This group has technical training and a knowledge of procedures.

   Faulty: Before operating the boiler, the fireman should both check the water level and he should be sure about the draft.

   Revised: Before operating the boiler, the fireman should check both the water level and the draft.

   Faulty: U.S. Highway 40 extends from the Coastal Plain, runs across the Piedmont, and into the mountains.

   Revised: U.S. Highway 40 extends from the Coastal Plain, across the Piedmont, and into the mountains.

2. Illogical shifts:

   Shift in tense: These effects were usually concentrated in the blood system of the animal. The count of red and white blood cells is affected greatly. The percentage of hemoglobin was also reduced.

   Revised: These effects were usually concentrated in the blood of the animal. The count of red and white blood cells was affected. The percentage of hemoglobin was also reduced.

Shift in person: First, a new filing system can be introduced. Second, you can train personnel to handle the present complicated system.

Revised: First, a new filing system can be introduced. Second, personnel can be trained to handle the present complicated system.

Shift in voice: The technicians began the tests on July 7, and the results were tabulated the following day.

Revised: The technicians began the tests on July 7 and tabulated the results on the following day.

Shift in mood: Class members should take meaningful notes on the lectures. Do not cram for quizzes that are to follow.

Revised: Class members should take meaningful notes on the lectures. They should not cram for the quizzes that are to follow.

Confused sentence structure: It was because of a natural interest that made me choose the profession of technical writing.

Revised: A natural interest in writing led me to choose the profession of technical writing.

parentheses (). *Parentheses* are used to enclose additional, explanatory or supplementary matter to help the reader in understanding the thought being conveyed. These additions are likely to be definitions, illustrations, or further information added for good measure. *Parentheses* are used also to enclose numbers of letters to mark items in a listing or enumeration.

per. *Per* is a preposition borrowed from the Latin, meaning *by, by means of, through,* and is used with Latin phrases that have found their way and use in English: *per diem, per capita, per cent (or percent), per annum.* *Per* has established itself by long range usage in business English.

per cent. The term *per cent* is two words, but more and more the term is being written as one word. The sign, %, is convenient and appropriate in tables, but its use in text should be avoided.

period. *Periods* are used to mark the end of complete declarative sentences and abbreviations. Periods are also used in a request, phrased as a question out of courtesy: Won't you let me know if I can be of further service.

phenomenon, phenomena. The plural form *phenomena* is frequently misused for the singular *phenomenon.*

prepositions. A *preposition* is a word of relation. It connects a noun, pronoun, or noun phrase to another element of the sentence. Certain prepositions are used idiomatically with certain words. For example:

knowledge of
interest in
hindrance to
agrees with (a person agrees to) a suggestion, agrees with (another)

agrees in principle to a suggestion
agrees to (a plan)
obedience to
responsibility for
fear of (high places)
means of
connected with

A number of years ago, it was fashionable for grammarians to put a stigma upon prepositions standing at the end of sentences. Actually, it is a characteristic of English idiom to postpone the preposition. Many technical writers still feel the pain inflicted in years gone by when some die-hard high school teacher slapped their wrists for ending a sentence with a preposition. Today, they still fall into very clumsy constructions to avoid it. Winston Churchill had the final word on final prepositions. Sir Winston, a famous stylist of his time, had little patience with rigid rules of grammar. When an assistant underlined a Churchillian sentence and noted solemnly in the margin, "Never end a sentence with a preposition," Sir Winston marginally noted back, "This is the sort of English up with which I will not put."

principal, principle. *Principal* is used both as a noun and an adjective. As a noun it has two meanings: (1) the chief person or leader; and (2) a sum of money drawing interest. As an adjective, *principal* means "of main importance" or "of highest rank or authority." *Principle* is always a noun and means fundamental truth or doctrine, or the basic ideas, motives, or morals inherent in a person, group, or philosophy.

pronoun. A *pronoun* refers to something without naming it. Its meaning in the reader's mind must be completed by a clear reference to some other word or group of words called its antecedent.

1. *Pronouns* must have definite antecedents or clearly understood antecedents.

Antecedent implied by not expressed:	The generator was overloaded and could not carry it.
Revised:	The generator could not carry the overload.
Or:	The overload was so large that the generator could not carry it.
Vague second-person reference:	The generator is not satisfactory when shunt-excited. You must have separate excitation.
Revised:	The generator is not satisfactory when shunt-excited. It must have separate excitation.

2. A *pronoun* agrees with its antecedent in number.

Plural:	Multi-stage amplifiers are the heart of radio, television, and almost all electronic equipment. *Their* circuits have

(not its circuit has) features of great practical importance.

Singular: The *antibody* is a protein and is produced as a response to the presence in the blood of foreign antigens; *it* passes (not *they*) into the blood through the lymphatic vessels.

3. Place relative *pronouns (who, that, which)* as close as possible to their antecedents.

Faulty: As vacuum tubes and amplifiers have become available which have higher powers and less noise, we have succeeded in developing instruments of improved ranges of operation.

Revised: As vacuum tubes which have higher powers and as amplifiers which have less noise become available, we have succeeded in developing instruments of improved ranges of operation.

punctuation. *Punctuation* is one means by which a writer can achieve clarity and exactness of meaning. Sloppy punctuation can distort meaning and confuse the reader. There are two principles governing the use of punctuation marks:

1. The choice and placement of punctuation marks are governed by the writer's intention of meaning.
2. A punctuation mark should be omitted if it does not clarify the thought.

There is a modern tendency to use open punctuation marks—that is, to omit all marks except those absolutely indispensible. A common mistake of the inexperienced writer is to overuse punctuation marks, especially the comma. A good working rule is to use only those marks for which there is a definite reason, either in making clear or in meeting some conventional demand of correspondence. To illustrate, notice the difference in meaning provided by the punctuation in the following two sets of sentences:

2. a) The professor said the student is a fool.
   b) "The professor," said the student, "is a fool!"
2. a) All of your resistors, which were defective, have been returned.
   b) All of your resistors which were defective have been returned.

The differences in meaning of sentences 1*a* and 1*b* because of the changes in punctuation in set one are quite obvious. The use of commas in sentences 2*a* makes the sentence say that "all of your resistors were defective, and I have returned them all." Sentence 2*b* without the commas tells the reader that "only those resistors which were defective have been returned." The ultimate test for any punctuation mark is "Is this punctuation needed to make the meaning of my words clearer?"

question mark (?) This punctuation mark is used to denote the end of a question. The *question mark* is placed inside a quotation mark only when the quoted matter itself is a question.

quotation marks ("). *Quotation marks* are used to enclose direct quota-tions, titles of articles and reports, and coined or special words or phrases.

said. In legal documents, the use of *said* as a demonstrative pronoun (this, that, these, those) has a long tradition of use (e.g., *said* Mr. Peterson, *said* dwelling). However, in business correspondence, it should be avoided as a cliché.

semicolon (;). The principal use of the *semicolon* is to separate inde-pendent clauses that are not joined by a conjunction or are joined by a conjunctive adverb or some other transitional term (e.g., therefore, how-ever, for example, in other words). A *semicolon* is also used to separate clauses and phrases in a series when they already contain commas.

> Among the articles offered for sale were a harpsichord, which was at least 200 years old; a desk, which had the earmarks of beautiful crafts-manship; and a table and four chairs, which were also antique.

shoptalk. *Shoptalk* is the specialized vocabulary used by specialists in their everyday work activity. It is usualy a jargon known only to the initiated and is both obscure and obstructive to those not "in the know." It is not limited to the shop occupations. Physicians, lawyers, and college professors have their own "talk" which obfuscates. *Shoptalk* is appropriate in formal channels of communication within particular occupations, but it is out of place in formal writing aimed at persons or groups outside of that specialized activity.

should, would. *Should* and *would* are used in statements that suggest some doubt or uncertainty about the statement that is being made. Many years ago, *should* was restricted to the first person, but usage now is so divided in the choice of these words that personal preference rather than rule is the guide today. However, consistency in usage in the same letter should be followed.

> I would greatly appreciate your granting me an interview.
> or:
> I should greatly appreciate your granting me an interview.

slow, slowly. Slow is both an adjective and an adverb; *slowly* is an adverb. As adverbs, the two forms are interchangeable, but *slow* is more forceful than *slowly*.

some, somebody, someone, somewhat, somewhere, etc. *Some* is used as a pronoun and an adjective.

> *Some* think otherwise (pronoun).
> He has *some* ideas on the problem (adjective).

Somebody (pronoun), *someone* (pronoun), *someday* (adverb), *somehow* (adverb) are written as one word.

split infinitives. Usage has won out. To split an infinitive is no longer viewed as a grammatical misdemeanor. Frequently the interpolation of a word between the parts of an infinitive adds clarity and emphasis. For example:

To carefully examine the evidence
To forcefully impeach
To seriously doubt
To better equip

tautology. Needless repetition of meanings differently expressed is an extreme form of wordiness. Examples are:

true facts
when gases combine together
consensus of opinion
first and foremost
same identical
basic fundamentals
initial beginning

than, then. *Then* is an adverb relating to time; *than* is a conjunction in clauses of comparison.

Then came the dawn.
Fortran is a more universal programming language *than* are any machine languages.

there is, there are. These expressions delay the occurrence of the subject in a sentence. The verb in the expression must agree in number with the real subject.

There is a difficult problem associated with the system.
There are two solutions to the problem.

Often these expressions can be omitted with no loss to the sentence.

A difficult problem is associated with the system.
Two solutions are available for the problem.
or:
The problem has two solutions.

toward, towards. The two forms are interchangeable.

underlining. In typewritten or handwritten copy *underlining* is used (in place of italics):

1. To indicate titles of books and periodicals.
2. For emphasis:
   Words that would be heavily stressed when spoken are *underlined.*
3. To indicate foreign words.

unique. The word means single in kind or excellence, unequalled; therefore, it cannot be compared. It is illogical to speak of (or to write): a more *unique* or most *unique* thing.

very. *Very* is an intensive word. It has become so overused that its intensive force is slight. If you are tempted to use *very* in a sentence or in the complimentary close, ask yourself what the word adds to the meaning you wish to convey. If the answer is negative, do not use it.

which. *Which* is a pronominal word, used in both singular and plural instructions. It refers to things and to groups of people regarded impersonally (The crowd, which was a large . . .).

who, whom. Who is the form used in the nominative case (as the subject) and *whom* is the form used in the objective case (when the form is the object of a verb or of a preposition).

He is the one *who* is responsible.
For *whom* the bell tolls.
Whom did he ask to serve as secretary?

your, you're. Your is a possessive pronoun.
Your company.

You're is a contraction for *you are.*

You're to leave before noon.

Selected Bibliography

General References

Abbreviations for Use on Drawings and in Text, New York: American National Standards Institute, 1972.

Baird, Russell N. and A. T. Turnbull, *Industrial and Business Journalism*, Philadelphia: Chilton Company, 1961.

Baldwin, J M., Editor, *Dictionary of Philosophy and Psychology*, New York: Peter Smith, 1949, 3 vols.

Bird, George L., *Article Writing and Marketing*, Revised Edition, New York: Holt, Rinehart and Winston, 1956.

Brittain, J. M., *Information and Its Users*, New York: Wiley Interscience, 1970.

Bush, George D., Editor, *Technology and Copyright: Annotated Bibliography and Source Materials*, Mt. Airy, Md.: Lomond Systems, 1972.

Estrin, Herman A., *Technical and Professional Writing, A Practical Anthology*, New York: Harcourt, Brace and World, Inc., 1963.

Evans, Bergen and Cornelia Evans, *A Dictionary of Contemporary English Usage*, New York: Random House, 1967.

Forsyth, D. P., *The Business Press in America*, 1750–1865, Philadelphia: Chilton Company, 1964.

Fowler, H. W., *Modern English Usage*, Revised Edition, Edited by Sir Ernest Gowers, New York: Oxford Press, 1965.

Hough, John N., *Scientific Terminology*, New York: Holt, Rinehart and Winston, Inc., 1953.

IEEE Standard Dictionary of Electrical and Electronic Terms, New York: Wiley-Interscience, 1972.

Kent, Allen and Harold Lancour, *Copyright, Current Viewpoints on History, Laws, Legislation*, New York: R. R. Bowker Company, 1972.

Klare, George R., *The Measurement of Readability*, Ames, Iowa: Iowa State University Press, 1963.

Kronick, David A., *A History of Scientific Periodicals*, New York: Scarecrow Press, 1962.

Mandel, Siegfried, *Dictionary of Science*, New York: Dell Publishing Company, Inc., 1969.

A Manual of Style, 12th edition, Chicago: University of Chicago Press, 1969.

Martin, R. C., and W. Jett, *Guide to Scientific and Technical Periodicals*, Denver: Swallow Press, 1963.

Mott, Frank Luther, *A History of American Magazines*, New York: D. Appleton and Company, 1930.

New York Times Style Book for Writers and Editors, New York: McGraw-Hill Book Company, 1962.

Nicholson, Margaret, *A Manual of Copyright Practices for Writers, Publishers and Agents*, New York: Oxford University Press, 1956.

Passman, Sidney, *Scientific and Technological Communication*, Oxford: Pergamon Press, 1969.

Perrin, Porter G., George H. Smith, and Jim W. Corder, *Handbook of Current English, 4th Edition*, Glenview, Ill.: Scott, Foresman and Company, 1969.

Robinson, David M., *Writing Reports for Management*, Columbus, Ohio: Charles E. Merrill Publishing Co., 1969.

Rogers, Walter T., *Dictionary of Abbreviations*, Detroit, Michigan: Gale Research Company, 1969.

Roget's Thesaurus of English Words and Phrases, Revised Edition, New York: St. Martins, 1964.

Schwartz, Robert, *Dictionary of Business and Industry*, New York: B. C. Forbes and Sons Company, 1954.

Scientific and Technical Communication, A Pressing National Problem and Recommendations for Its Solution, National Academy of Sciences-National Academy of Engineering, Washington, D. C.: Printing and Publishing Office, National Academy of Sciences, 1969.

Shidle, Norman G., *The Art of Successful Communication*, New York: McGraw-Hill Book Company, 1965.

Sigband, Norman B., *Communication for Management*, Glenview, Ill.: Scott, Foresman and Company, 1968.

Steen, Edwin B., *Dictionary of Abbreviations in Medicine and the Related Sciences*, 2nd Edition, London: Cassel, 1963.

Strunk, William, Jr., *The Elements of Style*, revised by E. B. White, New York: The Macmillan Co., 1959.

Turner, Rufus P., *Grammar Review for Technical Writers*, San Francisco: Rinehart Press, 1971.

Tweney, C. F. and E. C. Hughes, editors; *Chamber's Technical Dictionary*, Third Revised Edition with supplement, New York: The Macmillan Co., 1962.

U. S. Government Printing Office Style Manual, Washington, D. C.: U.S. Government Printing Office, 1973.

Van Nostrand's Scientific Encyclopedia, New York: D. Van Nostrand Co., Inc., 1958.

Vardaman, George T., Carroll C. Halterman, and Patricia Black Vardaman, *Cutting Communications Costs and Increasing Impacts, Diagnosing and Improving the Company's Written Documents*, New York: John Wiley and Sons, Inc., 1970.

Webster's New Collegiate Dictionary, Springfield, Mass.: G. and C. Merriam Co., Publishers, 1973.

Weisman, Herman M., *Information Systems, Services, and Centers*, New York: Becker and Hayes, Inc.; John Wiley and Sons, Inc., 1972.

Winchell, Constance, M., *Guide to Reference Books*, 8th Edition, Chicago: American Library Association, 1967.

Wittenberg, Phillip, *The Law of Literary Property*, New York: World Publishing, 1957.

Technical Writing and Editing

Brogan, John A., *Clear Technical Writing*, New York: McGraw-Hill Book Company, 1973.

Campbell, John M. and G. L. Farrar, *Effective Communications for the Technical Man*, Tulsa, Oklahoma: The Petroleum Publishing Co., 1972.

Clarke, Emerson and Vernon Root, *Your Future in Technical and Science Writing*, New York: Richard Rosen Press, 1972.

Coleman, Peter and Ken Brambleby, *The Technologist as Writer — An Introduction to Technical Writing*, New York: McGraw-Hill Book Company, Inc., 1972.

Damerst, William A., *Clear Technical Reports*, New York: Harcourt, Brace, Jovanovich, Inc., 1972.

Dodds, Robert H., *Writing for Technical and Business Magazines*, New York: John Wiley and Sons, Inc., 1969.

Gould, Jay R., *Opportunities in Technical Writing*, New York: Vocational Guidance Manuals, 1964.

Gray, Dwight E., *So You Have to Write a Technical Report*, Washington, D. C.: Information Resources Press, 1970.

Hays, Robert, *Principles of Technical Writing*, Reading, Mass.: Addison Wesley Company, 1965.

Hicks, Tyler G., *Writing for Engineering and Science*, New York: McGraw-Hill Book Company, Inc., 1961.

Houp, Kenneth W. and Thomas E. Pearsall, *Reporting Technical Information*, Second Edition, New York: Glencoe Press, 1973.

Huber, J. T., *Report Writing in Psychology and Psychiatry*, New York: Harper and Brothers, 1961.

Jordan, Stello, Editor, *Handbook of Technical Writing Practices*, (Two Volumes) New York: John Wiley and Sons, Inc., 1971.

Kapp, Reginald O., *Presentation of Technical Information*, New York: The Macmillan Co., 1957.

Maizell, Robert E., Julian Smith, and T. E. R. Singer, *Abstracting Scientific and Technical Literature, An Introductory Guide and Text for Scientists, Abstractors, and Management*, New York: Wiley-Interscience, 1971.

Mandel, Siegfried and David L. Caldwell, *Proposal and Inquiry Writing*, New York: The Macmillan Company, 1962.

Mills, Gordon H. and John A. Walter, *Technical Writing*, New York: Holt, Rinehart and Winston, Inc., 1970.

Mitchell, John H., *Writing for Technical and Professional Journals*, New York: John Wiley and Sons, Inc., 1968.

Noland, Robert L., *Research and Report Writing in the Behavioral Sciences*, Springfield, Illinois: Charles C. Thomas, Publisher, 1970.

Peterson, Martin S., *Scientific Thinking and Scientific Writing*, New York: Reinhold Publishing Corporation, 1961.

Rathbone, Robert R., *Communicating Technical Information*, Reading, Mass.: Addison-Wesley Publishing Company, 1966.

Rogers, Raymond A., *How to Report Research and Development Findings to Management*, New York: Pilot Books, 1973.

Singer, T. E. R., Editor, *Information and Communication Practice in Industry*, New York: Reinhold Publishing Corporation, 1958.

Tichy, H. J., *Effective Writing for Engineers-Managers-Scientists*, New York: Wiley-Interscience, 1966.

Turner, Rufus, P., *Technical Report Writing*, Revised Edition, San Francisco: Rinehart Press, 1971.

Ulman, Joseph N., Jr. and Jay R. Gould, *Technical Reporting, Third Edition*, New York: Holt, Rinehart and Winston, Inc., 1972.

Weil, B. H., Editor, *Technical Editing*, New York: Reinhold Publishing Corporation, 1958.

Weil, B. H., *The Technical Report: Its Preparation, Processing and Use in Industry and Government*, New York: Reinhold Publishing Corporation, 1954.

Woodford, F. Peter, Editor, *Scientific Writing for Graduate Students: A Manual on the Teaching of Scientific Writing*, New York: The Rockefeller University Press, 1968.

Graphics and Production

Arkin, Herbert and Raymond R. Colton, *Graphs: How to Make and Use Them*, Revised Edition, New York: Harper and Brothers, 1940.

Arnold, Edmund C., *Ink on Paper*, New York: Harper and Row, Publishers, 1963.

Bowman, William J., *Graphic Communication*, New York: John Wiley and Sons, Inc., 1968.

Caird, Ken, *Cameraready*, Pasadena, California: Cameraready Corp., 1973.

Clarke, Emerson, *A Guide to Technical Literature Production*, River Forest, Ill.: T.W. Publishers, 1961.

Croy, Peter, *Graphic Design and Reproduction Techniques*, Revised Edition, New York: Hastings House, Publishers, 1968.

Ettenburg, Eugene M., *Type for Books and Advertising*, New York: D. Van Nostrand Co., Inc., 1947.

Feinberg, Milton, *Techniques of Photojournalism*, New York: John Wiley and Sons, Inc., 1970.

Haughney, John D., *Effective Catalogs*, New York: John Wiley and Sons, Inc., 1968.

Hymes, David, *Production in Advertising and the Graphic Arts*, New York: Henry Holt and Co., 1958.

Illustrations for Publication and Projection, New York: American National Standards Institute, 1959.

Lewis, Richard A., *Annual Reports: Conception and Design*, Zurich, Switzerland: Graphis Press, 1971.

Lojko, Grace R., *Typewriting Techniques for the Technical Secretary*, Englewood Cliffs, N. J.: Prentice-Hall, Inc., 1972.

Magnon, George A., *Using Technical Art*, New York: Wiley-Interscience, 1970.

Marinaccio, Anthony, *Exploring the Graphic Arts*, Scranton, Pa.: International Textbook Company, Revised Edition, 1959.

Melcher, Daniel and Nancy Larrick, *Printing and Promotion Handbook*, New York: McGraw-Hill Book Company, Inc., Revised Edition, 1966.

Modley, Rudolph, and others, *Pictographs and Graphs: How to Make and Use Them*, New York: Harper and Brothers, 1952.

Nelms, Hemming, *Thinking With a Pencil*, New York: Barnes and Noble, 1964.

Price, Matlack, *Advertising and Editorial Layout*, New York: McGraw-Hill Book Co., 1949.

Turnbull, Arthur T. and Russell N. Baird, *The Graphics of Communication*, New York: Holt, Rinehart and Winston, 1964.

Correspondence

Lesikar, Raymond V., *Business Communication: Theory and Application*, Homewood, Illinois: Richard D. Irwin, 1972.

Menning, J. H. and C. W. Wilkenson, *Communicating Through Letters and Reports*, Homewood, Ill.: Richard D. Irwin, 1972.

Murphy, Herta M. and Charles E. Peck, *Effective Business Communication*, New York: McGraw-Hill Book Company, 1972.

Shurter, Robert L., *Written Communication in Business, Third Edition*, New York: McGraw-Hill Book Company, 1971.

Sigband, Norman B., *Effective Report Writing: For Business, Industry, and Government*, New York: Harper and Brothers, 1960.

Weisman, Herman M., *Technical Correspondence: A Handbook and Reference Source for the Technical Professional*, New York: John Wiley and Sons, Inc., 1968.

Communication

Borden, George A., *An Introduction to Human-Communication Theory*, Dubuque, Iowa: Wm. C. Brown Company, Publishers, 1971.

Bryson, Lyman, Editor, *The Communication of Ideas*, New York: Institute for Religious and Social Studies, 1948.

Cherry, Colin, *On Human Communication*, New York: John Wiley and Sons, Inc., 1957.

Dance, Frank E. X., Editor, *Human Communication Theory*, New York: Holt, Rinehart and Winston, Inc., 1967.

Doubleday Pictorial Library of Communication and Language, New York: Doubleday and Company, 1965.

Gordon, George N., *The Languages of Communication: A Logical and Psychological Examination*, New York: Hastings House, 1969.

Hartley, Eugene L. and Ruth E. Hartley, *Fundamentals of Social Psychology*, New York: Alfred A. Knopf, Inc., 1952.

Ogden, C. K. and I. A. Richards, *The Meaning of Meaning*, 8th Edition, New York: Harcourt, Brace and Company, 1947.

Probert, Walter, *Law, Language and Communication*, Springfield, Illinois: Charles C. Thomas Publisher, 1972.

Smith, Alfred G., *Communication and Culture*, New York: Holt, Rinehart and Winston, Inc., 1966.

Stewart, Daniel K., *The Psychology of Communication*, New York: Funk and Wagnalls, 1968.

Vardaman, George T. and Carroll C. Halterman, *Management Control Through Communication, Systems for Organizational Diagnosis and Design*, New York: John Wiley and Sons, Inc., 1968.

Weiss, Harold and James B. McGrath, Jr., *Technically Speaking*, New York: McGraw-Hill Book Company, Inc., 1963.

Wiksell, Wesley, *Do They Understand You?*, New York: The Macmillan Co., 1960.

Language and Semantics

Andrews, Edmund, *A History of Scientific English*, New York: Richard R. Smith, 1947.

Bloomfield, Leonard, *Language*, New York: Holt, Rinehart and Winston, Inc., 1933.

Chase, Stuart, *Power of Words*, New York: Harcourt, Brace and Company, 1954.

Gilman, William, *The Language of Science*, New York: Harcourt, Brace and World, Inc., 1961.

Hayakawa, S. I., *Language, Meaning and Maturity*, New York: Harper and Brothers, 1954.

Hayakawa, S. I., *Language in Thought and Action*, New York: Harcourt, Brace and Company, 1949.

Hughes, John P., *The Science of Language*, New York: Random House, 1962.

Huxley, Aldous, *Words and Their Meaning*, Los Angeles: The Ward Ritchie Press, 1940.

Jesperson, Otto, *Language: Its Nature, Development and Origin*, New York: The Macmillan Co., 1949.

Lee, Irving J., *Language Habits in Human Affairs*, New York: Harper and Brothers, 1941.

Mencken, H. L., *The American Language*, 4th Edition, New York: Alfred A. Knopf, Inc., 1936, Supplement 1, 1945, Supplement 2, 1948.

Sapir, Edward, *Language*, New York: Harcourt, Brace and Company, 1921.

Urban, Wilbur M., *Language and Reality*, New York: The Macmillan Co., 1951.

Walpole, Hugh R., *Semantics, The Nature of Words and Their Meaning*, New York: W. W. Norton and Company, Inc., 1941.

Whorf, Benjamin Lee, *Language, Thought and Reality, Selected Writings of Benjamin Whorf*, John B. Carroll, Editor, New York: John Wiley and Sons, Inc., and Massachusetts Institute of Technology Press, 1956.

Woodger, J. H., *Biology and Language*, Cambridge, England: Cambridge University Press, 1942.

Research and Its Methods

Altick, Richard D., *The Art of Literary Research*, New York: W. W. Norton and Company, Inc., 1963.

Beveridge, W. I. B., *The Art of Scientific Investigation*, London: William Heinemann, Ltd., 1951.

Braithewaite, R. B., *Scientific Explanation*, New York: Harper Torchbooks, Harper and Brothers, 1960.

Butterfield, Herbert, *The Origins of Modern Science*, Revised Edition, New York: Collier Books, 1962.

Cohen, Morris R. and Ernest Nagel, *An Introduction to Logic and Scientific Method*, New York: Harcourt, Brace and Company, 1934.

Conant, James B., *Science and Common Sense*, New Haven: Yale University Press, 1951.

Crumley, Thomas, *Logic: Deductive and Inductive*, New York: The Macmillan Co., 1947.

Dewey, John, *How We Think*, New York: D. C. Heath and Company, 1933.

Dewey, John, *Logic: The Theory of Inquiry*, New York: Henry Holt and Company, 1938.

Feibleman, James K., *Scientific Method The Hypothetico-Experimental Laboratory Procedure of the Physical Sciences*, The Hague, Netherlands: Martinus Nijhoff, 1972.

Finocchiaro, Maurice A., *History of Science as Explanation*, Detroit, Michigan: Wayne State University Press, 1973.

Galtung, Johan, *Theory and Methods of Social Science Research*, New York: Columbia University Press, 1969.

Games, Paul A. and G. R. Klare, *Elementary Statistics: Data Analysis for the Behavioral Sciences*, New York: McGraw-Hill Book Company, 1967.

Gee, Wilson, *Social Science Research Methods*, New York: Appleton-Century-Crofts, Inc., 1950.

Giere, Ronald N. and Richard S. Westfall, *Foundations of Scientific Method: The Nineteenth Century*, Bloomington, Indiana: Indiana University Press, 1973.

Good, Carter V. and Douglas E. Scates, *Methods of Research*, New York: Appleton-Century-Crofts, Inc., 1958.

Huff, Darrell, *How to Lie with Statistics*, New York: W. W. Norton and Company, Inc., 1954.

Keppel, Geoffrey, *Design and Analysis A Researcher's Handbook*, Englewood Cliffs, N. J.: Prentice-Hall, Inc., 1973.

Kish, Leslie, *Survey Sampling*, New York: John Wiley and Sons, Inc., 1965.

Mill, John Stuart, *A System of Logic*, London and New York: Longmans, Green and Co., Inc., 1948.

Noltingh, B. E., *The Art of Research: A Guide for the Graduate*, New York: Elsevier Publishing Company, 1965.

Northrup, F. S. C., *The Logic of the Sciences and the Humanities*, New York: The Macmillan Co., 1947.

Parten, Mildred, *Surveys, Polls and Samples*, New York: Harper and Brothers, 1950.

Payne, Stanley L., *The Art of Asking Questions*, Princeton, N. J.: Princeton University Press, 1955.

Pearson, Karl, *Grammar of Science*, London: Everyman's Library, J. M. Dent and Sons, Ltd., 1937.

Russell, Bertrand, *An Inquiry into Meaning and Truth*, New York: W. W. Norton and Company, Inc., 1940.

Sarton, George, *A History of Science*, Cambridge: Harvard University Press, 1952.

Schiller, F. C. S., *Logic For Use*, New York: Harcourt, Brace and Company, 1930.

Searles, Herbert L., *Logic and Scientific Methods*, New York: The Ronald Press Company, 1948.

Sinclair, W. A., *An Introduction to Philosophy*, New York: Oxford University Press, 1944.

Thompson, John Arthur, *An Introduction to Science*, New York: Holt, Rinehart and Winston, Inc., 1911.

Wallis, W. Allen and Harry V. Roberts, *The Nature of Statistics*, New York: Collier Books, 1962.

Wilson, E. Bright, Jr., *An Introduction to Scientific Research*, New York: McGraw-Hill Book Co., Inc., 1952.

Appendix
Examples of Various Forms
of Technical Writing
Covered in the Text

Student Employment Letters

203 West Lake Street
Fort Collins, Colorado 80521
December 12, 1974

Mr. E. A. Ferris
Administrator of Training
Combustion Engineering, Inc.
Prospect Hill Road
Windsor, Connecticut 06095

Dear Mr. Ferris:

Colorado State University has a central steam heating system for all the buildings on campus. One of the steam generators is a Combustion Engineering boiler erected in 1939. Since 1939, this boiler has performed satisfactorily with a minimum of maintenance. Modifications were recently completed for modernization of the heating plant and your boiler appears ready for another twenty years of service.

The performance of this boiler and the reputation of your company have prompted me to write you. If your company has an opening for a graduating mechanical engineer with previous mechanical experience, please consider my qualifications.

After attending college for one year, I joined the army to gain practical experience and maturity. While in the army, I attended school at Ft. Belvoir, Virginia for the maintenance of engineering equipment: bulldozers, cranes, etc. After equipment repair school, I was stationed in Germany where I had 2½ years of practical experience in the field repair of heavy equipment. This period also gave me a chance to mingle with people of other customs and learn to live and cooperate with a variety of personalities. Also during this time I was able to rise to specialist second class (E-5) which gave me experience in organizing and supervising the work of a group of men.

My tour of duty convinced me that I had been right in choosing mechanical engineering for a career, so I returned to Colorado State University to finish my studies for a B. S. M. E. degree. When I graduate, I will have taken such courses as thermodynamics, heat transfer, metallurgy, compressible fluids, fluid mechanics, nuclear engineering, statics, dynamics, strength of materials, and calculus.

I am a member of Sigma Tau, the national engineering honorary, and Phi Kappa Phi, national all-school senior and graduate honorary. My offices as treasurer of Sigma Tau and program chairman of our student section of A. S. M. E. have given me experience in responsibility and cooperation.

To help pay college expenses, I supplemented my scholarship and savings by summer work and part time work during the school year. All the jobs which I have held involved maintenance or repair of machinery. Last summer I worked as a junior engineer in the plant services section, Transport Division of Boeing Airplane Company. This job involved an efficiency and cost analysis of both steam and hot water heating systems, evaluation of economy suggestions and inspection of building repairs and maintenance. The job was interesting and gave me an insight into the depth of the mechanical engineer's field.

If my qualifications, as outlined on the enclosed resume and transcript, meet with your approval, may I have an interview at your convenience? I may be reached at the school address listed on the resume.

Respectfully,

John B. Wood

enc.

RESUME

Wood, John Bruce

School address:
203 West Lake Street
Fort Collins, Colorado
(no phone)

Permanent address:
1304 Pando
Colorado Springs, Colorado
MElrose 5-1473

Height: 5'11"   Weight: 205   Health: Excellent

Marital status: single

Memberships:

Professional Societies: A. S. M. E., Student Section

Honoraries: Sigma Tau, Honorary Engineering Fraternity
Phi Kappa Phi, All-school senior and graduate
honorary.

Clubs: Camvets, campus veterans' social club.

Special interests: mechanical repairs, skiing, hunting, reading, travel.

Education:

B. S. M. E.: Colorado State University; June, 1974

Service Schools: Certificate of Completion, Engineer,
Equipment Repairman; February 27,
1972

High School Diploma: Rifle Union High School, Rifle,
Colo.; June, 1969.

Experience:

Summer Work:

Summer — 1973: Junior Engineer, Plant Services Section,
Transport Division, Boeing Airplane Company. The job required
basic engineering knowledge of thermodynamics, strength of
materials, engineering drawing, principles of a-c motors, fluid
dynamics; past experience in machinery repair also helpful. Duties
included an efficiency and cost analysis for steam heating system
of main factory building and hot water system for the engineering
and administration buildings, also helping with engineering
analysis of existing chilled water system for air conditioning in
the engineering and administration buildings and coordinating
with water treatment consultants and plant facilities section concerning water treatment for the hot water system. Minor duties
were evaluation of employee suggestions and inspection of roof
maintenance and painting at Seattle field. An interesting and instructive summer.

Summer — 1972: Laborer and mechanic's helper, Colorado
State University Physical Plant. Job required strength, stamina,
normal dexterity, willingness to work unsupervised, driving skill,
and ability to operate and maintain crawler tractors, crane, grader,
ditcher, trucks and arc welders. Duties varied from laying sewer
pipe to operating a bucket loader; drove truck, both dump- and
semi- ; was responsible for scheduled maintenance and lubrication
of ditcher, grader, crawler tractors, and other construction equipment; operated bulldozer and pulled rollers, sheep's foot and
wobble wheel on small jobs. Weekend duties consisted of washing
and maintenance of motor pool trucks, cars and tractors.

Summer — 1969: Laborer, Gardner Construction Company,
Glenwood Springs, Colorado. Job required strength, stamina,
manual dexterity, and ability to read steel prints for concrete construction. Duties included placing and tying reinforcing rods for
concrete walls, raceways, valve boxes at the Colorado State Fish
Hatchery on Rifle Creek; later in the summer I was able to work
directly from steel prints without direct supervision. At times I
was also called on to return trucks from repair shop in town to job
site and operate air machinery. This job impressed upon me the
necessity of reliability and willingness.

Part-time Work:

April — 1970 to June — 1972: Service station attendant, Colorado State University Physical Plant, Fort Collins. Job required
manual dexterity; knowledge of maintenance and lubrication of
trucks, tractors, cars, construction equipment; reliability and responsibility. Duties included washing and helping grease all
physical plant vehicles on a monthly schedule, operation of C. S. U.

service station on Saturdays, minor tune-up and repair. No direct supervision during working hours.

January — 1966 to July — 1966: Operation of automatic screw machine, Woodward Governor Company, Ft. Collins. Job required manual dexterity, ability to use machinist's tools and micrometers and ability to work from detail drawings. Job consisted of set up and operation of four spindle, 2-inch automatic screw machine for primary machining operations on components of Woodward governors; maintenance of close tolerances frequently required. A very interesting and instructive period.

Personal references and confidential credentials will be supplied by the Colorado State University Placement Office on your request.

---

640 South Howes
Fort Collins, Colorado 80521
December 10, 1974

Battelle Memorial Institute
505 King Avenue
Columbus, Ohio 43201

ATTENTION: Personnel Director

Dear Sirs:

Research, in its most basic form, has always presented a challenge to me. As a boy interested in model airplanes, I read books on elementary aerodynamics. Then I tried to apply the principles I had learned to original flying model designs. At the age of fourteen, I got a part-time job in a grocery store so I could save enough money to buy a motorcycle. My motorcycle and I were seldom seen on the streets, however, because I always had it dismantled — trying to improve the fuel economy and performance of the engine.

Upon graduation from high school, the decision to continue my education in college was easy to make. I entered the Engineering College at Colorado State University with the intention of gaining a position in a research organization, such as yours, upon graduation as a Mechanical Engineer. During my school years, I have studied hard and maintained a good grade average. This past summer and fall I have been working in the Aerodynamics Lab at the University. While not directly engaged in research at this lab, I have had the opportunity of associating with many fine young research engineers and have learned much about wind tunnel techniques.

This June I will graduate with a Bachelor of Science degree in Mechanical Engineering. If you have an opening for such an engineer with a great interest in research, will you please examine the enclosed list of qualifications and grant me an interview? I will be in Columbus March 19 through March 23, 1975. Meanwhile, I may be reached at the above address.

Respectfully yours,

Douglas A. Strong

ds
Enc.

RESUME

Douglas A. Strong
640 South Howes
Fort Collins, Colorado
HU 4-0365
(address and telephone number until June 10, 1975)

Personal Data

Age: 22
Height: 5'11"
Weight: 140 lbs.
Health: Excellent
Marital Status: Married
Special interests: Motorcycles, tennis, fishing
Memberships: ASME, Pi Tau Sigma

Education

BSME, Colorado State University   June 10, 1975
    studied Statics, Dynamics, Thermodynamics, Principles of Elec-
    trical Engineering, Fluid Mechanics, Heat Transfer
Diploma, Grand Junction High School   June 7, 1971

Experience

August 1972 to present: Aerodynamics Lab, Colorado State University;
    did drafting and computations; designed carriage and transition
    sections for new wind tunnel; first experience with actual engineer-
    ing work.
Summer 1974: Great Western Sugar Company, Fort Collins, Colorado;
    loaded sugar bags into railroad cars; required physical stamina,
    ability to get along with many different kinds of people.
November 1972 to September 1973: American Gilsonite Company,
    Fruita, Colorado; worked as timekeeper and laborer; had some ex-
    perience as helper for pipefitters, machinists, carpenters, and boil-
    ermakers; became familiar with hardware used in refining gasoline
    and coke; job required knowledge of use of tools, responsibility.
Summer 1971, 1972: Seven-Up Bottling Company, Grand Junction,
    Colorado; worked as truck loader, driver-salesman, machine opera-
    tor, repaired machines; job required mechanical ability, physical
    stamina, sales ability.

References

Marvin Jackson, Seven-Up Bottling Company, Grand Junction Colo-
    rado
John Henderson, Head Engineer, American Gilsonite Company, Fruita,
    Colorado
Erich Plate, Aerodynamics Lab, Colorado State University, Fort Col-
    lins, Colorado

Proposal Letter

Department of Agricultural Engineering
Colorado State University
Fort Collins, Colorado 80521

December 8, 1974
(CEP 60 NAE 35)

Mr. W. D. Dreher, Manager
Wire Products Sales
The Colorado Fuel and Iron Corp.
Box 1920
Denver, Colorado 80201

Dear Mr. Dreher:

I have your letter inviting a proposal on testing nails for withdrawal resistance. We would be glad to conduct the required standard test on one or more types or sizes of nails. We would conduct the test on 20 specimens of each type or size of nail in accord with paragraph 8 of the "Gypsum Dry Wall Contractors International, Recommended Performance Standards for Nails for Application of Gypsum Wallboard" dated June 30, 1974.

We could begin the tests immediately on the signing of a contractual agreement between your firm and the Colorado State University Research Foundation. The costs given below include the preparation of a report and an overhead cost. Additional specimens can be tested for considerably less than the first one.

Our proposal for testing the first size, $1\frac{1}{4}$ inch No. 14 gage nail is:

1. Salaries:
   Professional Engineer, 0.6 month $1000.00
   Clerical, 8 hrs. at $1.50 .... 20.00
2. Equipment and Supplies
   Strain gage elements 30.00
   Lumber and supplies 30.00
3. Report duplication
   50 copies, 3 pages each 50.00
4. Annuities, 6% of Salaries 61.70
5. Overhead, 36% of Salaries 571.20
   Total (first test) $1762.90

Tests on additional types or sizes will be made for 40% of the cost of the first test, or $705.16.

We would be glad to have the opportunity to be of service to your firm and will look forward to hearing from you.

Yours truly,

Norman A. Evans
Head of Department

NAE/law

Description of a Process

How to Determine the Internal Resistance of a D.C. Voltage Source

A manufacturer does not always provide specific information on the internal resistance of a D.C. voltage source. You can determine the unknown resistance by making several simple voltage measurements using ordinary laboratory equipment.

The laboratory equipment you need for the measurements are:

A high impedance D.C. voltmeter

A calibrated variable resistor

Four insulated conductors

The steps followed are: (1) connecting the voltmeter, (2) measuring the open-circuit voltage, (3) connecting the variable resistor, (4) measuring the voltage across the variable resistor.

Connecting the Voltmeter

1. Estimate the magnitude of the internal resistance of the D.C. source, and choose a voltmeter with a resistance of at least twenty times this estimated value.

2. Connect one insulated conductor to the voltmeter terminal marked (+).

3. Connect one insulated conductor to the voltmeter terminal marked (KV). K is the voltage magnitude corresponding to full scale deflection of the meter. The scale chosen should be large enough to measure the total source voltage. See Figure 1.

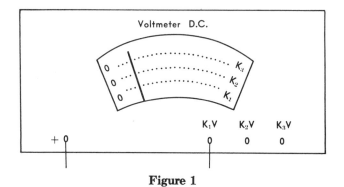

Figure 1

Measuring the Open-Circuit Voltage

1. Connect to one source terminal the free end of the conductor attached to the (KV) terminal of the voltmeter.

2. Touch the other source terminal with the free end of the conductor attached to the (+) terminal of the voltmeter, and observe the voltmeter needle. If the voltmeter needle moves to the right, the source terminal touched is positive. If the needle moves to the left, the terminal is negative.

3. Having determined the source polarity, connect the (−) terminal of the voltmeter to the positive terminal of the source and the (KV) terminal of the voltmeter to the negative terminal of the source.

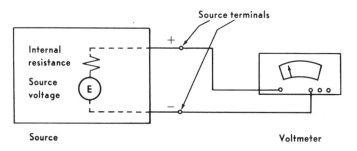

Figure 2

4. Read and note the voltage indicated on the appropriate voltmeter scale. The voltage thus read is effectively the source voltage. See Figure 2. Since the voltmeter resistance is so much larger than the internal resistance of the source, the voltage drop across the internal resistance is insignificant compared to the voltage drop across the meter. Also the large impedance limits the current, thus making the voltage drop across the internal resistance of the source even more insignificant.

Connecting the Variable Resistor

1. Do not change the voltmeter connections.
2. By means of the other two conductors, connect the variable resistor between the two source terminals. See Figure 3.

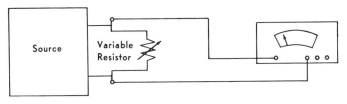

Figure 3

Measuring the Voltage Across the Variable Resistor

1. Adjust the variable resistor manually until the voltmeter indicates a voltage exactly one-half that indicated for the open-circuit condition previously noted.
2. Note the resistance of the variable resistor at that setting. That value of resistance is the internal resistance of the source. The high impedance voltmeter will draw only an insignificant current when in parallel with a low resistance, so therefore will indicate the actual voltage across the low value resistor. Since the sum of the voltages around a closed loop is zero, and since one-half of the source voltage now appears as a voltage drop across the variable resistor, the other half must be a voltage drop across the internal resistance of the source. With the same current through both resistors and equal voltage drops across both, the resistors must be equal in magnitude.

Sample Questionnaire

Dear Sir:

May we have just a few minutes of your time and the benefit of your experience in answering a few questions pertinent to the development of a Suggestion System Program?

Your assistance will be especially useful in helping us to plan effectively for a suggestion program. We would greatly appreciate your answering the brief survey which is attached and returning it to us in the attached postage-paid envelope. If you would like the tabulated results of this survey, please sign your name so that we may direct the results to you personally.

We shall appreciate your cooperation.

Cordially yours,

Please Answer All Applicable Questions

1. Age of Company: Less than 5 years._____, 5 to 10 yrs._____, more than 10 years._____.
2. Approximate number of employees: Under 50_____, 50-99_____, 100-250_____, 251-499_____, 500 or more_____.
3. Do you *have* or *have you had* a Suggestion System in your Company? Yes_____, No_____.
4. Is it in operation now? Yes_____, No_____.
5. If answer in 4 is no, please indicate reasons by checking:
   a. Too costly
   b. Lack of interest of employees
   c. Management indifference
   d. Supervisory resistance
   e. Reward system inadequate or unsatisfactory
   f. Other
6. If you do not have a Suggestion System, do you have alternate means of eliciting employee ideas: Please indicate_____
   _____.
7. How long have you had your Suggestion System in operation?_____
   _____.
8. Does your program include participation by Engineers_____, Foremen_____, Middle Management (Department Heads)_____, Upper Management (Division Heads)_____?
9. Types of Awards Offered: Cash_____, Certificates_____, Other _____?
10. What is your maximum award_____; minimum award_____; average size of award_____?
11. Who administers your Suggestion System?
    a) Personnel Office
    b) Suggestion System Director
    c) Committee
    d) Other_____
12. Do you consider that your suggestion system accomplishes its objectives? Yes_____, No_____.

COMMENTS:_____
_____
_____
_____

Sentence Outlines

PLASTICS IN AVIATION

Thesis Sentence: In the aircraft field, plastics are used for three general purposes, which, in order of their importance, are as follows: (1) construction of the air frame and wings and other major parts, (2) construction of minor parts not necessary to the structural integrity of the airplane, and (3) accessories such as control knobs, etc.

   I. The purpose of this paper is to inform the general reader of the importance of plastics and plastic products to the aircraft industry.

      A. Although it would be nearly impossible to tell of all the applications for which plastics are used in the aircraft industry, the entire field will be surveyed.

      B. There are several terms that should be defined in order that the data of the report can be fully understood and appreciated.

         1. Plastics are a group of organic, synthetic, or processed materials which may be molded or cast.

         2. Plastics are nonmetallic compounds synthetically produced (usually from organic compounds by polymerization) which can be molded into various forms and hardened for commercial use.

         3. A phenolic is a synthetic resin, usually made by reacting phenol with an aldehyde.

      C. The investigation of plastics in aviation will be made on the basis of the advantages of plastics over other materials and on the basis of the three general purposes of plastics in the field of aviation.

  II. The most important use of plastics in the aviation industry today is that of structural components.

      A. The more prominent plastic products which are used in this field are bonded plywoods, glass reinforced plastics, honeycomb core plastics, and foamed plastic.

      B. Some of the favorable characteristics which make these materials suitable for structural components are high tensile strength, good torsional oscillation strength, light weight, low coefficient of expansion, aerodynamic smoothness, and the ease with which they can be altered to fit the design.

      C. Some of the structural components for which the above materials are used are wings, bulkheads, air frames, and tail cones.

 III. Plastics are now trying to work their way into the engine and other parts which are not necessary to the structural integrity of the plane.

A. Some of the plastic products which are used in this phase of the aviation industry are Fiberglas reinforced plastics, corrugated laminated plastics, bonded honeycomb cores, and foamed plastics.

B. A few of the outstanding characteristics of these materials are light weight, good resistance to temperature, and low cost of production.

C. These materials are used for making such parts as engine compressor blades, disks, and housings; canopies, armament, dropable fuel tanks, doors, partitions, and control surfaces.

IV. The smaller parts and accessories of an aircraft which used to be troublesome to produce are now being made of plastics very satisfactorily.

A. Most of these small parts are made from polyester laminates and resin impregnated glass fiber pads.

B. Durability, light weight, appearance, and low cost of production are among the advantages of plastics over other materials for these products.

C. Such parts as battery containers, power distribution boxes, interior trim, cargo compartment liners, and instrument panels are among the many now being made from plastics.

V. Although plastics offer many advantageous characteristics and there seems to be a type of plastic for every purpose, it is unlikely that the airplanes of the near future will be made entirely of plastics or plastic products even though the plastics family will be well represented.

Student Reports

ADDITIONAL WATER SUPPLY FOR THE CITY OF FORT COLLINS

Thesis Sentence: The most feasible method of supplying Fort Collins with additional domestic water is one of the following:

1. a dam on the Cache La Poudre River and 100% metering of the city, or

2. a dam on the Cache La Poudre River and construction of an addition to the present water treatment plant, or

3. construction of a new treatment plant which draws water from Horsetooth Reservoir and which will be operated in conjunction with the present treatment plant.

I. The city of Fort Collins, like many other cities in the western United States, is presently faced with a water shortage problem which will become serious in a very few years if not alleviated.

A. The purpose of this investigation is to offer a solution to the water storage problem.

B. The people of Fort Collins first considered a municipal water system in 1875 and since that time the size of the system has been greatly increased.

   1. From 1883, when the original system was completed, until 1903 the city relied on a ditch taking water from the Poudre River for its water supply.

   2. In 1903 a diversion structure was built on the Poudre River itself, and in 1909 a treatment plant consisting of a small settling basin and 4.5 mgd capacity of rapid sand filters was completed.

   3. A 4.5 million gallon treated water storage reservoir was built on Bingham Hill in 1910 and in 1922 a two million gallon plain sedimentation tank was added at the treatment plant.

   4. A considerable enlargement to the system was undertaken and completed in 1925-1926.

     a. Rapid sand filters with a capacity of 6.0 mgd and new chlorinating equipment were added to the treatment plant.

     b. Soldier Canyon distribution reservoir, having a capacity of eight million gallons, was constructed and several new transmission and distribution pipelines were added to the system.

   5. A pretreatment works, consisting of a flash mixer, flocculator, and mechanical clarifier, was added to the treatment plant in 1948.

   6. In 1954 capacity of the plant was increased to 18.0 mgd through the addition of six new rapid sand filters and additional transmission and distribution pipelines.

C. This report is limited to the investigation of three possible methods of supplying Fort Collins with an additional water supply which will meet 1980 demands.

   1. One possibility considered was 100 percent metering of the city and construction of a dam on the Poudre River.

   2. Another method was to construct a dam on the Poudre River and enlarge the present treatment plant.

   3. A third alternative is to construct a new treatment plant at Horsetooth Reservoir to be used in conjunction with the present plant.

D. The collection of the data and information for this report was carried out by myself and two other students.

   1. Assisting in the collection of data were Bob Saulmon and Chester Smith, seniors in Civil Engineering at CSU.

2. Except for the collection of the data and the compilation of tables and graphs, this investigation and the conclusions drawn are entirely the work and opinions of the writer.

E. This problem will be approached by analyzing the city's present and future water requirements, then by considering each of the individual elements of the three alternatives mentioned above, and finally by drawing conclusions as to which of the alternatives is the most feasible.

II. The present Fort Collins water works is required to operate at capacity during peak consumption and will not be able to meet greater demands in the near future.

   A. The peak water demand during the summer of 1960 was 21.0 mgd for a population of 25,000, whereas the maximum capacity of the present water works is only 18.0 mgd.

      1. The peak demand was for a short period of time so it was possible to make up the deficiency by drawing water from the 12.0 million gallon storage tanks.

      2. If the peak consumption were to last for four consecutive days, there would be a water shortage.

   B. It is estimated that the population of Fort Collins will be about 40,000 by 1980, thus if present water consumption rates continue the city will require 28 mgd.

      1. One method of meeting this demand will be to provide additional treatment facilities capable of treating 12 mgd.

      2. A second method of meeting the demand is to reduce the consumption rate.

   C. The major deficiency of the present water works is lack of storage.

      1. The city now has no water storage available at all and if the Poudre River should become extremely low the city would have a very serious water shortage.

      2. Storage facilities are considered absolutely necessary to any future plans for Fort Collins.

III. One method of furnishing water storage is to build a dam and reservoir.

   A. It is estimated that at least 10,000 acre feet of storage must be provided if a dam is built.

   B. The most feasible dam location appears to be the Rockwell Ranch site, located on the South Poudre River just above its confluence with the main river.

      1. It is expected that a 10,000 acre foot reservoir at this site would be completely filled on an average of one year out of every five.

2. The total cost of an earth fill dam and reservoir would be about $1,700,000.

3. The dam would be constructed for the purpose of water storage only.

IV. Storage could also be obtained by taking water from Horsetooth Reservoir through the Soldier Canyon dam outlet.

    A. The city of Fort Collins presently owns 6,000 shares of Horsetooth water.

        1. This 6,000 shares provides an average of 4,200 acre feet of water each year, which is probably sufficient to supply Fort Collins until 1980.

        2. The water contains a large amount of algae which would have to be eliminated in the treatment process.

    B. At the present time this water can only be used from April to October so a treatment plant would probably be used as a peaking plant in the summertime.

V. Additional treatment plant facilities will be necessary in order to meet the future demand if the present consumption rate continues.

    A. If a dam were built on the Poudre River, an addition to the present treatment plant and additional transmission pipelines would be necessary.

        1. A filter gallery containing eight rapid sand filters would supply 12.0 mgd at a cost of about $280,000.

        2. A 27-inch steel pipeline with a capacity of 12.0 mgd could be constructed for about $1,240,000.

    B. A complete new treatment plant and short transmission pipeline will be required if Horsetooth Reservoir water is used.

        1. The cost of a complete treatment plant capable of treating 12.0 mgd will be approximately $1,500,000.

        2. A short transmission pipeline leading from Horsetooth to the treatment plant and from the treatment plant to Soldier Canyon reservoir will cost approximately $20,000.

VI. The only method of reducing the consumption rate of water, other than rationing, is to install water meters.

    A. At the present time only industrial concerns and large water users are metered in Fort Collins.

    B. It is estimated that if Fort Collins were converted to 100% meters the rate of water consumption would be decreased by at least one-third.

        1. Assuming a consumption rate of 28.0 mgd in 1980, based on present use, the installation of meters would reduce this to 18.7 mgd.

2. The present treatment plant can supply this amount of water, thus meters would postpone the construction of additional treatment facilities until about 1980.

C. It will be necessary to install about 7,600 meters costing $608,000 at the present time and it is estimated that 4,500 meters costing $540,000 will be required by 1980.

VII. The basis for selecting one of the above alternatives will be the feasibility and adaptability of the project as well as the total cost.

A. The cost of the project will be the main item for consideration by those concerned.

1. The total cost of building a dam and placing the city on water meters is estimated to be about $2,848,000.

2. Construction of a dam and an addition to the present treatment works will cost approximately $2,900,000.

3. If Horsetooth Reservoir water is used, the initial cost of the facilities will be about $1,520,000.

B. The feasibility and adaptability to the present system must also be considered before deciding on any one alternative.

1. Alternative 1 is entirely feasible; however, the residential sections of Fort Collins have never been metered and there will probably be considerable objection to metering.
   a. For this reason it is recommended that the city water remain on a flat rate as long as possible.
   b. This alternative should be kept in mind for future use if the need for water arises again.

2. Alternative 2 is easily adaptable to the present water system, however it may not be feasible if it is found that the reservoir will not fill as often as expected.

3. Alternative 3 is also adaptable to the present system and seems to be entirely feasible.

C. From the above considerations, it is recommended that alternative 3, construction of a system to permit use of Horsetooth water, be undertaken as soon as possible.

1. The initial cost of alternative 3 is considerably less than the other alternatives.

2. If, in the future, more water is required, it can be purchased at Horsetooth, whereas the other two alternatives would require a considerable expenditure to obtain additional water rights.

3. If it becomes necessary to use this water year round the law requiring use for only six months out of the year could probably be changed.

4. This alternative will also result in fewer transmission pipelines to be maintained.

636½ South Loomis
Fort Collins, Colorado 80521
May 26, 1974

Mr. John R. Doe
City Engineer
City of Fort Collins
Fort Collins, Colorado 80521

Dear Mr. Doe:

In compliance with the terms of my employment by the city of Fort Collins, I submit this report concerning the problem of furnishing Fort Collins with an additional water supply which will meet its needs until 1980.

I wish to express my appreciation to the Fort Collins Water Department for their assistance in the completion of this report.

Respectfully submitted,

M. Dean Skalla

Enclosure

Report on

ADDITIONAL WATER SUPPLY FOR THE CITY OF
FORT COLLINS

Submitted to
Mr. John R. Doe
City Engineer
Fort Collins, Colorado

By
M. Dean Skalla
Civil Engineering Student
Colorado State University

Fort Collins, Colorado
May 26, 1974

ABSTRACT

The purpose of this report is to investigate the Fort Collins domestic water situation and make recommendations as to the most feasible method of supplying the city with additional water. The report is based on the city's estimated water needs in 1980.

The investigation was carried out by studying methods of furnishing Fort Collins with adequate water storage, and a waterworks system capable of meeting the 1980 demands. Several methods of meeting these two requirements were studied and then the individual elements were combined into three alternatives, each of which will solve the problem.

The three alternatives were (1) construction of Rockwell Dam on the Little South Poudre River and enlargement of the present waterworks system, (2) construction of Rockwell Dam and 100 percent metering of the distribution system, and (3) construction of a waterworks system, which will be operated in conjunction with the present system, to use Horsetooth Reservoir water.

It was found that the cost of alternatives 1 and 2 was considerably more than the cost of alternative 3. Therefore, it is recommended that the City of Fort Collins adopt the program set forth in alternative 3 at the earliest possible date.

APPENDIX A

FIGURES

APPENDIX B

TABLES

ADDITIONAL WATER SUPPLY FOR THE CITY OF FORT COLLINS

INTRODUCTION

Significance and Purpose of Investigation

Like many other cities in the western part of the United States, the city of Fort Collins is presently faced with a water shortage problem which, if not alleviated, will become quite serious in a very few years. The purpose of this report is to offer several possible solutions to the problem and to make recommendations as to which plan the city should adopt.

History of the Fort Collins Water-Works System

The people of Fort Collins first considered a municipal system in 1875 and since that time the system has been steadily improved and its size greatly increased.

The original water works was completed in 1883 and consisted of a small pumping plant which pumped water from a ditch supplied by the Cache La Poudre River. Many difficulties were experienced with the original system and, as a result, a diversion structure was built on the Poudre River itself in 1903. From the intake, water was transported to the city through wood stave pipes of various sizes.

In 1909 a small treatment plant consisting of a settling basin and rapid sand filters was built; the following year a 4.5 million gallon treated water storage reservoir was constructed on Bingham Hill. A two million gallon plain sedimentation tank was added to the treatment plant in 1922.

Due to increased consumption, a considerable enlargement to the water-works system was undertaken in 1925-1926. The treatment plant was enlarged through the addition of 6.0 mgd (million gallons per day) capacity of rapid sand filters and a new chlorination system. The wood stave transmission pipelines were replaced with a 24-inch steel transmission line from the treatment plant to the city, and a major extension of city distribution lines was completed. Another treated water storage reservoir with a capacity of 8.0 mgd was constructed at Soldier Canyon.

A pretreatment works, consisting of flash mixer, flocculator, and mechanical clarifier, was added to the treatment plant in 1948. In 1954 the capacity of the plant was increased to 16.5 mgd through the addition of six new rapid sand filters, a 27-inch steel transmission main, and additional distribution lines.

In the spring of 1961 a small addition to the transmission lines was added. This increased the capacity of the system to 18.5 mgd. The water-works system then consisted of the following elements, with capacities as indicated:

    A. Treatment plant, 20.0 mgd

    B. Pipelines from treatment plant to storage reservoirs, 18.5 mgd

    C. Pipelines from storage reservoirs to city, 30.0 mgd

    D. Treated water storage reservoirs, 12.0 mgd

Such is the state of the Fort Collins water-works system upon which this report is based.

Scope of the Investigation

This report is limited to the investigation of three possible methods of supplying Fort Collins with an additional domestic water supply which will meet the city's demands in 1980. These alternatives are:

Alternative 1. Construction of a dam and reservoir on the Little South Poudre River and enlargement of the present treatment plant.

Alternative 2. Construction of a dam and reservoir on the Little South Poudre River and 100 percent metering of the city's water distribution system.

Alternative 3. Construction of a new treatment plant, which will treat Horsetooth Reservoir water, to be used in conjunction with the present treatment plant.

Personnel Involved in the Investigation

The collection of the data and information for this report was carried out by two other students and the writer. Assisting in the collection of The data and the compilation of graphs and tables were Robert Saulmon and Chester Smith, seniors in Civil Engineering at Colorado State University. Except for the aforementioned assistance, this investigation, the conclusions drawn, and the recommendations made are entirely the work and opinions of the writer.

Organization of the Report

This investigation will be approached by first discussing the city's present water rights and its present and future water requirements. The individual elements of each of the three alternatives will then be considered and conclusions drawn from an analysis of these considerations. Finally, recommendations will be made as to which of the alternatives is the most feasible.

FORT COLLINS WATER RIGHTS

The city of Fort Collins now has rights to water from two sources, the Poudre River and Horsetooth Reservoir. The city has direct flow decrees on 19.93 sec. ft. of Poudre River water which can be diverted at the present treatment plant site. These decrees will furnish a maximum of 13.0 mgd. Tabulated information on the individual decrees is shown in Table 1.

Fort Collins also has direct flow rights to 8.00 sec. ft. of No. 1 decree and 3.10 sec. ft. of No. 3 decree. However, this water cannot be diverted at the treatment plant site, so it has been used only for irrigation. The city is now trying to change the point of diversion through court action. In the past it has been difficult to change point of diversion under Colorado water law; therefore this report is written under the assumption that Fort Collins will be unsuccessful in its attempt. The city also has direct flow rights to 1200 ac. ft. of ditch water each year. This water is used only for irrigation.

In addition to the Poudre River water, Fort Collins also owns 6,000 shares of Horsetooth Reservoir water. In the past, these rights have yielded an average of 4,260 ac. ft. of water each year. The outlet from Horsetooth Reservoir to the Poudre River is several miles below the present treatment plant so the water cannot be directly used for a domestic water supply. However, in the past some of this water has been traded for Poudre River water, which can be diverted through the treatment plant.

The Horsetooth Reservoir rights are the only storage rights now owned by Fort Collins. All rights which the city owns on the Poudre River are for direct flow only, and none of the water can be stored. Since Horsetooth water cannot be routed through the treatment plant the city has no storage at all that is avaiable as a direct domestic supply. As long as there is sufficient water flowing in the Poudre River, it will be possible to trade Horsetooth water for Poudre water when the need arises. However, if the water in the Poudre were to become extremely low, the city would have no other water source to turn to. This is a definite possibility and point out the danger involved in operating a water system without storage facilities. Therefore, adequate storage facilities are considered absolutely necessary to any future plan for supplying Fort Collins with additional water.

Present and Future Water Requirements of Fort Collins

The past records of the Fort Collins treatment plant show a rapid rise in the amount of water used the past few years (see Fig. 1). If this trend continues the water-works system will soon be unable to meet the demand.

The present water works is now required to operate at capacity during peak consumption periods in the summer months, and even now it cannot keep up with the demand. During the summer of 1960, the peak water demand reached 21.0 mgd for a population of 25,000, whereas the maximum capacity of the present water works is only 16.5 mgd. However, the maximum demand was for short periods of time so it was possible to make up the deficiency by drawing water from the clear water storage tanks. The tanks were then refilled at night when the demand was not so great.

Even with the proposed installation of additional transmission lines, which will raise the capacity to about 18.5 mgd, it is doubtful if the plant can meet the demand for more than three or four more years at the most. By 1980 it is estimated that the population of Fort Collins will be about 40,000 (see Fig. 2). If the city's present consumption rates are applied to this population figure, it is estimated that the 1980 water demand will be about 28 mgd (see Fig. 1).

One method of meeting this demand will be to provide additional water treatment facilities capable of treating enough water to satisfy the 1980 demands. A second method is to reduce the consumption rate of the city to such a point that the present water-works system will still be adequate in 1980. Methods of obtaining adequate storage facilities and meeting the 1980 water demand will be discussed in the following sections.

STORAGE POSSIBILITIES

Rockwell Dam and Reservoir

One method of furnishing water storage to the city of Fort Collins is to build a dam on the Poudre River or on one of its tributaries. It is estimated that at least 10,000 ac. ft. of storage must be provided if a dam is built.

An excellent site for a dam is on the Poudre River near its confluence with Elkhorn Creek. However, a dam on this site would preferably be a large undertaking capable of supplying hydroelectric power, and its cost would be far more than the city of Fort Collins can afford. The site is now being studied by the Bureau of Reclamation, and it is possible that they will build a dam there in the future. If the dam is constructed, Fort Collins could probably obtain water from the project; however, the city's needs are immediate, so no further investigation was carried out on this plan.

The most feasible location for a city-constructed dam appears to be the Rockwell Ranch site, located on the Little South Poudre River. The site is located eight miles above the mouth of the Little South, approximately thirty miles west of Fort Collins (see Fig. 5). The main reason for choosing this site is that the city owns the land. This will mean a considerable savings in the cost of the project if it is carried out. A dam constucted at this site would be for storage purposes only, since it would not be large enough for hydroelectric facilities and the flood control benefits would be negligible.

If a dam were constructed at this site there is some question as to whether or not there is enough water available to keep the reservoir adequately full. A study of the decree rights on the Poudre River indicates that the river is far overdecreed. This is possible because most of the water users do not divert all of the water which their decrees call for. In an average or dry year almost all of the Poudre water is used and there has been very little, if any, available for storage. However, in years of more than average rainfall there is a considerable amount of water wasted. This is shown by comparing river flow past the Greeley gaging station in average and wet years. In an average year the flow past the Greeley station averages about 25,000 ac. ft., whereas in a wet year the flow averages about 60,000 ac. ft. Thus it appears that in some years there may be as much as 35,000 ac. ft. of water which could be stored.

It is this excess flow in wet years upon which Fort Collins must depend to fill a reservoir. Through the use of a flow duration curve (see Fig. 3), it is expected that an 11,500 ft. reservoir on the Poudre River itself will be completely filled on an average of one year out of every five. The Little South Poudre supplies about 10 percent of the total flow in the Poudre River. By capturing all the flow in this tributary, and this can only be done during periods of high flow in the Poudre River itself, it is probable that a reservoir on the Little South will also completely fill on an average of one year out of every five. It is expected that between periods of complete fillage the reservoir will capture sufficient water to ensure the city of Fort Collins of an adequate supply.

A study was made and a report written in 1960 on the Rockwell Dam site by the Engineering Consultants Inc., of Denver, Colorado. All the information concerning cost and size of the dam and drawings of the dam site were taken from that report.

In that report several possible dam and reservoir sites were studied. An analysis of these possibilities by the writer of this paper indicated that a dam which will provide 11,500 ac. ft. of storage was the most feasible and economical. The requirement that at least 10,000 ac. ft. of storage be provided was the primary item influencing this decision. The relationship between the cost of a dam and storage provided by the dam is shown in Fig. 4.

The high cost of constructing a concrete dam at the Rockwell site will rule it out; therefore, the dam will of necessity be an earthfill structure. The dam will be 104 ft. high and require about 681,000 cu. yds. of fill material. The fill material is readily accessible and will present no problems. Diversion through the dam will be handled by twin 6' x 6' concrete conduits. The outlet gates will pass 400 cfs which is the estimated maximum river discharge which will occur each year. The spillway will be an unlined open channel cut in the left embankment of the dam. Dimensions for the dam are shown in Fig. 6.

The total cost of constructing a dam at the Rockwell Ranch site will be about $1,614,000 or $140 per ac. ft. of storage. The cost will consist of the items shown in Table 2. It is estimated that the project will take about two years to complete because of the short working season.

Horsetooth Reservoir

Another source of water storage for Fort Collins is Horsetooth Reservoir. As previously mentioned, the city owns 6,000 shares of Horsetooth water which provide an average of 4,260 ac. ft. of water each year. The only Horsetooth water which the city now uses is that which is traded for Poudre River water during peak demand periods in the summer. This is done because the Poudre water can be diverted through the treatment plant, whereas this is impossible with Horsetooth water, as was shown above. This practice is carried out only when the city's Poudre River rights are insufficient to meet the demand. The amount of water traded each year is only a small percentage of the Horsetooth water owned by the city. Although this procedure has been used quite often, it is unsatisfactory because of the danger that the flow in the Poudre River may become so low that there will be no water to divert. Another drawback to trading Horsetooth water is the difficulty experienced in negotiating the trades. Many of the farmers, with whom the trades must be made, seem to feel that they will not receive a fair exchange of water so they are reluctant to trade.

Because of the danger of water freezing in canals, the Bureau of Reclamation presently allows water to be taken from Horsetooth only from April to October. The peak demand on the city's water supply is also experienced during this period, usually from June to October, so it will be beneficial to operate a water-works system at Horsetooth as a peaking operation. However, if the present plant could not meet demands at other times, Fort Collins would then be allowed to use Horsetooth water at all times. This would be acceptable because the water would be transported in pipes, not in open canals, and there would be no danger of the water freezing.

It is estimated that the 4,200 ac. ft. of storage provided by Horsetooth Reservoir will be more than sufficient to meet Fort Collins 1980 water demands if it is used in conjunction with Poudre River water. The city now has access to 3.5 billion gallons of Poudre River water each year and it is estimated that the demand in 1980 will be about 4.4 billion gallons. Thus the city must obtain an additional 0.9 billion gallons. Horsetooth Reservoir will supply 1.37 billion gallons. If more water is needed in the future, it will probably be possible to purchase more Horsetooth water.

The water in Horsetooth is readily accessible to the city of Fort Collins. A diversion structure was provided at Soldier Canyon Dam at the time of its construction especially for the city's future use. Since the city already owns water in Horsetooth, there will be no initial cost for the storage.

One drawback to using Horsetooth water is that the water contains a considerable amount of algae. High concentrations of algae cause an undesirable odor and taste in the water. This problem will have to be carefully considered when designing a treatment plant. It will be discussed further in a later section. Otherwise, the Horsetooth water will present no special problems.

ADDITIONAL WATER-WORKS SYSTEMS

As previously pointed out, the present Fort Collins water system is incapable of meeting the water demands during periods of peak consumption. Thus, if present consumption rates continue, it will be necessary to construct additional treatment facilities and transmission lines, as well as to supply more water. Although the present water works can supply a maximum of 18.5 mgd, the system is more efficient if operated at 16 mgd or less. Therefore, based on a population of 40,000 and a consumption rate of 28 mgd, an additional water system must have a capacity of 12 mgd to meet the estimated 1980 demands. The present clear water storage facilities and transmission lines from the reservoirs to the city will still be adequate in 1980.

The two alternatives discussed above for furnishing Fort Collins with additional water supply and storage will require different treatment and transmission facilities. These are discussed below.

Water-Works System for Proposed Rockwell Reservoir

The water from Rockwell Reservoir will enter the Poudre River above the diversion to the present treatment plant. Thus the additional water can also be treated at this plant if the facilities are expanded. The present pretreatment works are adequate to handle 28 mgd, but the rapid sand filters and transmission lines have a capacity of only 13.5 mgd. There is ample space at this location for additional construction. The treatment plant site is shown in Fig. 8.

Additions to the present system will consist of a filter gallery containing eight rapid sand filters and a transmission line from the treatment plant to the city. This new system will have a capacity of 12 mgd and the total cost will be about $1,520,000 as shown in Table 3. It is estimated that the project will take about one year to complete.

Horsetooth Reservoir Water-Works System

The city of Fort Collins has been reluctant in the past to use Horsetooth Reservoir water as a domestic supply. The primary reason for this reluctance is that Hoorsetooth water contains a high concentration of algae, which has caused considerable difficulties when treated by conventional methods. The algae can be removed by ordinary rapid sand filters, but the filters must be backwashed so often that this has proved uneconomical. The city of Loveland, Colorado, has found that the algae, along with a considerable amount of turbidity, can be removed from the water by microstrainers. They found that when the microstrainers were used the rapid sand filters required backwashing every 48 hours; whereas, when the microstrainers were not used, the filters had to be backwashed as often as every four hours.

The Denver water board has also experienced difficulties due to algae in the city's water and has conducted an intensive investigation into the possibilities of using microstrainers. Tests conducted on lake water have shown that microstraining reduced the microorganisms by about 94 percent and the turbidity by about 31 percent. In comparison, rapid sand filters reduced the microorganisms by about 88 percent and the turbidity by about 68 percent. (12:13) The tests also showed that microstraining required less wash water. As a result of these tests, the Denver water board feels that microstrainers will furnish adequate filtration of their water. Since the initial costs and maintenance costs of microstrainers are considerably less than that of rapid sand filters, Denver is now installing them as a primary treatment process. Rapid sand filters will not be used at all.

In view of the foregoing discussion, it appears that microstrainers could also be used to remove the algae from Horsetooth water. However, the use of microstrainers as the only form of filtration is highly questionable. If Horsetooth water is used for domestic use, it is recommended that microstrainers be used in conjunction with rapid sand filters. The microstrainers would be used for initial filtration and the rapid sand filters for final filtration. This is the procedure adopted at the Loveland plant, and it has given very satisfactory results.

As previously mentioned, the water demand in 1980 is estimated to be about 28.0 mgd. If the Horsetooth treatment plant is designed for a maximum output of 12.0 mgd, the present plant must supply 16.0 mgd. It is estimated that this output will provide the most economical service from the plant. However, the maximum amount of water which can be taken from the Poudre River is 13.0 mgd. It would be uneconomical to operate the present plant at a maximum of 13.0 mgd when it could operate 16.0 mgd per day because then the Horsetooth plant would have to be larger. Thus, on days of maximum demand, it may still be necessary to trade a small amount of Horsetooth water for Poudre River water.

The size and cost of a plant for treating Horsetooth water will be considerably less than that of a plant treating river water. This is because the water is relatively free of turbidity so flocculation and sedimentation tanks will not be required. The most suitable location for the treatment plant is the 12 acre plot of land upon which Soldier Canyon Reservoir is located. This site, which is about one-quarter of a mile east of Soldier Canyon Dam, is shown in Fig. 9. The layout of the treatment plant is shown in Fig. 10.

In addition to the treatment plant, a short transmission line will be required from the outlet works in Soldier Canyon to the plant and from the plant to Soldier Canyon Reservoir. The total cost of the system is estimated to be about $567,240, as shown in Table 4. It is estimated that the project will require one year to complete.

REDUCTION OF WATER CONSUMPTION

In the past history of Fort Collins, it has never been necessary to ration or meter domestic water; as a result the city has a very high water consumption rate. The average daily consumption during the year 1960 was 300 gallons per capital. At times during the summer, the daily consumption rate has exceeded 500 gallons per capita.

The only way to reduce the water consumption rate is to either ration the water or meter it. Water rationing is usually very unpopular, and Fort Collins will resort to it only as an emergency measure.

Water Meters

Part of the Fort Collins water system is now metered, but most of it is on a flat rate basis. All commercial customers or large water users, such as Colorado State University, are metered. Residential customers are charged a flat rate. The University and other large water users will probably use about the same amount of water no matter how they are charged for it. Thus it is the residential water users who will control the rate of consumption.

Several other cities have found it necessary to convert from a partially metered system to 100 percent meters. A study of the results of these conversion projects showed that meters reduced the consumption anywhere from 25 to 300 percent. E. W. Steel states that metering all services of a city should reduce the consumption by about 50 percent without meters (11:17). After the original installation of meters, there is a tendency for the consumption rate to gradually increase over a period of time.

If Fort Collins were to convert to 100 percent meters, it is estimated that the initial reduction in water consumption will be at least 40 percent. By 1980 the reduction will be less, probably about 35 percent. Without meters, the 1980 consumption was estimated at about 28.0 mgd, thus if meters are installed this will be reduced to about 18.5 mgd. This is the capacity of the present water-works system, so 100 percent metering of the system will postpone the construction of additional water supply facilities until about 1980.

Other cities have found that the conversion to meters works the best when the city itself purchases, installs, and maintains the meters. The costs given below are based on such a plan.

To convert Fort Collins to 100 percent meters will require the installation of 7,600 meters now and an additional 4,500 by 1980. The wabling disk type meter, costing about $80 in place, is the most common. The installation of 7,600 meters of this type will cost the city about $608,000. It is estimated that the additional 4,500 meters will cost about $540,000, based on an estimated average price for 1970. Thus the total cost of installing all meters required by 1980 will be about $1,148,000. The engineering fees are considered negligible for this project. A small allowance for contingencies is included in the total cost figure.

CONCLUSIONS

As previously mentioned, the city of Fort Collins feels it is absolutely necessary that any future water-works expansion include storage facilities. This means that an addition to the present treatment plant or

100 percent metering of the city water cannot be considered as alternative plans in themselves. Next to supplying adequate storage, the most important items to be considered when selecting a plan are the cost, the feasibility, and the adaptability of the project. These will be discussed below. The maintenance costs of all the alternatives will be ignored because they will be about the same for each alternative and will have no bearing on the recommendations.

Alternative 1: Construction of Rockwell Dam and enlargement of the present water-works system.

Costs involved in this plan include the cost of the dam, the addition to the treatment plant, and a transmission line from the plant to the city. The total cost is estimated to be about $3,134,000 (see Table 5).

The feasibility of this plan is somewhat questionable because of the uncertainty of the reservoir filling as often as expected. After the dam is completed, Horsetooth Reservoir water can be traded for Poudre River water to initially fill the reservoir. If, after the initial filling, the reservoir will not capture enough water to supply the city, it would be impractical to keep trading Horsetooth water to keep the reservoir full instead of using Horsetooth water directly. Otherwise this plan is feasible.

This alternative will adapt very well to the present water-works system. The new treatment facilities would be located on the same site as the present one, keeping the plant centralized. The new pipeline would discharge into the present storage tanks from which the present pipelines to the city will handle the additional water.

Thus this alternative has two serious drawbacks which must be carefully considered before undertaking the project. These drawbacks are the relatively high cost and the uncertainty of filling the reservoir. Because of the high cost it is doubtful if the city of Fort Collins can afford to proceed with the project at the present time.

Alternative 2: Construction of Rockwell Dam and 100 percent metering of the distribution system.

The costs of this project will be separated into an initial cost, consisting of the costs of the dam and 7,600 meters, plus the cost of 4,500 meters distributed over a 19-year period. The total cost of the project is estimated to be about $2,762,000 (see Table 5), of which $2,222,000 is the initial cost. The cost of the additional meters is very uncertain because of the difficulty in estimating the future cost of meters.

This project appears to be feasible; however, since the residential sections of Fort Collins have never been metered, there will probably be considerable objection to this project. Many cities have had to resort to court action before they could install meters. For this reason the present city government favors a flat-rate system as long as it is possible.

Less water will be required for this alternative than was required for alternative 1 so it is estimated that Rockwell Reservoir would easily store enough water to meet the 1980 demand. This alternative is the most adaptable of the three to the present system. The present water works would be adequate until about 1980 as it now exists.

The high cost and the reluctance by the city to use a metered system are the main deterrents to this alternative. Again it is estimated that the cost is considerably more than Fort Collins can afford. However, the city may someday find it absolutely necessary to use meters to conserve water.

Alternative 3: Construction of a new treatment plant which will treat Horsetooth Reservoir water.

The only costs involved in this plan are the cost of a treatment plant and a short transmission line. The total cost is estimated to be about $567,240 (see Table 5).

The project appears to be entirely feasible. The city now owns the land required for the treatment plant, and it owns enough water in Horsetooth Reservoir to meet the 1980 demands. One great advantage of this alternative is that the city will be able to purchase additional Horsetooth water when the demand exceeds the supply in the future.

This project is also easily adaptable to the present water system. The effluent from the treatment plant can be piped directly to Soldier Canyon Reservoir, from which there is adequate piping to transport the water to the city.

The cost of this alternative is considerably less than that of the other two alternatives (see Table 5), and it is estimated that the city of Fort Collins can afford to proceed with the project.

The only objection to this alternative seems to be the high concentration of algae in the water, which has been the main reason for not using the water previously. The use of microstrainers should eliminate this problem.

RECOMMENDATIONS

After consideration of the foregoing discussion and conclusions it is recommended that:

1. To satisfy future water requirements and also to supply storage facilities, provisions be made to use Horsetooth Reservoir as a direct domestic water supply.
2. The present water-works system be expanded to a capacity of 28.0 mgd by building a treatment plant at Soldier Canyon Dam capable of supplying 12.0 mgd.
3. The program be carried out at the earliest possible date in order to alleviate the present situation.
4. Water meters can be considered as a means of reducing consumption if, in the future, the city finds it is impossible to supply enough water to meet current consumption demands.

The primary reason for recommending alternative 3 is that its cost is considerably less than that of the other two alternatives. The possibility of purchasing additional water from Horsetooth Reservoir in the future has also greatly influenced the decision. On the whole alternative 3 is considered more feasible than the other two alternatives.

APPENDIX A

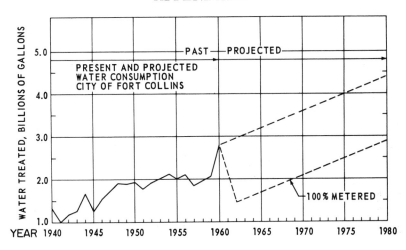

Figure 1

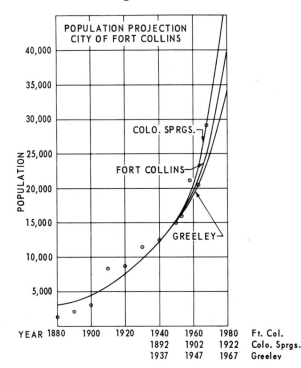

Figure 2

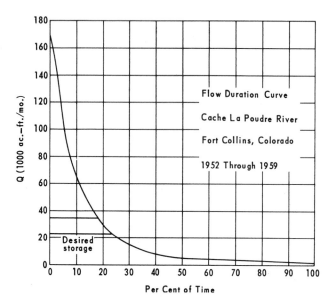

Figure 3

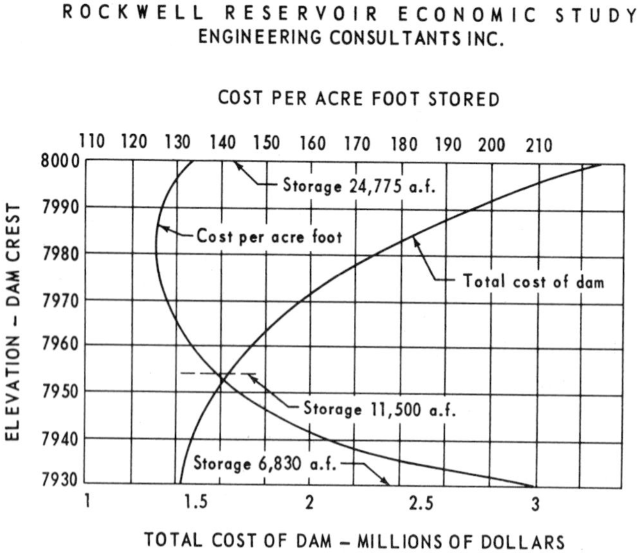

ROCKWELL RESERVOIR ECONOMIC STUDY
ENGINEERING CONSULTANTS INC.

COST PER ACRE FOOT STORED

Figure 4

GENERAL LOCATION

OF

PROPOSED ROCKWELL DAM

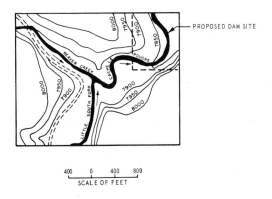

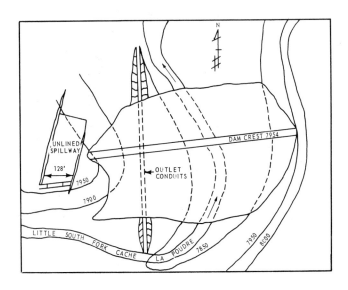

Figure 5

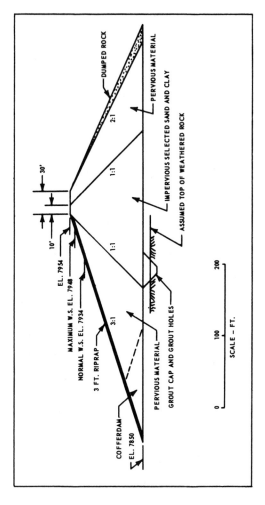

ENGINEERING CONSULTANTS INC.
CITY OF FORT COLLINS, COLORADO

ROCKWELL DAM
TYPICAL SECTION
STORAGE — 11,500 ACRE FEET

Figure 6

CITY OF
FORT COLLINS

WATER TREATMENT PLANT

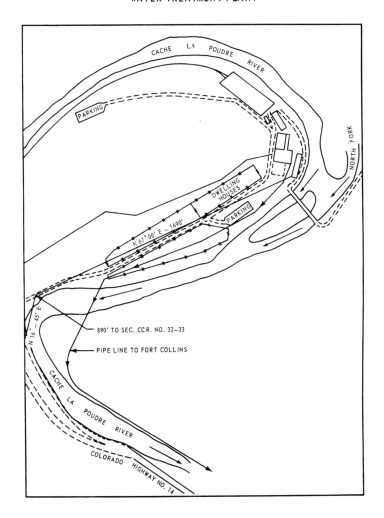

Figure 7

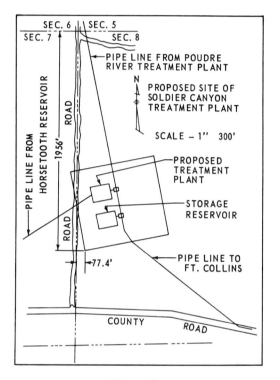

Figure 8

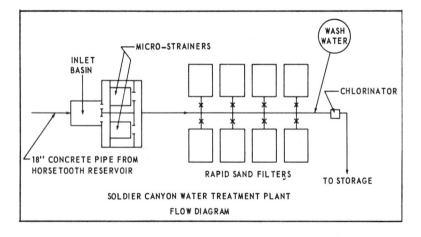

Figure 9

APPENDIX B

Table 1. Fort Collins Water Rights on the Poudre River.

Decree Priority No.	Flow in Sec. Ft.	Average Number of Days Right is Usable	Average Yield in Acre Feet
1	3.50	365	2534
5	2.15	212	904
6	7.00	350	4860
12	2.78	183	1009
14	4.50	270	2410
Total	19.93		11,724

Table 2. Breakdown on Rockwell Dam Costs.

Item	Cost
Dam Embankment	$693,540
Outlet Works	313,610
Spillway	287,260
Miscellaneous	108,500
Contingencies and Engineering Fees	211,090
Total	$1,614,000

Table 3. Breakdown of Costs on Water-Works System to Use for Rockwell Reservoir Water.

Item	Cost
12 miles—27″ Steel Pipe and Connections	$1,128,000
8—1.5 mgd Rapid Sand Filters	254,000
Contingencies and Engineering Fees	138,000
Total	$1,520,000

Table 4. Breakdown of Costs on Horsetooth Water-Works System.

Item	Cost
8—1.5 mgd Rapid Sand Filters	$280,000
2—10 Ft. Microstrainers	150,000
1—25,000 Gallon Steel Tank	23,000
1—Chlorinator Unit	21,240
2—Operators' Houses	15,000
2,500 Ft.—18″ Reinforced Concrete Pipe	26,380
Contingencies and Engineering Fees	51,562
Total	$567,240

Table 5. Breakdown of Costs on the Three Alternatives for Supplying Fort Collins with Additional Water.

	Item	Cost
Alternative 1.	Rockwell Dam	$1,402,910
	Transmission Pipeline	1,128,000
	Treament Facilities	254,000
	Contingencies and Engineering Fees	349,090
	Total	$3,134,000

Alternative 2.	Rockwell Dam	$1,402,910
	7,600 Initial Meters	608,000
	4,500 Additional Meters by 1980	540,000
	Contingencies and Engineering Fees	211,090
	Total	$2,762,000
Alternative 3.	Treatment Facilities	$515,678
	Contingencies and Engineering Fees	51,562
	Total	$567,562

BIBLIOGRAPHY

1. Beck, R. W. and Associates, *Survey and Report, Comprehensive Program, Water Supply—Improvements—New Financing—Rates for the City Water Utility System*, December 1, 1953, 90 pp.

2. *Colorado-Big Thompson Project*, U. S. Department of Interior, Bureau of Reclamation, April 1957, Vol. II, 212 pp.

3. Engineering Consultants, Inc., *Preliminary Report on Feasibility and Cost, Rockwell Dam*, Denver, Colorado, December, 1960, 68 pp.

4. Evans, George R., "Review of Experiences with Microstrainer Installations," *American Water Works Association*, Vol. 49, No. 5, May, 1957.

5. "Four Pipe Alternatives Priced for Water Transmission Line," *Engineering News Record*, 164, April 14, 1960, p. 111.

6. Linsley, Ray K., Jr., Max A. Kohler, and Joseph L. H. Paulhus, *Hydrology for Engineers*, New York: McGraw-Hill Book Company, Inc., 1958, 340 pp.

7. Loyd, D. F., "Cost Comparison of Unmetered and Metered Systems at Idaho Falls, Idaho." *American Water Works Association*, 52, April 1960, pp. 433-36.

8. Owen, L. M., "Meters Solve a Water Crisis: Costine, Maine," *Water and Sewage Works*, 106, May 1959, pp. 191-92.

9. "Rapid City Water Treatment Plant," *Engineering News Record*, 165, November 10, 1960, p. 73.

10. Rynders, A., "Demand Rates and Metering Equipment of Milwaukee," *American Water Works Association*, 52, October 1960, pp. 1239-43.

11. Steel, Ernest W., *Water Supply and Sewerage*, New York: McGraw-Hill Book Company, Inc., 1960, 645 pp.

12. Turre, George J., "The Use of Microstraining for Marston Lake at Denver," *American Water Works Association*, September, 1959.

13. "Water and Sewer Main Job for Wisconsin Subdivision," *Engineering News Record*, 160, January 2, 1958, p. 63.

422 West Laurel
Fort Collins, Colorado
November 30, 1974

Professor Weisman
Department of English
Colorado State University
Fort Collins, Colorado 80521

Dear Professor Weisman:

In compliance with your term assignment to complete a technical paper, I hereby submit my report on the Possible Origin of the Intrusive Bodies in Big Thompson Canyon.

This report includes field work, description and location of the subject, laboratory research, and library research.

From research, laboratory, and field work, I found that the intrusive is an offshoot from a mass called Silver Plume Granite.

It is recommended that for additional study, thin sections of the intrusive should be used, thereby more accurately determining the percentage composition.

Sincerely yours,

Gerald L. Owens

Enclosure
GLO/lo

Preliminary Report
on the
Possible Origin of the Intrusive Bodies
in
Big Thompson Canyon

Submitted to
Professor H. M. Weisman
of
Colorado State University
Instructor
Technical Writing

By
Gerald L. Owens
Fort Collins, Colorado
November 30, 1974

Abstract

The intrusive body which lies at the base of Palisade Mountain on Highway 34 west of Loveland, Colorado, is an igneous mass that has been intruded into a metamorphic country rock. The country rock, which is essentially mica schist, covers a wide area with no other intrusives in evidence. How does this one small body of granite come to be in this location, and is it related in any way with known bodies of granite in the front range?

From the investigation, as to mineral constituents and structural correlation as compared with a known outcrop of Silver Plume Granite, it was found to be an offshoot from a main mass of Silver Plume Granite. The granite was intruded into pre-cambrian country rock during pre-cambrian times. Locally, the granite changes color due to varying percentages of mineral constituents. Individual crystal sizes depend on the rate of cooling in conjunction with the amount of mineralizers present.

For further study, it is recommended that thin-section studies be made of the intrusive in order to determine more accurately the percentage composition of the body.

Table of Contents

INTRODUCTION

West of the city of Loveland, Colorado, in the front range of the Rockies, lies the Canyon of the Big Thompson River. The river, while the mountains were being thrust upward, cut a narrow, deep canyon through the ancient metamorphic rock. At the base of Palisade Mountain, where construction work on Highway 34 exposed the flanks of the mountain, lies an area in which igneous material has been intruded.

An occurrence of this nature is not uncommon in a mountainous region. However, the fact that there are no other igneous bodies within miles led to the question of origin. Where did this lava originate? What were the factors of its intrusion into the metamorphic country rock? Could this body be correlated in any way with known igneous bodies in the front range? If there is a connection, would comparison of constituents of the two prove that they are of the same age?

Significance of Problem

The problem and its solution is of most interest to geology students, for it will show them how important correlation and scientific investigation is to the study of geology. As in every unusual geologic feature, there is a subtle challenge to every geologist that says, "Come, find me out." The intrusive has, many times in the past, issued this challenge; but, due to other obligations, few students have tried to work it out. To benefit those who have the desire to know, but who do not have the time to spare, this report is submitted in hopes that it will give them an insight into the processes of such an investigation.

Scope

Limited available laboratory time has necessitated investigation of only two major areas. One area deals with field observations in respect to correlation of the known and unknown bodies; and the other, in laboratory analysis of the hand specimens.

Historical Background

There have not been any professional papers written on this particular intrusive body, but there are professional papers about the Silver Plume Granite formation. According to my advisor, Mr. Campbell of the Geology Department, there have been a few papers written about the intrusive by students in the past.

Definitions

The definition of terms in relation to description is a necessity for the non-technical as well as technical reader.

Magma—A magma is a naturally occurring molten rock mass in or on the earth which is composed of silicates, oxides, sulfides, and volatile constituents such as boron, fluorine, and water (1:177).

Igneous Rocks—Igneous rocks are those rocks which have formed in or on the earth's crust by solidification of molten lava (5:121).

Intrusive Body—An intrusive body is an igneous rock mass which has solidified within the earth's crust (1:31).

Acidic—Acidic refers to rocks which are very high in silica content (2:321).

Metamorphic Rocks—Metamorphic rocks are rocks that were originally of igneous or sedimentary origin. Tremendous pressure and heat generated during mountain building transformed them into new minerals with a foliated outward appearance (3:35).

Country Rock—A country rock is defined as any rock that is penetrated by an intrusive, igneous body (5:92).

Stock—A stock is an intrusive igneous body less than 40 square miles in area (3:77).

Batholith—A batholith is an intrusive igneous body more than 40 square miles in area (3:78).

Dike—A dike is a rock of liquid origin that intruded along faults and slips of a formation and cut across bedding planes (3:78).

Organization

The manner in which material in this paper is presented will bear some explanation. In order for the reader to more fully understand the processes and geologic theories, it was necessary to include, at some points in the body of the report, certain evaluations of the material presented.

ACKNOWLEDGMENTS

I sincerely thank the following people who have contributed much to the value of this report: Professor Weisman, of the English Department of Colorado State University, who set forth the principles of technical report writing; Mr. Campbell, of the Geology Department of Colorado State University, who aided me in my research work; Lyn Owens, my wife, who checked the paper for grammatical and mechanical errors and also typed the final draft.

ANALYZING THE PROBLEM

The intrusive body in Big Thompson Canyon offered no tangible clues as to its origin. That it was a granite rock intruded into a metamorphic country rock was apparent, but how would this help? The only method to follow was to find out whether or not there were any known intrusive bodies in the area; and if there were, could they be correlated in any way?

From notes taken on a G-7 field trip, it was found that there was a Silver Plume Granite outcrop near Masonville, Colorado. Although this was approximately 12 airline miles distant from the intrusive, it was the only one in the area that was positively known.

From this start it was decided that samples should be taken from both bodies and analyzed. Also, any material in the library that pertained to igneous rock in the front range should be studied.

With the thought that correlation between the two bodies might answer the question, work was begun.

LOCATION AND DESCRIPTION OF INTRUSIVE

In order for this report to be of value to those who wish to conduct subsequent studies or check the validity of the existence of the intrusive, directions will be given as accurately as possible.

Location of the Intrusive

Starting from the south city limits of Fort Collins, Colorado, proceed south to Loveland, Colorado. At the junction of U. S. Highway 287 with State Highway 34, turn west on Highway 34. A general

Figure 1

store is located near the entrance of Big Thompson Canyon called the "Dam Store." Check the mileage on the car's odometer at this point and proceed approximately eight miles west up the Canyon (see Appendix 1). On the north side of the highway stands a small white sign which reads "Palisade Mountain Elevation 8,258 Ft." (See Figure 1). At this point the small bodies which make up the intrusive extend parallel to the highway for a distance of 150 yards. The main intrusive is 15 feet to the northeast of the sign in Figure 1 and is readily identified by its clean, gray color.

Description of the Intrusive

The larger of the four small intrusive bodies will be described as it is representative of the other three.

The intrusive extends vertically from the level of the highway approximately 25 feet to its uppermost contact with the metamorphic country rock. Horizontally, from contact to contact, the distance is 40 feet (see Figure 1, Appendix II).

Externally, the intrusive is a light-gray, fine-grained granite which lies in perfect contact with the dark-gray, foliated, micaceous schist (see Figure 2, Appendix II). Upon closer examination, small black books of biotite mica are seen to be incorporated into the mass. Cigar-shaped portions of the country rock, ranging from two to six inches long, and one block three feet by six feet long, are evident in the mass, thus indicating that the intrusive cooled before complete melting had occurred. On a smaller intrusive, to the west, a pegmatic zone surrounds the intrusive at its contact with the country rock. The zonal growth of the large crystals grades inward for two feet where it again assumes the smaller size of crystals of the mass proper. This zone of pegmatite granite was the result of mineralizers, or volatile constituents, which were trapped in this area. The fluidity of the cooling mass was higher than normal for this particular intrusive; and, subsequently, crystal growth was quite rapid as proved by the large size of the individual crystals.

LOCATION AND DESCRIPTION OF THE SILVER PLUME GRANITE OUTCROP

For the reasons mentioned in "Location and Description of Intrusive," the following directions for finding the Silver Plume Outcrop will be given as accurately as possible.

Location of Outcrop

Starting from the south city limits of Fort Collins, Colorado, proceed south on U. S. Highway 287 for three miles and then turn west on County Road Number 186 which goes to Masonville, Colorado. Upon reaching the location of the South Horsetooth Dam, check the odometer reading on the car and proceed for approximately four miles. At this point the road will be rising to the top of a narrow canyon. At the crest of the hill on the north side of the road lies a reddish-brown, granite mass which has been partially exposed by construction work. This is the outcrop of Silver Plume Granite (see Appendix I).

Description of the Outcrop

The outcrop is approximately 15 feet high by 50 feet long from east to west. Erosion has worn down the metamorphic rock into which the granite was intruded, but now a soil mantle covers it on the topside to a depth of several inches. Considerable chemical and mechanical weathering has taken place as the mass is fractured and crumbly in places. Also, a reddish film of iron has been deposited by circulating ground water.

Upon closer examination of a freshly chipped surface, crystal sizes appear to be fairly uniform with an occasional crystal approaching 15 millimeters in length by 10 millimeters in width. Color is pinkish-gray due to the flesh-colored feldspar and the black biotite crystals.

SPECIMEN SELECTION AND FIELD TRIMMING

Specimen Selection

Specimens were taken from the intrusive and the known outcrop of Silver Plume Granite with an eye towards specimens that would be representative of the whole mass, as well as specimens that indicate unusual facets of its formation. Specimens were taken from the centers of the two bodies and also along the contacts with the country rock.

Field Trimming

It is sometimes necessary to use a three- to four-pound striking hammer to obtain rocks from a solid mass, or even to break up large chunks of rock. However, there is generally enough rock material lying on the ground from which a specimen can be chosen. If possible, rocks should be chosen that approximate the 3″ x 5″ x 1″ size that hand specimens should be. In any event, a rock should be chosen that has at least one right angle and/or two flat sides.

A geology pick and a pair of heavy work gloves were used to chip the specimens down to size (see Figure 2). While holding the rock in one gloved hand, the geology pick was used to strike a glancing blow, making sure that the chips would fly away from the operator. This operation may be understood more readily with a diagram (see

Figure 2   **Figure 3**

Figure 3). As indicated by the drawing, an imaginary line should be visualized down the center of the rock. Small fragments indicated by the dotted lines should be chipped off. The blows should be struck down and away so as not to crack the specimen in half. When one side has been trimmed off flat, the specimen should be turned over and chipped on the other side. Considerable patience is needed as these rocks are brittle and fracture very easily.

MICROSCOPIC ANALYSIS OF SPECIMENS

The Intrusive Specimen

The specimen from the intrusive was observed under fifty power magnification. The predominate mineral was orthoclase feldspar with its white color and irregular grains showing many of the faces with nearly perfect right-angle cleavage. The next mineral, in order of abundance, was quartz. This mineral is the clear, colorless variety which has no crystal form as it was the last mineral to crystalize out in the cooling intrusive. It filled in all pore spaces left by the other crystals. Biotite mica, with its black, splendent luster and flexible folia, was regularly spaced in the matrix. The relatively small size of the crystals made them unnoticeable beyond a few feet. Infrequent, but quite large, crystals of plagioclase feldspar were observed. These were identified by their gray-white color and twinning striations on the cleavage surfaces. As evidenced by the high percentages of quartz, this rock can be termed acidic (see Table 1, Appendix III).

The Silver Plume Specimen

The specimen from the Silver Plume outcrop was also observed under fifty power magnification. The orthoclase feldspar was the most abundant mineral found. It varied from white to flesh-colored with euhedral, crystal faces. The quartz, next in abundance, which filled in the remaining spaces, was rose-colored to colorless. The color could have been imparted to the quartz by a minute quantity of cobalt. The biotite mica present was in small, black masses which were quite close together. A very small quantity of white plagioclase feldspar, showing excellent twinning striations, was present. This specimen represents a magma which was high in silica content. Therefore, this rock is also termed acidic (see Table 2, Appendix III).

COMPARISON OF FIELD AND LABORATORY DATA

Similarity of Occurrence in Field

Comparison of the types of country rock penetrated by the respective bodies showed that they are of the same age and the same rock type. The country rock in both localities is a pre-cambrian, metamorphic rock called "mica schist." Both of the bodies are granite, varying only in percentages of minerals present, as well as in color. This follows quite well with T. S. Lovering's professional paper entitled *Geology and Ore Deposits of the Front Range*, U. S. Geol. Surv. Professional Paper, 223, page 28. The following is the section written on the Silver Plume Granite:

> In the front range are a large number of small stocks and batholiths generally intruded at the same time. . . . Most of the stocks and batholiths are pinkish-gray, medium-grained, slightly porphyritic biotite granite, composed chiefly of pink and gray feldspars, smoky quartz, and biotite mica; but muscovite is present in some facies. . . . The percentage of biotite varies from place to place. . . .

The intrusive bodies are very small when compared to the size of a stock (3:77), or a batholith (3:78). They do, however, compare favorably with the definition for a dike (3:78). For the most part, dikes are rather small in a cross-sectional area but are known to extend up to 100 miles in length (3:78). For both of the bodies, which the author will call dikes, the rock that the molten lava passed through would have some bearing on the composition. This could be one of the reasons why there is a difference in color and percentage of composition. From this, then, it is logical to assume that both bodies are offshoots from a larger mass. They probably followed fractures or faults caused by crystal movements during the time of the pre-cambrian, metamorphic country rock.

Comparison of Laboratory Data

The specimen of the intrusive, in comparison with that of the Silver Plume Granite, shows that they are very close in chemical composition. The intrusive specimen had 60% orthoclase in comparison with the Silver Plume's 40%. Quartz in the intrusive averaged 20%, while in the Silver Plume it was 30%. For biotite mica, the percentage was 15% in the intrusive and 25% in the Silver Plume. Both of the bodies had 5% plagioclase. For a better visual comparison, see Appendix III.

Conclusion

The intrusive bodies in Big Thompson Canyon are of the same geologic age as that of the Silver Plume Granite. They both were intruded into a pre-cambrian, metamorphic country rock during pre-cambrian times. The country rock, into which the granite was intruded, is a mica schist which is quite common in the front range.

Chemically, the intrusive and the Silver Plume Granite are the same. Although there is some variation in color and percentages of the constituents, this is within the somewhat broad range given for the Silver Plume Granite (4:28). Locally, the percentage can change due to an increase or decrease in amount of constituents. Color variations are due to minute inclusions of a metallic ion or molecule derived from the walls of the conduit or feeder pipe.

Recommendations

It is recommended by the author that a large number of Silver Plume Granite specimens be analyzed. These should come from different outcrops. Also, to more accurately determine the constituents and their percentages, thin-section studies of the specimens should be made.

Appendix I

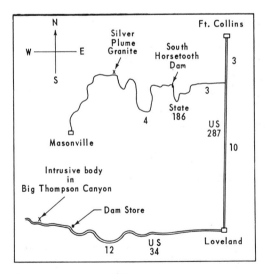

Figure 1
Illustrations Showing Work Sites (Not drawn to scale)

APPENDIX II

Figure 1

Photo showing relative size of intrusive in Big Thompson Canyon

Figure 2

Photo showing contact between country rock and granite intrusive

APPENDIX III

Tables of Comparative Constituents

Table 1

Specimen From Intrusive

Mineral	Percentages	Color
Orthoclase Feldspar	60%	White
Quartz	20%	Clear, glassy
Biotite Mica	15%	Black
Plagioclase Feldspar	5%	Gray-white

Table 2

Specimen From Silver Plume

Mineral	Percentages	Color
Orthoclase Feldspar	40%	Flesh-colored to white
Quartz	30%	Rose to clear, glassy
Biotite Mica	25%	Black, vitreous
Plagioclase Feldspar	5%	White

BIBLIOGRAPHY

1. Bowen, N. L. (1922), "The Reaction Principle in Petrogenesis," *Journal of Geology*, Vol. 30, pages 177-198.

2. Grout, F. F. (1941), "Formation of Igneous Looking Rocks by Metasomatism. A Critical Review and Suggested Research," *Bul. Geol. Soc. Am.*, Vol. 52, pages 1525-1526.

3. Longwell, Chester R.; Flint, Richard Foster, *Introduction to Physical Geology*, New York, John Wiley and Sons, 1955, 432 pages.

4. Lovering, T. S., *Geology and Ore Deposits of the Front Range*, U. S. Geol. Surv., Professional Paper, 223, page 28.

5. Pirrson, Louis V.; Knolph, Adolph, *Rocks and Rock Minerals*, New York, John Wiley and Sons, 1957, 365 pages.

Student Popular Science Article

HOW AIRPLANES FLY
by
David M. Graham[1]

Today's airplanes vary in weight from a few hundred pounds to many thousands of pounds. It is truly amazing, and seemingly a miracle, that these heavy machines are able to stay up in the air and even travel through it at tremendous speeds without falling from the sky. However, the fact that a large airplane can fly at a height of thousands of feet is due to a fairly simple principle. Jet planes are held up by the same principle, but they are capable of forward flight for a different reason than propeller planes. We will learn in this article how the planes that are propeller-driven operate.

First, how does an airplane keep from falling down once it gets up into the sky? The airplane wings' main function is to hold the airplane up. These wings are not just slabs of metal, all flat and stiff. They have a very definite shape which helps them lift the airplane.

Notice in Figure 1 that the bottom of the wing is fairly flat while the top is curved. The front of the wing is thick, and the rear, or trailing edge, is very thin. When the airplane is flying, air rushes over the wings at high speed. Because the top of the wing is curved, the air passes over

---

[1]Written as a student assignment for a publication like *Boy's Life*.

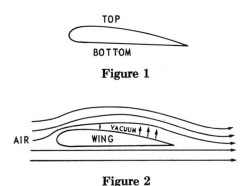

Figure 1

Figure 2

the top of the wing much faster than under the bottom of the wing. The air passing over the top, shown by the lines in Figure 2, causes a vacuum, or suction, which tends to lift the wing. The air going under the wing doesn't create such a vacuum. So, the airplane, which is attached to the wing, is lifted instead of pushed down. If both surfaces of the wing were equally flat, the air rushing over the surfaces woud push on both sides with an equal pressure. This wouldn't help the airplane stay up in the air. That's why the top is curved—so the air will help lift the airplane while it is flying. This vacuum, which lifts the plane, works by trying to suck the wing up to fill the space where there is less air just above the wing. The suction, like the suction from a vacuum cleaner, is called lift.

The airplane propeller pulls the aircraft through the air. The blades of the propeller are just like small wings, the front of the propeller blade is curved and the trailing edge is thin and flat. The side of the blade that corresponds to the top of the wing is facing towards the front of the airplane. The rear side of the blade corresponds to the under-side of the wing. As the propeller turns, the air rushing past the curved part of the blade creates a vacuum just as it did on the wing. This vacuum pulls the propeller forward and the airplane with it (figure 3). The action of the

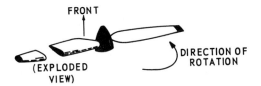

Figure 3

propeller is very much like the action of a screw being driven into a piece of wood. As the screw turns, it pulls itself into the wood like the propeller pulls the airplane through the air. The propeller, with the engine, is really the most important part of the airplane. When the airplane is just taking off from the ground, the propeller has to pull the airplane forward until it is going fast enough to give lift to the wings.

Once the airplane is up in the air, how is it steered so it won't just fly in a straight line and smack into a mountain or something? Don't worry, the airplane can be steered. There are three different controls which enable the airplane to go up and down, and turn to the right or the left. These are called control surfaces and their specific names and locations are: the ailerons, which are located on the trailing edges of the wings; the elevators, which are part of the small wings on the rear of the airplane; and the rudder, which is located on the tail-fin above the little wings.

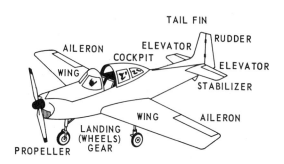

Figure 4

The control surfaces are moved by means of a steering wheel in the cockpit. This wheel turns just like the one in a car, but it can also be pushed forward and backward. Turning the wheel to the right makes the airplane turn, or bank, to the right. A left bank is made by turning the wheel in that direction. Pushing the wheel forward makes the airplane point down towards the ground. Pulling it out makes the airplane climb. Some of the older airplanes and most of the big airliners have a different system of moving the airplane around. The controls are the same but, instead of having a single wheel that does everything, these airplanes have a small steering wheel on a "stick" which operates just the elevators and the rudder. On the cockpit floor, there are two big pedals, like the ones in an automobile, which move only the ailerons. Now let us find out what happens when the airplane controls are moved.

The ailerons are long, thin, door-like pieces. They fit on the trailing edge of the wings and can be moved up or down. If the aileron lies straight out, and even with the rest of the wing, the air rushing by is not disturbed and the airplane flies straight. Now, suppose the aileron pointed down. The air that normally whizzed under the wing would strike the aileron with a lot of push. This causes the wing to be forced up. The same thing happens when the aileron points up instead of down. The air hits the aileron and forces the wing down. The ailerons are so connected that when one of them is pointed down, the other points up. Then the rushing air pushes one wing up and pushes the other down.

On the rear of the airplane, called the tail-section, are what looks like three small wings. Two are level like the large wings in front, and one points straight up. This one is called the tail-fin, while the level ones are called stabilizers. The elevators on the stabilizers operate just like the ailerons with one exception—both elevators move in the same direction, instead of opposite directions as with the ailerons. These control surfaces make the airplane either climb or dive. When the elevators point up, the air pressure forces the tail down, like with the ailerons, and points the airplane up so it can climb higher. Just the opposite happens when the elevators point down.

The remaining control surface, the rudder, steers the airplane just like the rudder on a boat. The air does the job instead of water. If the rudder points to the right, the airplane turns to the right, and to the left if the rudder points to the left.

Now, let's climb in and make an actual flight. First, the engine is turned on. A knob on the dashboard controls the gas just like the accelerator pedal in a car. We push the knob in to make the propeller turn faster and pull the airplane forward. Out on the runway, we push the knob all the way in and we speed down the runway faster and faster. The rushing wind creates lift on the wings and the airplane leaves the ground. We're flying! We climb high into the air so we can test our controls. You may have noticed, we had to pull back on the wheel to get the plane into the air, and we kept the wheel pulled back until we were far enough above the ground. Now that we're high enough, let's make a right turn. We slowly turn the wheel to the right and the airplane seems to tip over to the right. The right aileron points up and the left one points down. Pulling back on the wheel just slightly helps the airplane go through the turn by raising the elevators which pushes the tail down and to the left. We do the opposite movements when we want to make a left turn. But, we still have to pull back slightly on the wheel to help push the tail around where we want it. Since we've already seen how the airplane climbs, let's put it into a dive. We push the wheel forward slowly and the plane flies towards the ground. If we were to push the wheel forward too fast or too far, the airplane would point down too far and get out of control. Then we might crash. Now, as the wheel goes forward, the elevators are pointing down and pushing the tail up because of the air pressure. To pull out of the dive, we just pull back on the wheel.

It's time to take the airplane back for a landing. We turn the airplane so we can see the runway straight ahead. We push the wheel in slowly to fly the airplane closer to the ground. We keep down until we are very close by the time we're over the end of the runway. As soon as we can feel the wheels touch the ground, we pull out the gas knob to slow the propeller down. But, we keep the wheel pushed in slightly. In slowing down, the tail section will settle by itself as the air can no longer hold it up. And, we've landed safely.

That is how an airplane works. With a little practice, flying becomes easier than driving a car. But we have to start out in the airplane with much more caution than we use when driving a car. After all, an airplane doesn't travel as close to the ground as does the car.

Full Scale Test on a Two–Story House Subjected to Lateral Load

Felix Y. Yokel, George Hsi, and
Norman F. Somes

Center for Building Technology
Institute for Applied Technology
National Bureau of Standards
Washington, D.C. 20234

U.S. DEPARTMENT OF COMMERCE, Frederick B. Dent, *Secretary*

NATIONAL BUREAU OF STANDARDS, Richard W. Roberts, *Director*

Issued March 1973

Contents

FULL SCALE TEST ON A TWO-STORY HOUSE SUBJECTED TO LATERAL LOAD *

Felix Y. Yokel, George C. Hsi, and Norman F. Somes

Tests were carried out on a single-family, detached house to determine its deflection characteristics under lateral loads. The house was a two-story building of conventional wood-frame construction. Two series of tests were conducted. The first of these was to determine the stiffness of the house
when subjected to a simulation of wind loading. The second was to determine the dynamic response of
the house to a single impulse load.

The report presents the results of these tests from which the following primary conclusions were
derived:

(1) The measured second-story drift of the building was considerably less than that derived using
present design criteria for medium- and high-rise buildings as applied to most areas of the United
States.

(2) Only a small portion of the distortion of the exterior walls was transmitted to the interior gypsum board finish material.

(3) The upper-ceiling diaphragm experienced significant in-plane deformation. On the other hand,
the floor/ceiling diaphragm at the lower ceiling level tended to act as a rigid diaphragm and to translate
as a rigid body when the building was subjected to lateral load.

(4) The natural frequency of the structure was approximately 9 Hz and damping averaged approximately 6 percent of critical damping varying from 4 to 9 percent.

Key words: Building damping; drift; dynamics; earthquake; frequency; housing; lateral resistance;
racking; stiffness; structural deflections; vibration; wind load; wood-frame construction.

1. Introduction

There is currently in the United States a strong trend
toward the production of housing by industrialized
methods. These methods frequently involve the innovative use of materials, structural concepts, or manufacturing procedures. It is reasonable that the occupant of
housing, produced by these methods, should anticipate
at least that level of structural performance which
society has come to expect from conventional housing.

Conventional housing in the United States has
evolved over several hundred years and, although this
evolution has been based upon achieving a satisfactory
level of performance, the final result has frequently
gone unquantified. Thus, for example, the occupant of
a conventional house would probably consider it to be
adequately stiff but not know how stiff it is or how stiff
he implicitly requires it to be. Surprisingly, there is
very little published information on the response of
conventional housing to static and dynamic loads and
the information available [1,2][1] does not necessarily

*Research sponsored by the Department of Housing and Urban Development, Washing-
ton, D.C. 20410.
[1]Numbers in brackets refer to literature references.

apply to housing built in the United States. How then
are the criteria to be established by which to evaluate
innovative products?

In the case of structural response to lateral load,
there are design criteria for medium- and high-rise
buildings limiting their drift (lateral displacement at
any story level) under certain wind and seismic loads.
These design criteria are not normally used to design
low-rise buildings. Provisions presently used to assure
minimum stiffness of low-rise buildings against lateral
load [3] stipulate minimum stiffness of shear walls.
Because of the complex interaction between structural
elements and the contribution of partitions and
cladding to lateral stiffness, these provisions do not
provide a basis for predicting the response of conventional low-rise buildings to lateral load.

The purpose of the investigation reported herein was
to measure the lateral drift of a conventional wood-
frame building under simulated wind load to determine
whether the drift limitations required in the design of
medium- and high-rise structures are applicable to low-
rise housing units.

It was realized that the drift displacement of
buildings under static lateral load is not necessarily the

only independent variable affecting the adequacy of their real-life performance in a high wind. However, in absence of a full understanding of all the variables in this problem, information on the properties of conventional housing known to be acceptable to occupants provides the only available guidance for the evaluation of innovative, untried systems.

A secondary objective of the investigation was to determine the dynamic response characteristics of the structure in order to permit a more accurate calculation of the effects of dynamic lateral loads such as earthquake loads.

2. Scope of Investigation

2.1. Selection of House

Discussions were held with one of the largest producers of conventional housing in the United States. This builder agreed to make available a recently completed house located in the Washington, D.C. suburban subdivision of Bowie, Md. In terms of construction details, the house was considered to be typical of much of the conventional two-story housing in this country. The site provided adequate space in which to set up equipment for the application of horizontal loads to the exterior of the house. Figure 2.1 shows a general view of the subdivision.

FIGURE 2.1 *General view of the Bowie, Md. subdivision.*

2.2. Choice of Loading

It was decided to carry out two basic experiments. The first of these was to determine the stiffness of the house when subjected to a simulated wind load. The second was to determine the dynamic response of the house when subjected to an impulsive load. Because of the difficulty in applying a simulated wind load in a distributed manner over the building elevation, a series of concentrated forces were applied horizontally at two levels against the rear elevation of the building. These levels corresponded approximately to the levels of the centerlines of the upper-story floor joists and the underside of the lower chords of the roof trusses, respectively. Forces were applied first at one level and then at the other. The separate effects of these forces were combined to compute a total effect using the principle of superposition. Justification for applying this principle is provided by the approximately linear response of the house to the loading. Four rams were used to apply the forces at each level and these were spaced so as to achieve a reasonably uniform distribution of load along the rear wall.

2.3. Measurements

In the first experiment, the measurements included loads, vertical deflection, horizontal drift of floors and walls, and racking distortions of walls. In the second experiment, the basic measurement was that of the deflection amplitude/time relationship for the vibration of the building.

3. Description of Test Structure

3.1. Building Tested

A front view of the building is shown in figure 3.1. It is a two-story wood-frame structure with a partial brick veneer front at the lower story. The front entrance is from a portico having a 4-in-thick concrete floor slab resting on compacted fill. The fill is retained by an 8-in-thick hollow concrete block masonry wall. The portico slab abuts the front wall of the house with its top surface 2 ft 10 in above the first-floor level.

Figures 3.2 and 3.3 show floor plans and elevations, respectively. The lower story contains a family room, bedroom, bathroom, and a garage. The upper story has an L-shaped living and dining room, a kitchen, three bedrooms, and two bathrooms.

The lower[2] floor consists of a 4-in-thick concrete slab

[2]For brevity, the term "lower" will be used to describe items within the lower story and the term "upper" will be correspondingly used for items within the upper story.

FIGURE 3.1   *Front view of the building.*

on grade. Walls are of wood-stud construction covered on all interior surfaces of the building with 3/8-ir.-thick gypsum board (conforming to ASTM C36) [4] and on all exterior surfaces with a 1/2-in-thick gypsum sheathing (conforming to ASTM C79) [5]. Stud sizes[3] and spacings are shown in figure 3.2. The brick veneer is a single wythe of 4 in nominal thickness. The veneer

[3]All sizes of wood members in the figure as well as in the text hereafter are given as nominal sizes in inches, in accordance with SPR 16-53 [6].

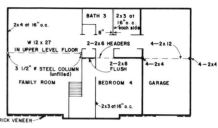

LOWER LEVEL FLOOR PLAN

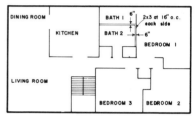

UPPER LEVEL FLOOR PLAN

FIGURE 3.2   *Floor plans of the building.*

FIGURE 3.3   *Elevations of the building.*

is tied to the stud wall by galvanized corrugated metal ties, provided in every 7th course and spaced 32 in on center. Exterior-wall studs are braced at all building corners with 1 × 4 let-in bracing installed at a 45° angle to the horizontal in accordance with the provisions in the FHA Minimum Property Standards [7]. Exterior siding is asbestos shingle or 3/8-in-thick beveled wood siding with a 6-in exposure, as shown by the elevations in figure 3.3.

The structural framing of the upper floor consists of 2 × 8 wood joists spaced 12 in on center, supported by bearing walls and intermediate supports as shown in figure 3.2. The lower- and upper-story ceilings are 1/2-in-thick gypsum board (conforming to ASTM C36). The upper-floor subflooring is 5/8-in-thick plywood covered in all areas by resilient vinyl floor tile.

The upper ceiling and the roof are supported by trussed roof rafters made of 2 × 4 wood members and spaced at 24 in on center. Roofing consists of 1/2-in-thick plywood covered by asphalt shingles.

3.2. Special Structural Features Affecting Response to Lateral Load

Lateral resistance to movement of the building in the direction in which load was applied is increased by the backfilled portico abutting the front wall. However, the four portico columns shown in figure 3.1 are dowelled to the 4-in-thick portico slab by single 1/2-in-diameter bolts, and probably do not contribute materially to the rigidity of the building. The 7-ft floor-to-ceiling height, which is relatively low for this type of building, also tends to increase lateral resistance to movement. The brick veneer front, which could influence resistance to movement in the long direction of the building, does not substantially affect rigidity in the direction in which load was applied, namely, normal to the plane of the veneer.

4. Test Arrangement

4.1. Loading System

The loading arrangement is shown in figure 4.1. Two 10-ton forklifts were used to hold the loading assembly in its desired position and to resist the reaction forces. Each forklift carried a 5-kip concrete block to counter-balance the overturning moment caused by the applied horizontal load and to increase frictional resistance to sliding on the ground. A loading assembly was bolted to each concrete block. The assembly consisted of two

vertical 5-ft-long W8 × 31 beams[4] and one horizontal 14-ft-long W8 × 31 beam which supported two rams spaced 12 ft on center. Figure 4.2 shows the concrete blocks during the attachment of the vertical beams. In figure 4.3, a horizontal beam is positioned in its correct location by the forklift. The horizontal beams were attached to the vertical beams by clips which permitted leveling and vertical adjustment of their position. Slotted bolt holes permitted some horizontal adjustment of the positions of these beams; however, in general, the correct horizontal positioning of the rams was achieved by correct positioning of the forklifts.

[4]Nominal size of wideflange steel beams in accordance with AISC [8].

FIGURE 4.1 *Loading arrangement.*

FIGURE 4.2 *Mounting of loading assembly.*

Figure 4.4 shows one of the rams carried by a horizontal beam.

The loading system and the horizontal position of the loads are shown schematically in figure 4.5. Each of the four single-acting hydraulic rams had a load capacity of

FIGURE 4.3   *Positioning of horizontal loading beam.*

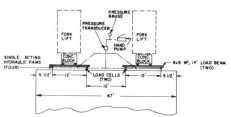

FIGURE 4.5   *Loading system.*

FIGURE 4.4   *Loading ram.*

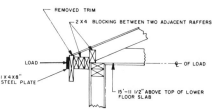

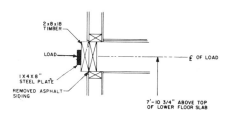

FIGURE 4.6   *Load transfer to the building.*

10 tons, a 2-in-diameter piston and a 6-in stroke. Oil pressure was applied by a single hand pump to all four rams through a manifold and hoses arranged in parallel. Load was monitored by two separate and redundant systems: a pressure transducer connected to the hydraulic system; and two load cells, each with an operating range of 5 kip, monitoring two of the ram loads as shown in the figure. Compressive loads were applied to the rear wall of the building. Loads were monitored electronically. Data in this report are referenced to readings of one of the load cells as described in the appendix.

The manner of load transfer to the building as well as the vertical positioning of the loads are illustrated in figure 4.6. Load was applied at the underside of the horizontal roof-truss members and at the level of the centerline of the second-story floor joists. At the upper level, load was applied through an 8 × 4 × 1-in steel plate to a 2 × 8 wood member which in turn transferred the load to 2 × 4 members nailed between two successive roof trusses. This method of load transfer was chosen to prevent local load concentrations and accompanying permanent damage and also to minimize the removal of siding. At the lower level, load was applied through an 8 × 4 × 1-in steel plate to a 2 × 8 wood member which in turn transferred the load through the rim joist to the floor joists.

In the experiment to determine the dynamic response of the building, a 12-in-long piece of 3/4-in-diameter steel pipe was inserted between the ram and

FIGURE 4.7 *Reaction blocks at rear of forklift.*

FIGURE 4.8 *Reaction block in front of forklift.*

the loading plate in one of the loading points at the lower level. After a predetermined load was applied, the pipe was removed by a sharp hammer blow.

Some difficulties developed during the testing because of muddy ground conditions which reduced the friction force on the forklift wheels. Wooden blocking had to be used in the front and the rear of the forklifts to increase lateral load resistance (see figs. 4.7 and 4.8). This arrangement did not affect the accuracy of load positioning.

ALL LVDT'S NEAR THE SIDE WALLS, EXCEPT THE LVDT'S AT THE CENTER VERTICAL PLANE, WERE SET 5.5" BELOW THE CEILING AND 9" AWAY FROM THE WALL.

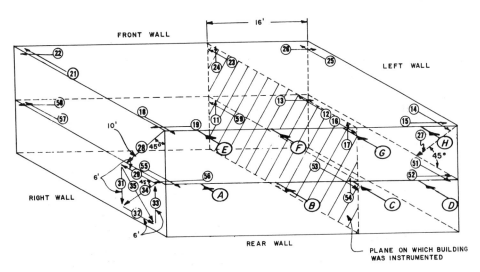

FIGURE 4.9 *Location of transducers and loading points.*

4.2. Displacement-Measurement System

Displacement transducers were used to monitor movement of the building and racking distortion of walls. Of the total of 32 displacement transducers used, the readings from 30 had a resolution of approximately 1×10^{-3} in. The two transducers used to monitor dynamic response were more sensitive. The resolution of the graphical records of the dynamic response is approximately 5×10^{-4} in. All transducers were mounted in the interior of the building where the temperature could be maintained at approximately 70 °F using the conventional heating control system provided in the house.

The location of the transducers is shown in figure 4.9 wherein the circled numbers identify each displacement transducer. The arrows show the displacement vectors measured. Points of load application are identified by letters. The letters and numbers shown in figure 4.9 are used throughout this report to designate the locations of loads and displacements. Instruments were installed near the corners of the building and in the plane, shown cross-hatched, which is parallel to the side walls and located 16 ft from the exterior face of the left side foundation wall. The location of the measurement points relative to the interior wall and ceiling surfaces is noted in the figure. It also can be seen that the rear part of the right side wall in the lower story was instrumented to measure racking distortion. Similar measurements were made on the rear part of the right and left side walls in the upper story by instrumenting a single diagonal.

Typically, transducers were mounted on stands consisting of 2-in-diameter steel pipes welded to a base of a steel plate as shown in figure 4.10 for a typical transducer installation in the upper story. Figure 4.11 shows a closeup view of two transducers. Contact with the measurement point was maintained by the spring pressure.

A typical setup for measuring racking distortion of walls is shown in figure 4.12. Each displacement transducer was attached to one end of a 3/4-in-diameter brass or aluminum tube. The plunger of the transducer impinged on an aluminum plate which was glued to the wall surface and its movement recorded the change of length of wall over which the tube and transducer was extended.

In order to maintain stable temperature and protect the equipment against inclement weather, it was necessary to measure deformations inside the building. This precluded the use of transducer supports which would reference upper-story measurements to the ground floor. Since upper-story instruments were supported on the second-story floor, total displacements had to be

FIGURE 4.10 *Transducer stand.*

FIGURE 4.11 *Displacement transducers.*

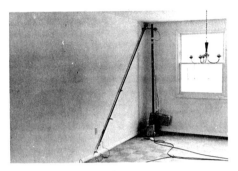

FIGURE 4.12 *Instrumentation of wall diagonal.*

FIGURE 4.13 *Data logging system.*

derived by allowing for the displacement of the upper-story floor. It was not feasible to measure, and correct for, transducer-support rotation which was judged too

small to be measurable with the available instrumentation.

4.3. Data Acquisition

Data from 32 displacement transducers, two load cells and one pressure transducer were electronically scanned, converted into digital form, and transmitted to a teletype unit. The teletype unit produced a printout which could be examined in the field and, in addition, punched the data on paper tape. The data were later transferred from the paper tape to magnetic tape for electronic computer processing.

The data acquisition system had a 100-channel capacity and was capable of logging approximately one reading per second. Data were scanned after each increment of loading and after unloading. The test schedules did not permit readings of deflection recovery to be made several hours after unloading.

Data on dynamic response from either of the two more-sensitive displacement transducers were recorded by a strip-chart recorder. The deflection-time history was recorded on a horizontal scale of 50 mm/second and a vertical scale of 20 mm per 0.01-in displacement. The data logging system shown in figure 4.13, from left to right, consisted of a digital scanning system, teletype, and a strip-chart recorder.

4.4. Visual Inspection

Visual observations of the interior and exterior surfaces of the building were made before and after each test.

5. Test Program

The static tests are summarized in table 5.1. Four

TABLE 5.1. *Summary of static tests*

(See figure 4.9 for location of loads)

Test No.	Date	Loading points	Loading sequence kip	Conditions	Comments
1	2/5/71 pm	A, B, C, D	0– 5.63–0.14 0.14– 8.00–0 0– 7.93–0	Cloudy 35 °F	Loading was limited by frictional resistance of forklifts.
2	2/6/71 am	A, B, C, D	0–10.00	Sunny 40 °F	Horizontal resistance was increased by blocking of forklifts.
3	2/7/71 am	E, F, G, H	0– 2.00–0 0– 4.00–0 0– 6.00–0 0– 7.23–0	Rainy 36 °F	Do.
4	2/7/71 pm	A, B, C, D	0–10.00–0	Rainy 36 °F	Do.

TABLE 5.2. *Summary of dynamic tests*

(See figure 4.9 for location of loads and transducers)

TABLE 5.2. *Summary of dynamic tests*

(See figure 4.9 for location of loads and transducers)

All tests performed on 2/8/72

Test No.	Load kip	Location of load	Location of transducer	Notes
5.1	2.0	C	53	
5.2	2.0	C	53	
5.3	2.0	C	53	Not a clean hit.
5.4	2.0	C	53	
5.5	1.5	C	53	
5.6	1.0	C	53	
5.7	1.0	C	53	
5.8	2.0	B	53	
5.9	2.0	B	53	

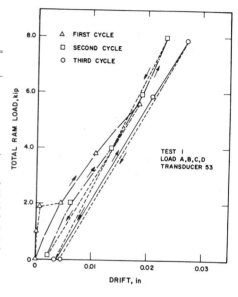

FIGURE 6.1 *Response to lateral load in Test 1.*

tests were performed. Test 1 consisted of three cycles of load applied at the lower level. In this test, the load was limited due to sliding of the forklifts. In Test 2, a single cycle of load was applied at the lower level after increasing the horizontal resistance of the forklifts by reaction blocks. In this cycle, the load exceeded the maximum applied in Test 1. In Test 3, four cycles of load were applied at the upper level. In Test 4, a single cycle of load was applied at the lower level to determine residual drift after slackness in the building had been taken up by previous testing.

The dynamic tests are summarized in table 5.2. The response in each test was recorded by a single trans-ducer. In order to determine the response for various relative locations of loads and transducers, the test had to be repeated many times. Upper story transducers could not be used to measure vibrations because of ex-citation of the transducer support.

6. Discussion of Results

6.1. Overall Building Response in Tests 1 through 4

6.1.1. Residual Drift Displacements

Figure 6.1 shows a plot of total horizontal load against drift, measured at transducer 53, which was located at the lower-ceiling level near the point of largest dis-placement. The plot is for Test 1, in which three cycles of load were applied. In the unloaded position after the first loading cycle, the residual drift displacement was 0.002 in. This was approximately 12 percent of the drift due to the maximum load applied in this cycle. Sub-sequent increments of residual deflection were smaller, indicating that residual deflection was reduced by reduction of slackness in the system and possibly by response characteristics similar to strain hardening.

The total residual displacement in this test was 0.004 in, which is approximately 15 percent of the maximum displacement observed in this test (0.027 in).

Figure 6.2 shows a similar plot for Test 4, where the total ram load was increased to 10 kip. In this case, the residual displacement was 0.003 in, or approximately 8 percent of the 0.039-in total displacement in the test.

Figure 6.3 is a plot of Test 3 for transducer 16, which was located at the upper ceiling level near the point of the largest displacement. In this test, four cycles of load were applied at the upper-ceiling level. The load applied in this test was rather high (a 2.82 kip total ram load would be equivalent to the portion of a total wind load of 15 psf tributary to these points of load applica-tion. The total ram load actually applied was 7.23 kip.). The total residual displacement from all four cycles was 0.018 in, which is about 27 percent of the total max-imum 0.068-in displacement measured. However, if the last loading cycle is considered separately, it caused an additional residual displacement of 0.007 in, which is approximately 13 percent of the 0.057-in maximum ad-ditional displacement in this loading cycle. The first loading cycle in this test, where the 0.004-in residual displacement was almost 40 percent of the total 0.011-in displacement, indicates that additional residual displacements will decrease after preloading of the structure.

In the first load cycle in Test 1 and the second load cycle in Test 3, the deflection at maximum load exceeded the deflection at the same load level in the sub-

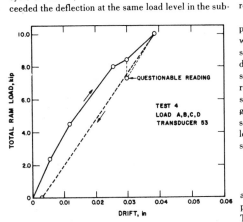

FIGURE 6.2 *Response to lateral load in Test 4.*

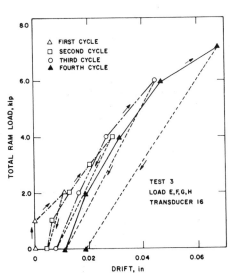

FIGURE 6.3 *Response to lateral load in Test 3.*

sequent load cycle. This response was probably caused by differences in the rate of loading. Miscellaneous adjustments in these initial load cycles substantially increased the time required for load application and data readout.

The test sequence did not allow time to measure displacement recovery several hours after unloading, which probably would have resulted in recovery of some of the residual displacements previously discussed. It is also reasonable to assume that at the smaller loads actually acting in service conditions, residual displacements would represent an even smaller percentage of total displacements. Thus, in general, it can be concluded from the previous discussion that after preloading, up to the level of the preload, the structural response to lateral load was substantially elastic.

6.1.2. Reproducibility of Data

Reproducibility of data is illustrated in figures 6.4 and 6.5. Figure 6.4 shows a plot of displacement at point 53 against lower-level loading for cycles 2 and 3 in Tests 1, 2, and 4. Data from all these four tests are reasonably consistent. Similarly, figure 6.5 shows displacements at transducer 16 for cycles 2 and 3 of Test 3, which are in good agreement. Similar trends were observed for other transducers measuring displacement

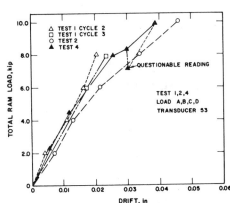

FIGURE 6.4 *Reproducibility of structural response to lower-level loading.*

in the direction of the applied load.

Thus, it can be concluded that after the application

of one or two load cycles, structural response to lateral load was predictable and reproducible.

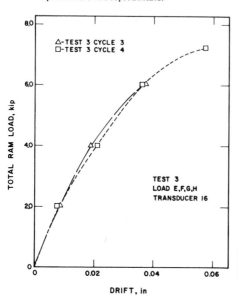

FIGURE 6.5 *Reproducibility of structural response to upper-level loading.*

6.2. Drift Under Wind Load

Since, after pre-loading, structural response to lateral load was approximately linear and measured load-displacement characteristics in various tests were reasonably consistent, results of upper- and lower-level loading can be combined by superposition to approximately predict the effect of simultaneously applying lower- and upper-level load. In this way, the effect of a load similar to that resulting from wind pressure, but statically applied, can be approximately predicted.

The wind load simulated by superimposing lower- and upper-level ram loads was computed in accordance with ANSI Standard A58.1-1955 [9]. Accordingly, no resultant load was assumed to act on the roof (see table A5 in A58.1-1955) and an evenly distributed load was assumed to act on the projected wall area in the direction of the ram load. Thus, it was assumed that the lower ram loads would simulate the wind load on a wall area equal to 376 ft², extending from midheight of the lower story to midheight of the upper story, 8 ft high

and 47 ft long. The upper ram loads were assumed to simulate the wind load acting on the wall area equal to 188 ft², extending from midheight of the upper story to the roof. It was further assumed that the area between the midheight of the lower story and grade level would be resisted at the foundation level and thus would not contribute to drift. Using this approach, a design wind load of 15 psf is simulated by the total ram loads of 2.82 kip at the upper level and 5.64 kip at the lower level, acting simultaneously.

Figures 6.6 through 6.9 show drift versus wind load for all the points where drift in the direction of the applied load was measured. The figures were derived by superimposing displacements caused by lower-level loading measured in Test 2 and displacement caused by upper-level loading equal to the average of the measurements in the second and third cycles in Test 3.

6.3. Resistance to Lateral Load

6.3.1. Lateral Displacements

Figure 6.10 shows the displacement at the upper-ceiling level relative to the surface of the upper floor. This drift was measured in the last loading cycle of Test 3, for which lateral load was applied at the upper-ceiling level. Drift was derived for a total ram load of 2.82 kip which simulates approximately the contribution of the wind load to the upper-ceiling level when a total wind load of 15 psf acts normal to the long walls of the building.

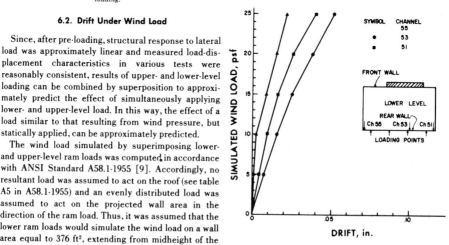

FIGURE 6.6 *Drift at upper-floor level.*

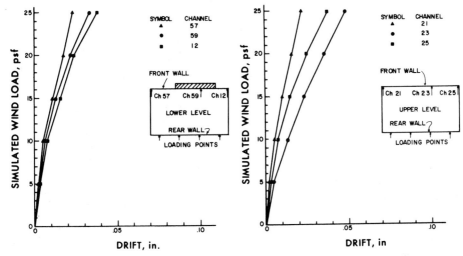

FIGURE 6.7 *Drift at upper-floor level.*

FIGURE 6.9 *Drift at upper-ceiling level.*

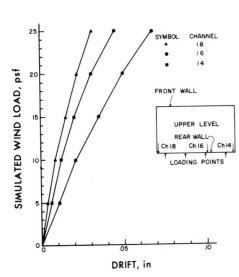

FIGURE 6.8 *Drift at upper-ceiling level.*

UPPER LEVEL

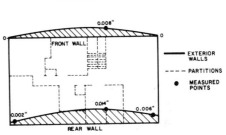

FIGURE 6.10 *Displacement of upper ceiling relative to upper floor under upper-level loading, simulating 15 psf wind load.*

The figure shows significant in-plane deformation of the upper-ceiling diaphragm. At the upper rear corners, the exterior side walls moved in the direction of the applied load; however, no translation was observed at the two front corners. The framing of these walls provides let-in bracing from each rear upper corner, at a 45° angle to the vertical, down to the upper-floor level.

Since there is incomplete rigidity at the connections, this brace could conceivably resist the horizontal thrust without participation of the 2 × 4 member at the top of the wall studs. This mechanism could account for the mode of deformation of these walls.

Figure 6.11 shows the total drift of the upper- and lower-story ceilings due to a lateral load applied at the upper-ceiling level (Test 3, last loading cycle). The total translation of the upper ceiling was derived by adding the measured lower-ceiling translation to the measured upper-ceiling translation. This is the best approximation that can be obtained from the data since measurements at the upper-ceiling level were not directly referenced to the ground level. No correction was made for possible rotation of the frames supporting the transducers, which was too small to be measured with the available instrumentation. It can be seen from figure 6.11 that while substantial in-plane deformation of the upper-ceiling diaphragm can be observed, drift at the lower-ceiling level approximates a rigid body translation. This leads to the conclusion that at the upper level, interior partitions experienced a much greater racking deformation than the side walls. This

mechanism probably caused the interior partitions to transfer a substantial part of the horizontal load, exerted at the upper-ceiling level, to the upper-floor level.

Figure 6.12 shows drift at the upper- and lower-ceiling levels under a simulated wind load of 15 psf normal to the long walls. The translations shown in the figure were computed by superposition, adding translations caused by applying ram loads at the upper-ceiling level (last loading cycle, Test 3) to those caused by ram loads applied at the lower-ceiling level (Test 2). Wind pressures corresponding to the ram load were computed as in section 6.2. The building as a whole translated and rotated slightly. The rotation is probably attributable to the unsymmetrical distribution of interior partitions and is an order of magnitude smaller than the overall drift deformations. Even though the drift near the center of the building at the lower-ceiling level was greater than that at the side walls on the same level, it cannot be concluded that this difference is attributable to overall in-plane deformation of the floor-ceiling assembly, since the translation near the front wall on the same level corresponds approximately to a rigid body displacement.

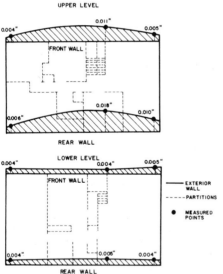

FIGURE 6.11   *Total drift under upper-level loading simulating 15 psf wind load.*

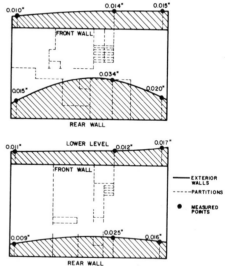

FIGURE 6.12   *Drift under combined upper- and lower-level loading simulating 15 psf wind load.*

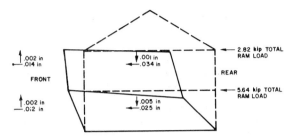

6.3.2. Vertical Displacements

Figure 6.13 shows lateral and vertical displacements near the center of the building (refer to fig. 4.9 for the position of displacement measurements), caused by a simulated 15 psf wind load. In general, the vertical displacements were one order of magnitude smaller than the horizontal displacements. As a result, no significant conclusions can be drawn from the vertical measurements.

6.3.3. Wall Distortion

Figure 6.14 shows a plot of displacements at the lower-ceiling level (points 55 and 57) corresponding to the extreme right of the building, both front and rear. The figure also shows a plot of the change in length of two diagonals across part of the right side wall instrumented at 45° angles to the horizontal (points 34 and 35). The gage points for the diagonal transducers were attached to the interior gypsum board and, therefore, the changes in length of these diagonals are a measure of the racking distortion of the interior gypsum board. It can be seen that the change in length of the diagonals was approximately 10 percent of the drift measured at corresponding load levels. If the gypsum board had participated fully in the racking distortion of the wall, the length change of the instrumented diagonals would correspond to approximately 60 percent of the drift measured at the upper wall corner. Thus, subject to the qualifications stated below (sec. 6.3.5), the important conclusion can be drawn that only a small portion of the racking force was transmitted to the gypsum board. It is, therefore, reasonable to assume that the let-in bracing provided the major portion of the resistance to the lateral force.

6.3.4. Lateral Displacement Normal to the Applied Load

Lateral displacements normal to the applied load were measured by transducers 56, 52, 13, and 58 at the lower level and by transducers 19, 15, 26, and 22 at the upper level. In general, these displacements were one

FIGURE 6.13 *Lateral and vertical displacements near the center of the building (refer to fig. 4.9 for location of measurements).*

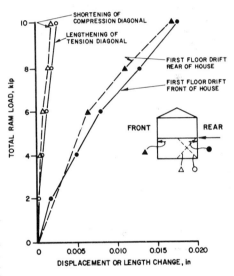

FIGURE 6.14 *Wall distortion.*

order of magnitude smaller than the drift displacements in the direction of loading and no significant conclusion can be drawn from these measurements. The following table shows the displacements corresponding to a simulated 15 psf wind load computed by superposition using Test 2 and the third cycle of Test 3. Drift displacement in the direction of the applied loads at the same locations are given for comparison.

Story	Transducer	Displacement normal to load in	Drift displacement in
Lower	56	0.002 inward	0.009
	52	0.001 inward	0.016
	13	0.000	0.017
	58	0.001 inward	0.011
Upper	19	0.002 inward	0.015
	15	0.002 inward	0.020
	26	0.003 inward	0.017
	22	0.007	0.010

6.3.5. Possible Effects of Slippage in Joints

The interpretation of structural response to lateral load presented in section 6.3 is based on the premise that joints were capable of transmitting displacements to structural elements without significant slippage. Thus, it was assumed in 6.3.1 and 6.3.3 that the displacements measured at the inside face of the rear wall of the building were imparted to the side walls. While this assumption seems reasonable, there is no evidence by actual measurements that the displacements of the rear walls were fully transmitted to the floor or ceiling diaphragms, and that the translation of these diaphragms near the side walls was actually transmitted to the side walls without slippage at the joints. The conclusions with respect to side wall distortion presented in 6.3.1 and 6.3.3 must therefore be qualified, since they are based on a premise that cannot be proven.

6.4. Dynamic Response

The plots on the left side of figure 6.15 show the results of various tests. Figure 6.15(a) shows the superimposed results of four tests, 5.1, 5.2, 5.3, and 5.4, in which a 2,000-lb force was removed at point C and response was measured by gage No. 53 (see fig. 4.9) which is close to point C. The width of the curve is partly due to higher-frequency electrical noise of the recording equipment, but also indicates the range of variation between the results of the four tests.

The range of variation between individual tests is relatively small, indicating that dynamic response was reproducible in successive tests. The response curves indicate that the method of removing the lateral force in the building apparently did not permit the structure to freely vibrate from the initial deflected position. The apparent deflection position at load release is shown in the figure as approximately 1/4 that of static displacement. The irregular response in the first cycle, subsequent to the load release, was reproducible in all the tests and is probably attributable to rebound of the spring-loaded gage. Similar plots are shown in figure 6.15(b) for tests 5.8 and 5.9, wherein a 2,000-lb force was removed at point B and response was measured at gage 53 which is located at a distance of 12 ft from point B, and in figure 6.15(c) for tests 5.6 and 5.7, wherein a 1,000-lb force was removed at point C and response was measured at gage 53 in the vicinity of point C.

Even though it is demonstrated that data are reproducible, the following must be taken into account in assessing the validity of conclusions drawn from the tests: resolution of displacement time-history records taken at the lower-ceiling level is marginal due to the low magnitude of displacement resulting from the method of excitation used (maximum peak-to-peak displacement of the free vibration was approximately 0.003 in). In view of this small displacement amplitude, it is possible that vibrations in the region of deflection measurement may not accurately reflect the vibration of the overall structure. In addition, transmission of structural motion to the transducers and seating in the structure may have adversely affected the resolution of these low magnitude vibration measurements.

Analysis of the superimposed records shown in the plots in figure 6.15(a), (b), and (c) shows a high frequency caused by electrical noise and a lower dominating frequency of approximately 9 Hz. Frequencies derived from these plots are reasonably consistent with each other. The frequency of 9 Hz appears reasonable for a structure of this type. The plots on the right side of figures 6.15(a), (b), and (c) show average deflection response and estimated envelopes of deflection decay from which the percentage of critical damping was computed by the following equations:

$$\text{percent critical damping} = \nu \times 100$$

$$2\pi\nu = \ln x_1/x_2$$

where:

$$\nu = \text{critical damping ratio}$$

$$x_1/x_2 = \text{two successive displacement peaks}$$

$$\ln = \text{natural logarithm}$$

Damping computed from the plots in figures 6.15(d), (e), and (f) averages 6 percent of critical damping with an approximate range from 4 to 9 percent. Damping ratios would probably increase for vibrations with larger displacement amplitude.

In summary, it can be concluded on the basis of the

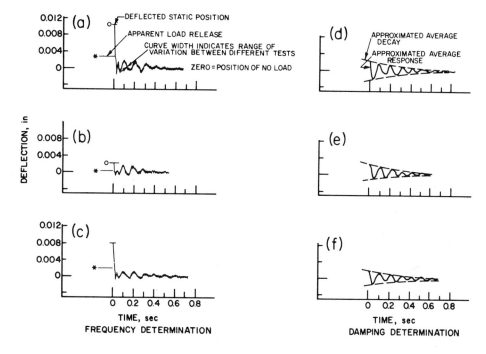

FIGURE 6.15 *Dynamic response characteristics.*

measurements and subject to the stated qualifications that the natural frequency is approximately 9 Hz and that damping averages 6 percent of critical damping, ranging from 4 to 9 percent.

6.5. Comparison of Building Response With Present U.S. Design Practice

Figure 6.16 shows a plot of drift at the lower-ceiling level against simulated wind pressure normal to the long wall. The solid line connecting the points designated by circles shows the drift measured in the fourth loading cycle and was derived by superposition of the results of Test 2 and the last loading cycle of Test 3. Drift is shown for the location of the largest displacement observed at this level (gage 53). Superposition was used, as discussed in section 6.2, assuming that a wind force of the magnitude shown on the vertical scale of the plot acted normal to the entire long-wall area, and that no resultant wind pressure acted on the projected roof area.

The dash-dotted vertical line represents the drift limit of $h/500$, where h is the height above grade level, which is normally used in the United States in the design of medium- and high-rise buildings for a wind of a 50-year mean recurrence interval.

To the right of the plot, wind loads computed for various areas in the United States and for different exposure conditions are shown. Columns A, B, and C were computed using the ANSI-A.58.1 Standard now proposed for adoption. Columns D were computed using ANSI Standard A.58.1-1955 which is presently in force and which does not distinguish between various exposure conditions. Note that, except for the Miami, Fla. region, the measured drift of this building, after slackness in the system was removed in three preloading cycles, is considerably less than that derived using present design criteria for medium- and high-rise buildings. No conclusion can be drawn for Miami, since the design wind loads required for this region exceed the loads actually applied in the test.

While the previously discussed plot provides information on drift that is anticipated when the building is subjected to repeated cycles of windload in one direction, consideration should also be given to the magnitude of drift in the first load cycle and to the total magnitude of lateral displacement, including residual displacement, when several load cycles are applied in one direction.

At least part of the residual displacement would be recovered if the direction of the lateral load is reversed. The first reverse cycle of lateral load could cause a total displacement which would include the cumulative effect of recovery of part of the residual displacement in the initial direction of loading, and the initial load cycle in the reverse direction.

Some indication of the magnitude of this total displacement can be derived from figure 6.16. The points designated by squares show the magnitude of drift that the building is estimated to experience in the first load cycle. The points designated by triangles show the total displacement relative to the initial position of the building estimated for the third load cycle. The points were derived using the results of Tests 1 and 3.

It can be concluded from the plot of these data that even the maximum drift that could reasonably be expected in the first cycle of load reversal under windloads up to 20 psf would be considerably less than h/500.

7. Conclusions

The following conclusions can be drawn on the basis of an analysis of the test results:

(1) The measured second-story drift of the building was considerably less than that derived using present design criteria for medium- and high-rise buildings as applied to most areas of the United States.

(2) Only a small portion of the distortion of the exterior walls was transmitted to the interior gypsum board finish material. Thus, the let-in bracing probably made the major contribution to the resistance to racking loads. This conclusion is subject to the qualification stated in section 6.3.5.

(3) The upper-ceiling diaphragm experienced significant in-plane deformation. Correspondingly, the racking distortion of the upper-story interior partitions exceeded that of the exterior side walls.

(4) The floor-ceiling diaphragm at the lower-ceiling level tended to act as a rigid diaphragm and to translate as a rigid body when the building was subjected to lateral load.

(5) The measured structural response indicates that the natural frequency of the structure is approximately 9 Hz and the damping averages approximately 6 percent of critical damping varying from 4 to 9 percent. The validity of these conclusions is somewhat in

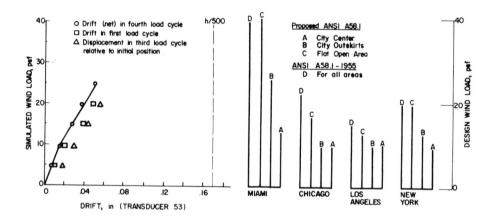

FIGURE 6.16 Comparison of measured building response with design criteria for medium- and high-rise buildings.

question because of the marginal resolution of the displacement time history records.

(6) Slackness in the structure caused relatively large residual displacements in the initial cycles of lateral loading. Response characteristics in subsequent cycles of loading were substantially elastic and reasonably consistent and reproducible, as were the measurements of dynamic response.

8. Acknowledgment

The authors wish to acknowledge the cooperation extended by Mr. Calvin B. Morstein, Vice President, and other members of the staff of Levitt Technology Corporation in furnishing the house for test. The authors also acknowledge the cooperation of Mr. William J. Werner, Project Manager, Operation Breakthrough staff, Department of Housing and Urban Development, and the assistance of Mr. Thomas H. Boone, Technical Representative, NBS, in providing liaison with Levitt Technology Corporation. Much credit is due to the following NBS staff and laboratory personnel who carried out their work in frequently unpleasant winter conditions and in accordance with difficult time schedules: Mr. Lloyd Davis, Mr. Frank Rankin, Mr. Randolph Williams, Mr. James Raines, Mr. Charles Bulik, Mr. Lymus Payton, Mr. James Warfield, Mr. Edward Tuma, and Mr. William Appleton.

9. References

[1] Short, A. and Simms, L. G., Survey of Loading Tests on Some Post-War House Prototypes, The Journal of the Institute of Structural Engineers (London, England, Feb. 1949).

[2] ASTM Special Technical Publication No. 210, Symposium on Full Scale Tests on House Structures, American Society for Testing Materials (Baltimore, Md., July 1957).

[3] FHA Technical Circular 12, A Standard for Testing Sheathing Materials for Resistance to Racking, Federal Housing Administration, Underwriting Division (Washington, D.C., Oct. 1949).

[4] ASTM C36-68, Standard Specifications for Gypsum Wallboard, 1970 Annual Book of ASTM Standards, Cement, Lime, Gypsum, Part 9, American Society for Testing Materials (Philadelphia, Pa., Nov. 1970).

[5] ASTM C79-67, Standard Specifications for Gypsum Sheathing Board, 1970 Annual Book of ASTM Standards, Cement, Lime, Gypsum, Part 9, American Society for Testing Materials (Philadelphia, Pa., Nov. 1970).

[6] Simplified Practice Recommendation 16-53, Lumber, American Standards for Softwood Lumber, U.S. Department of Commerce (Washington, D.C., Jan. 1953).

[7] Federal Housing Administration, Minimum Property Standards for one and two living units, FHA No. 300, U.S. Department of Housing and Urban Development, Federal Housing Administration (Washington, D.C. 20410, June 1969).

[8] American Institute of Steel Construction, Manual of Steel Construction, Seventh Edition (New York, N.Y., 1970).

[9] A58.1-1955, American National Standards Institute, Building Code Requirements for Minimum Design Loads in Buildings and Other Structures, National Bureau of Standards (Washington, D.C., Sept. 1955).

Brief History of Measurement Systems

with a Chart of the Modernized Metric System

"Weights and measures may be ranked among the necessaries of life to every individual of human society. They enter into the economical arrangements and daily concerns of every family. They are necessary to every occupation of human industry; to the distribution and security of every species of property; to every transaction of trade and commerce; to the labors of the husbandman; to the ingenuity of the artificer; to the studies of the philosopher; to the researches of the antiquarian, to the navigation of the mariner, and the marches of the soldier; to all the exchanges of peace, and all the operations of war. The knowledge of them, as in established use, is among the first elements of education, and is often learned by those who learn nothing else, not even to read and write. This knowledge is riveted in the memory by the habitual application of it to the employments of men throughout life."

JOHN QUINCY ADAMS
Report to the Congress, 1821

Weights and measures were among the earliest tools invented by man. Primitive societies needed rudimentary measures for many tasks: constructing dwellings of an appropriate size and shape, fashioning clothing, or bartering food or raw materials.

Man understandably turned first to parts of his body and his natural surroundings for measuring instruments. Early Babylonian and Egyptian records and the Bible indicate that length was first measured with the forearm, hand, or finger and that time was measured by the periods of the sun, moon, and other heavenly bodies. When it was necessary to compare the capacities of containers such as gourds or clay or metal vessels, they were filled with plant seeds which were then counted to measure the volumes. When means for weighing were invented, seeds and stones served as standards. For instance, the "carat," still used as a unit for gems, was derived from the carob seed.

As societies evolved, weights and measures became more complex. The invention of numbering systems and the science of mathematics made it possible to create whole systems of weights and measures suited to trade and commerce, land division, taxation, or scientific research. For these more sophisticated uses it was necessary not only to weigh and measure more complex things—it was also necessary to do it accurately time after time and in different places. However, with limited international exchange of goods and communication of ideas, it is not surprising that different systems for the same purpose developed and became established in different parts of the world—even in different parts of a single continent.

The English System

The measurement system commonly used in the United States today is nearly the same as that brought by the colonists from England. These measures had their origins in a variety of cultures—Babylonian, Egyptian, Roman, Anglo-Saxon, and Norman French. The ancient "digit," "palm," "span," and "cubit" units evolved into the "inch," "foot," and "yard" through a complicated transformation not yet fully understood.

Roman contributions include the use of the number 12 as a base (our foot is divided into 12 inches) and words from which we derive many of our present weights and measures names. For example, the 12 divisions of the Roman "pes," or foot, were called *unciae*. Our words "inch" and "ounce" are both derived from that Latin word.

The "yard" as a measure of length can be traced back to the early Saxon kings. They wore a sash or girdle around the waist—that could be removed and used as a convenient measuring device. Thus the word "yard" comes from the Saxon word "gird" meaning the circumference of a person's waist.

Standardization of the various units and their combinations into a loosely related system of weights and measures sometimes occurred in fascinating ways. Tradition holds that King Henry I decreed that the yard should be the distance from the tip of his nose to the end of his thumb. The length of a furlong (or furrow-long) was established by early Tudor rulers as 220 yards. This led Queen Elizabeth I to declare, in the 16th century, that henceforth the traditional Roman mile of 5,000 feet would be replaced by one of 5,280 feet, making the mile exactly 8 furlongs and providing a convenient relationship between two previously ill-related measures.

Thus, through royal edicts, England by the 18th century had achieved a greater degree of standardization than the continental countries. The English units were well suited to commerce and trade because they had been developed and refined to meet commercial needs. Through colonization and dominance of world commerce during the 17th, 18th, and 19th centuries, the English system of weights and measures was spread to and established in many parts of the world, including the American colonies.

However, standards still differed to an extent undesirable for commerce among the 12 colonies. The need for greater uniformity led to clauses in the Articles of Confederation (ratified by the original colonies in 1781) and the Constitution of the United States (ratified in 1790) giving power to the Congress to fix uniform standards for weights and measures. Today, standards supplied to all the States by the National Bureau of Standards assure uniformity throughout the country.

The Metric System

The need for a single worldwide coordinated measurement system was recognized over 300 years ago. Gabriel Mouton, Vicar of St. Paul in Lyons, proposed in 1670 a comprehensive decimal measurement system based on the length of one minute of arc of a great circle of the earth. In 1671 Jean Picard, a French astronomer, proposed the length of a pendulum beating seconds as the unit of length. (Such a pendulum would have been fairly easily reproducible, thus facilitating the widespread distribution of uniform standards.) Other proposals were made, but over a century elapsed before any action was taken.

In 1790, in the midst of the French Revolution, the National Assembly of France requested the French Academy of Sciences to "deduce an invariable standard for all the measures and all the weights." The Commission appointed by the Academy created a system that was, at once, simple and scientific. The unit of length was to be a portion of the earth's circumference. Measures for capacity (volume) and mass (weight) were to be derived from the unit of length, thus relating the basic units of the system to each other and to nature. Furthermore, the larger and smaller versions of each unit were to be created by multiplying or dividing the basic units by 10 and its multiples. This feature provided a great convenience to users of the system, by eliminating the need for such calculations as dividing by 16 (to convert ounces to pounds) or by 12 (to convert inches to feet). Similar calculations in the metric system could be performed simply by shifting the decimal point. Thus the metric system is a "base-10" or "decimal" system.

The Commission assigned the name *metre* (which we now spell meter) to the unit of length. This name was derived from the Greek word *metron*, meaning "a measure." The physical standard representing the meter was to be constructed so that it would equal one ten-millionth of the distance from the north pole to the equator along the meridian of the earth running near Dunkirk in France and Barcelona in Spain.

The metric unit of mass, called the "gram," was defined as the mass of one cubic centimeter (a cube that is 1/100 of a meter on each side) of water at its temperature of maximum density. The cubic decimeter (a cube 1/10 of a meter on each side) was chosen as the unit of fluid capacity. This measure was given the name "liter."

Although the metric system was not accepted with enthusiasm at first, adoption by other nations occurred steadily after France made its use compulsory in 1840. The standardized character and decimal features of the metric system made it well suited to scientific and engineering work. Consequently, it is not surprising that the rapid spread of the system coincided with an age of rapid technological development. In the United States, by Act of Congress in 1866, it was made "lawful throughout the United States of America to employ the weights and measures of the metric system in all contracts, dealings or court proceedings."

By the late 1860's, even better metric standards were needed to keep pace with scientific advances. In 1875, an international treaty, the "Treaty of the Meter," set up well-defined metric standards for length and mass, and established permanent machinery to recommend and adopt further refinements in the metric system. This treaty, known as the Metric Convention, was signed by 17 countries, including the United States.

As a result of the Treaty, metric standards were constructed and distributed to each nation that ratified the Convention. Since 1893, the internationally agreed-to metric standards have served as the fundamental weights and measures standards of the United States.

By 1900 a total of 35 nations—including the major nations of continental Europe and most of South America—had officially accepted the metric system. Today, with the exception of the United States and a few small countries, the entire world is using predominantly the metric system or is committed to such use. In 1971 the Secretary of Commerce, in transmitting to Congress the results of a 3-year study authorized by the Metric Study Act of 1968, recommended that the U.S. change to predominant use of the metric system through a coordinated national program. The Congress is now considering this recommendation.

The International Bureau of Weights and Measures located at Sevres, France, serves as a permanent secretariat for the Metric Convention, coordinating the exchange of information about the use and refinement of the metric system. As measurement science develops more precise and easily reproducible ways of defining the measurement units, the General Conference of Weights and Measures—the diplomatic organization made up of adherents to the Convention—meets periodically to ratify improvements in the system and the standards.

In 1960, the General Conference adopted an extensive revision and simplification of the system. The name *Le Système International d'Unités* (International System of Units), with the international abbreviation SI, was adopted for this modernized metric system. Further improvements in and additions to SI were made by the General Conference in 1964, 1968, and 1971.

The following "Chart of the Modernized Metric System" is adapted from a National Bureau of Standards Special Publication 304A (Revised October 1972), which is sold by the Superintendent of Documents, U.S. Government Printing Office, Washington, D.C.

THE MODERNIZED metric system

The International System of Units—SI

is a modernized version of the metric system established by international agreement. It provides a logical and interconnected framework for all measurements in science, industry, and commerce. Officially abbreviated SI, the system is built upon a foundation of seven base units, plus two supplementary units, which appear on this chart along with their definitions. All other SI units are derived from these units. Multiples and submultiples are expressed in a decimal system. Use of metric weights and measures was legalized in the United States in 1866, and since 1893 the yard and pound have been defined in terms of the meter and the kilogram. The base units for time, electric current, amount of substance, and luminous intensity are the same in both the customary and metric systems.

Symbol	When You Know	Multiply by	To Find	Symbol
in	inches	ᴬ25.4	ᴬmillimeters	mm
ft	feet	ᴬ0.3048	meters	m
yd	yards	ᴬ0.9144	meters	m
mi	miles	1.609 34	kilometers	km
	square yards	0.836 127	square meters	m²
	acres	0.404 686	ᶜhectares	ha
yd³	cubic yards	0.764 555	cubic meters	m³
qt	quarts (lq)	0.946 353	ᴰliters	l
oz	ounces (avdp)	28.349 5	grams	g
lb	pounds (avdp)	0.453 592	kilograms	kg
°F	Fahrenheit temperature	ᴬ5/9 (after subtracting 32)	Celsius temperature	°C
mm	millimeters	0.039 370 1	inches	in
m	meters	3.280 84	feet	ft
m	meters	1.093 61	yards	yd
km	kilometers	0.621 371	miles	mi
m²	square meters	1.195 99	square yards	yd²
ha	hectares	2.471 05	acres	
m³	cubic meters	1.307 95	cubic yards	yd³
l	ᴰliters	1.056 69	quarts (lq)	qt
g	grams	0.035 274 0	ounces (avdp)	oz
kg	kilograms	2.204 62	pounds (avdp)	lb
°C	Celsius temperature	ᴬ9/5 (then add 32)	Fahrenheit temperature	°F

ᴬexact

ᴮ for example, 1 in = 25.4 mm, so 3 inches would be (3 in) (25.4 mm/in) = 76.2 mm

ᶜhectare is a common name for 10 000 square meters

ᴰliter is a common name for fluid volume of 0.001 cubic meter

Note: Most symbols are written with lower case lette.s; exceptions are units named after persons for which the symbols are capitalized. Periods are not used with any symbols.

Prefixes and Multiples and Submultiples

	Prefixes	Symbols	Multiples and Submultiples
	tera (ter a)	T	1 000 000 000 000 = 10¹²
	giga (ji ga)	G	1 000 000 000 = 10⁹
	mega (meg a)	M	1 000 000 = 10⁶
	kilo (kil o)	k	1 000 = 10³
	hecto (hek to)	h	100 = 10²
	deka (dek a)	da	10 = 10¹
			Base Unit 1 = 10⁰
	deci (des i)	d	0.1 = 10⁻¹
	centi (sen ti)	c	0.01 = 10⁻²
	milli (mil i)	m	0.001 = 10⁻³
	micro (mi kro)	μ	0.000 001 = 10⁻⁶
	nano (nan o)	n	0.000 000 001 = 10⁻⁹
	pico (pe ko)	p	0.000 000 000 001 = 10⁻¹²
	femto (fem to)	f	0.000 000 000 000 001 = 10⁻¹⁵
	atto (at to)	a	0.000 000 000 000 000 001 = 10⁻¹⁸

The SI unit of area is the **square meter** (m²).

The SI unit of volume is the **cubic meter** (m³). The liter (0.001 cubic meter), although not an SI unit, is commonly used to measure fluid volume.

The SI unit for pressure is the **pascal** (Pa).
$$1Pa = 1N/m^2$$

The SI unit for work and energy of any kind is the **joule** (J).
$$1J = 1N \cdot m$$

The SI unit for power of any kind is the **watt** (W).
$$1W = 1J/s$$

Standard frequencies and correct time are broadcast from WWV, WWVB, and WWVH, and stations of the U.S. Navy. Many short-wave receivers pick up WWV and WWVH, on frequencies of 2.5, 5, 10, 15, and 20 megahertz.

SEVEN BASE UNITS

meter - m
LENGTH

1 METER

1 650 763.73 WAVELENGTHS

ONE WAVELENGTH

"Kʳ ATOM"

An interferometer is used to measure length by means of light waves

The meter (common international spelling, metre) is defined as 1 650 763.73 wavelengths in vacuum of the orange-red line of the spectrum of krypton-86.

kilogram - kg
MASS

U.S. PROTOTYPE KILOGRAM NO. 20

The standard for the unit of mass, the kilogram, is a cylinder of platinum-iridium alloy kept by the International Bureau of Weights and Measures at Paris. A duplicate (No. 20) is in the custody of the National Bureau of Standards and serves as the mass standard for the United States. This is the only base unit still defined by an artifact.

The SI unit of force is the **newton** (N). One newton is the force which, when applied to a 1 kilogram mass, will give the kilogram mass an acceleration of 1 (meter per second) per second.
$$1N = 1kg \cdot m/s^2$$

1N

ACCELERATION of 1m/s²

1 kg

second - s
TIME

TRANSITION REGION (CAVITY) OSCILLATING FIELD

DETECTOR

CESIUM SOURCE

DEFLECTION MAGNET   DEFLECTION MAGNET

OSCILLATOR

NBS ATOMIC-TIME-SCALE SYSTEM

The second is defined as the duration of 9 192 631 770 cycles of the radiation associated with a specified transition of the cesium-133 atom. It is realized by tuning an oscillator to the resonance frequency of cesium-133 atoms as they pass through a system of magnets and a resonant cavity into a detector.

Schematic diagram of an atomic beam spectrometer or "clock." Only those atoms whose magnetic moments are "flipped" in the transition region reach the detector. When 9 192 631 770 oscillations have occurred, the clock indicates one second has passed.

The number of periods or cycles per second is called frequency. The SI unit for frequency is the **hertz** (Hz). One hertz equals one cycle per second.

The SI unit for speed is the **meter per second** (m/s).

The SI unit for acceleration is the **(meter per second) per second** (m/s²).

ampere - A
ELECTRIC CURRENT

The ampere is defined as that current which, if maintained in each of two long parallel wires separated by one meter in free space, would produce a force between the two wires (due to their magnetic fields) of 2×10^{-7} newton for each meter of length.

The SI unit of voltage is the **volt** (V).
$$1V = 1W/A$$

The SI unit of electric resistance is the **ohm** (Ω).
$$1\Omega = 1V/A$$

---

kelvin - K
TEMPERATURE

The kelvin is defined as the fraction 1/273.16 of the thermodynamic temperature of the triple point of water. The temperature 0 K is called "absolute zero."

On the commonly used Celsius temperature scale, water freezes at about 0°C and boils at about 100°C. The °C is defined as an interval of 1 K, and the Celsius temperature 0°C is defined as 273.15K.

The Fahrenheit degree is an interval of 5/9°C or 5/9K; the Fahrenheit scale uses 32°F as a temperature corresponding to 0°C.

The standard temperature at the triple point of water is provided by a special cell, an evacuated glass cylinder containing pure water. When the cell is cooled until a mantle of ice forms around the reentrant well, the temperature at the interface of solid, liquid, and vapor is 273.16K. Thermometers to be calibrated are placed in the reentrant well.

---

mole - mol
AMOUNT OF SUBSTANCE

The mole is the amount of substance of a system that contains as many elementary entities as there are atoms in 0.012 kilogram of carbon 12.

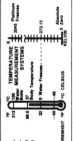

When the mole is used, the elementary entities must be specified and may be atoms, molecules, ions, electrons, other particles, or specified groups of such particles.

The SI unit of concentration (of amount of substance) is the **mole per cubic meter** (mol/m³).

---

candela - cd
LUMINOUS INTENSITY

The candela is defined as the luminous intensity of 1/600 000 of a square meter of a blackbody at the temperature of freezing platinum (2045K).

The SI unit of light flux is the **lumen** (lm). A source having an intensity of 1 candela in all directions radiates a light flux of 4π lumens.

A 100-watt light bulb emits about 1700 lumens.

TWO SUPPLEMENTARY UNITS

radian - rad
PLANE ANGLE

The radian is the plane angle with its vertex at the center of a circle that is subtended by an arc equal in length to the radius.

steradian - sr
SOLID ANGLE

The steradian is the solid angle with its vertex at the center of a sphere that is subtended by an area of the spherical surface equal to that of a square with sides equal in length to the radius.

NBS Policy for Usage of SI Units

The International System of Units (SI)[1] was defined and given official status by the General Conference of Weights and Measures in 1960. NBS adopted the SI for use by its staff in 1964,[2] and issued statements interpreting this policy in 1968[3] and 1971.[4] While the basic policy remains unchanged, its restatement below may be useful at this time. This discussion references NBS Special Publication 330, 1972 Edition,[1] for SI units and units that may be used with SI rather than listing them.

It is the policy of the Bureau to support the International System of Units (SI) and encourage its adoption. During a transition period, exceptions to the SI are allowed in the interests of effective communication with the expected audiences for Bureau technical writings. These exceptions are permitted only during the transition period; they are not to be used as a basis for perpetuating non-SI usages.

Numerical data are used in NBS publications in two distinct ways: as descriptive data and as essential data. NBS policy for the transition period accepts different treatment of these two classes of data, although they may be presented in the same text; each may cite appropriate units.

Descriptive data describe arrangements, environments, noncritical dimensions and shapes of apparatus, and similar measurements not entering into calculations or expression of results.

Essential data express, lead up to, or help to interpret the quantitative results of the activity that is being reported.

Descriptive Data

Descriptive data should be expressed in the most useful and convenient manner. Translation into SI is not required. The units best understood by the intended audience are the most appropriate. Where non-SI units are used, the author is encouraged to add SI equivalents in parentheses but usage within a paper should be consistent on this point. Commercial gauge designations or other standard nomenclatures, e.g., drill sizes, are acceptable. As SI units become more commonly used for commercial products, use of SI units in descriptive data should conform.

Essential Data

The transition period to general usage of SI units is expected to be dependent on the field of science or technology and on the particular unit itself. In those fields in which use of SI units would create a serious impediment to communication, the essential data may be expressed in the units customarily used in the relevant field of technology. SI equivalents should be added in parentheses, or in parallel columns in tables. If graphs are used as the primary or sole means of presenting essential data, the coordinates may be divided according to customary usage, but a secondary set of coordinate markings in SI units should be included. The top and right-hand side of the graph are often appropriate for this purpose. If graphs are used only to indicate trends, or as supplements to tables, units customary in the field are acceptable without SI translation. NBS authors should, however, use the SI as soon as the level of SI usage in the related field renders it an efficient communication device. Familiarity with SI units is recommended to all NBS staff.

Essential data shall be expressed in SI units, or in units approved by the International Committee of Weights and Measures (CIPM) for use with the SI (Tables 1-9 of NBS Special Publication 330). Values in other units may be added, in parentheses, where it is felt that this information will improve the communication between authors and readers.

The CIPM has authorized, for a limited time, the use of the following common units with the SI:

nautical mile	bar
knot	standard
ångström	atmosphere
are	gal
hectare	curie
barn	röntgen
	rad

Centimeter-Gram-Second (CGS) units with special names should not be used with SI units; use of the following units should be avoided (except, of course, during the transition period):

erg	oersted
dyne	maxwell
poise*	stilb
stokes*	phot
gauss	

The CIPM has deprecated the use of the following names:

fermi (use femtometre)
metric carat
torr
kilogram-force
calorie
micron (use micrometre)
X unit
gamma (use nanotesla)
gamma (use microgram)
lambda (use microliter)

Reference data used generally in science and technology should be expressed in SI units with appropriate indication of conversion factors into traditional units or with parallel columns of converted values. Standard reference data applicable primarily to scientific interests (e.g., tables of X-ray atomic energy levels) should be expressed in SI units. Non-SI units may be included as parallel entries.

Where the general usage in any field does not recognize SI units, NBS authors should employ units comprehensible to their readers, but should try to increase reader familiarity with SI units as rapidly as possible. Again, the device of parallel columns (familiar plus SI units) is recommended.

In the lists given in NBS SP-330 (e.g., Table 3), the short names for compound units, such as "coulomb" for "ampere-second," exist for convenience, and their use is not compulsory. For example, communication sometimes benefits if the author expresses magnetic flux in volt-seconds, instead of using the synonym webers, because of the descriptive value implicit in the former unit phrase.

The analysis, interpretation, or application of essential data may involve angles and the relative values of natural functions (sine, tangent, log sin, etc.). In these cases, the angles may be expressed in degrees rather than in radians.

Natural Units

In some cases, it is convenient to express quantities in terms of fundamental Lorentz invariant constants of nature. This situation arises in theoretical work; experimental data are rarely determined directly in terms of these constants. The use of these constants as "natural units" is acceptable. The author, however, should state clearly which natural units are being used; such broad terms as "atomic units" should be avoided when there is danger of confusion.

Typical examples of natural units are:

Unit	Symbol
elementary charge	e
electron mass	m_e
proton mass	m_p
Bohr radius	a_0
electron radius	r_e
Compton wavelength of electron	λ_C
Bohr magneton	μ_B
nuclear magneton	μ_N
velocity of light	C
Planck's constant	h or $\hbar$

Special Considerations

1. It is acceptable to use the Celsius temperature (symbol t) defined by the equation

$$t = T - T_0$$

where T is the thermodynamic temperature, expressed in kelvins, and $T_o =$ 273.15 K by definition. The Celsius temperature is in general expressed in degrees Celsius (symbol °C). The unit "degree Celsius" is thus equal to the unit "kelvin," and an interval or a difference of Celsius temperature may also be expressed in degrees Celsius.

2. In situations where the use of the symbol K for thermodynamic temperature would lead to confusion with other symbols, it may be replaced by °K.

3. Logarithmic measures such as pH, dB, and Np are acceptable.

4. On recommendation of the CIPM, the names of multiples and submultiples of the kilogram are formed by adding prefixes to the word gram.

5. Words and symbols should NOT be mixed; if mathematical operations are indicated, only symbols should be used. For example, one may write "joules per mole," "J/mol," "J mol^{-1}," but not "joules/mole," "joules mol^{-1}," etc.

6. In keeping with the recommendations of various international bodies and domestic professional societies, NBS encourages the use of the spelling "metre" in communications whose principal audience is the technical community. This would include students receiving technical training at the college level. However, in order to minimize possible resistance to the use of the metric system by nontechnical groups due to introduction of an unfamiliar spelling, NBS communications addressed to such audiences (including students through high school) should use the spelling "meter." Uniform international spelling of the word "litre" is of less significance than "metre" since "litre" is not a base unit of SI, and the staff is encouraged to continue using the spelling "liter." Both spellings for both words are considered to be acceptable.

---

[1] A complete description of SI, together with guidelines for its implementation established by the International Committee of Weights and Measures, appears in the International System of Units, translated by C. H. Page and P. Vigoureux, NBS Spec. Pub. 330, 1972 Edition.

[2] NBS Adopts International System of Units, Nat. Bur. Stand. (U.S.), Tech. News Bull. 48, No. 4, p. 61 (1964).

[3] NBS Interprets Policy on SI Units, Nat. Bur. Stand. (U.S.), Tech. News Bull. 52, No. 6, p. 121 (1968).

*Policy for NBS Usage of SI Units, Nat. Bur. Stand. (U.S.), Tech. News Bull. 55, No. 1, p. 18 (1971).

Course Outline:
Technical Report Writing

1. What Technical Writing Is
   Its Role and History
   Definition
   Problems

2. Scientific Method and Approach
   Historic Origins of Science
   The Problem Concept
      Types of Problems
   The Hypothesis
   Scientific Method

3. Technical Correspondence
   Psychology of Approach
   Format Considerations and Mechanical Details
   The Memorandum
   The Employment Letter
   Inquiries and Responses

4. Report Writing — Reconstruction of an Investigation
   What Reports Are
   Types and Forms

                    Drawings
                    Diagrams
                    Exploded Views
                    Maps
                    Graphs
                    Charts
                    Curves

8.   Mechanics and Grammar
                    Technical Style
                    Readability
                    Documentation
                    Handling of Footnotes and References

Major Assignments

1. Biographical letter

2. Memorandum justifying choice of problem for 2500-3000 word research report. (exercise 12, chapter 2.)

3. Work schedule and listing of tasks and timetable for accomplishing tasks. (exercise 7, chapter 3.)

4. Block diagram of problem of investigation and its components. (exercise 11, chapter 3.)

5. Application letter for a summer job.

6. Application letter, with resume, for the job desired after graduation.

7. Letter of inquiry and reply (related to the investigation for research report being written).

8. Book review. (Book should deal with some aspect of term research report [500 words].)

9. Write one of the following, as applicable to your report: (a) an expanded definition; (b) description of a mechanism; (e) explanation of a process; include illustrations as needed.

10. Sentence outline of term report.

11. Introduction to report.

12. Term research report, 2500-3000 words.

13. Oral presentation of term report.

Course Outline: Advanced Elements of Technical Writing

1. Semantics and the Process of Communication
   Nature of Language
   Human Communication
   Semantics
   Problems in Achieving Meaning
   Role of Writing in Scientific Method
2. Special Expository Techniques — Definition
   Nature of Definitions
     Formal Definitions
     Informal Definitions
   Methods of Developing Definitions
     Synonym and Antonym
     Illustration
     Stipulation
     Operational Definitions
     Expanded Definitions
       Amplification
       Example
       Explication
       Derivation

History
Analysis
Comparison and Contrast
Distinction
Elimination
Combination of Methods

3. Special Expository Techniques — Description
   Uses of Technical Description
   Outlining
   Major Divisions of the Written Description
   Illustrations
   Style

4. Special Expository Techniques — Explanation of a Process
   Uses of Explanation of Technical Processes
   Outlining
   Major Divisions
   Illustrations
   Style

5. Special Expository Techniques — Analysis
   Methods of Analysis
      Classification
      Partition
   Synthesis
   Interpretation

6. Professional and Technical Articles
   The Scientific Paper in Professional Journals
      Types
      Study of the Professional Journals
      Organization of the Scientific Professional Paper (Article)
      Style
   Communications to the Editor
      The Research Communication
      "Letting off Steam"
   The Technical Article
      The Technical Periodical
      Organization of the Article
      Style
   The Semitechnical Article
      Sources of Publication
      Organization
      Style

7. Popularization — Writing for the General Reader

8. Outlining and Organization

9. Problems in Writing the Paper or Article

10. Graphic Presentation in Technical Articles

11. Technical Style

12. Writing Query Letters to Editors

Major Assignments

1. Locate, describe, and analyze a scientific paper or book in which the author demonstrates use of critical analysis — showing discriminatory powers, judgment — particularly his consideration of the consequences of an interpretation, his capacity to see deficiencies in data or interpretations, and his discernment of trends in data and concepts; his perception of relationships and powers of synthesis.

2. Write a review of a book dealing with a scientific matter. Your review should place and identify the book written, the general field of its subject, and for whom it is meant; analyze its purpose and its area of usefulness; describe and demonstrate how successful or unsuccessful the author was in accomplishing his purpose. Your review, therefore, should provide comment on the authority of the author; the topics covered in the book; the completeness of the coverage; accuracy and recency of the information and its validity. Comment on its style. Finally, evaluate the worth of the book.

3. Choose any three common words from the following: "wave," "work," "compound," "twist," "web," "even," "baffle," "point," "table," "shell," "truss," "dip," "horn," "nut," "line." Use each word in four sentences, illustrating a different meaning in each sentence. Weave each of the four sentences of each set of three into a unified piece of one or two paragraphs, providing enough amplification so that you are confident the reader has received the meaning you intended.

4. Write the descriptions called for in exercise 1, chapter 8. Under each description, provide a diagram similar to that shown in figure 2 of chapter 8. Next, note the diagrams in figure 5a and 5b, chapter 8. Diagram the amount of "loss" there is in each description.

5. Write three expanded definitions of about 200–300 words each of any of the terms in exercise 2, or three similar terms in your own field of interest, using the expository devices of explication, example, analogy, comparison and contrast, analysis, history, operational, descriptive details, or cause and effect.

6. Choose a simple mechanism or organism with which you are familiar. Prepare an outline similar to that in chapter 10. Before describing or outlining it, make one or as many drawings of the subject as necessary to familiarize yourself with it. Then write the description; use headings for each major division and subdivision of your subject. Include whatever illustrations or drawings you need to supplement the description.

7. Write a detailed explanation of an experiment you have performed. In writing your exposition, use the first person, active voice, and answer the following questions:
   a) What was the experiment you performed?
   b) Why did you perform the experiment?
   c) What tools or instruments did you use?
   d) What materials did you use and where did you get them?
   e) How did you proceed in the process of performing the experiment? What steps did you follow?
   f) What was the result?

8. Locate a report, a professional paper, a semitechnical article, and a popular article, all dealing with the same subject. Discuss the organization, style of writing, opening, terminal section, and viewpoint of the four types of writing. Comment on the writer's success in transmitting his information to the intended readers of each of these types.

9. Choose a technical concept which you feel sufficiently informed about. Construct an analogy which will make this concept more intelligible to the uninformed reader.

10. Write a Letter or Communication to the Editor of a professional journal in your field. Your letter should express your views on a "critical" issue in your field, or express praise or criticism of a previous article or view appearing in the journal.

11. Choose a topic of interest to you and one which you have great familiarity with, based on past or present research; as the major assignment for the course, develop an article of about 2000-3000 words on one of any of the following types: a paper for a professional journal; a semitechnical article for a technical periodical; or a popularized article on a technical subject. Study the articles in the journal or magazine you plan to submit your article.

12. Write a query letter to a magazine about its possible interest in an article you wish to submit to it. Include a short outline together with statements on your competency for dealing with the subject and its appropriateness to the particular magazine.

13. Develop a detailed outline for the paper or article you are writing in exercise 11, above. (The outline should be written prior to and in preparation for the article.)

INDEX

INDEX